I0222893

Disaster upon Disaster

Catastrophes in Context

General Editors:
Gregory V. Button, University of Michigan at Ann Arbor
Anthony Oliver-Smith, University of Florida
Mark Schuller, Northern Illinois University

Catastrophes in Context aims to bring critical attention to the social, political, economic, and cultural structures that create disasters out of natural hazards or political events and that shape the responses. Combining long-term ethnographic fieldwork typical of anthropology and increasingly adopted in similar social science disciplines such as geography and sociology with a comparative frame that enlightens global structures and policy frameworks, *Catastrophes in Context* includes monographs and edited volumes that bring critical scrutiny to the multiple dimensions of specific disasters and important policy/practice questions for the field of disaster research and management. Theoretically innovative, our goal is to publish readable, lucid texts to be accessible to a wide range of audiences across academic disciplines and specifically practitioners and policymakers.

Volume 2
Disaster upon Disaster: Exploring the Gap between Knowledge, Policy, and Practice
Edited by Susanna M. Hoffman and Roberto E. Barrios

Volume 1
Contextualizing Disaster
Edited by Gregory V. Button and Mark Schuller

Disaster upon Disaster

*Exploring the Gap between Knowledge,
Policy, and Practice*

**Edited by
Susanna M. Hoffman and Roberto E. Barrios**

berghahn
NEW YORK · OXFORD
www.berghahnbooks.com

First published in 2020 by
Berghahn Books
www.berghahnbooks.com

© 2020 Susanna M. Hoffman and Roberto E. Barrios

All rights reserved. Except for the quotation of short passages
for the purposes of criticism and review, no part of this book
may be reproduced in any form or by any means, electronic or
mechanical, including photocopying, recording, or any information
storage and retrieval system now known or to be invented,
without written permission of the publisher.

Library of Congress Cataloging-in-Publication Data
Names: Hoffman, Susanna M., editor. | Barrios, Roberto E., editor.
Title: Disaster upon disaster : exploring the gap between knowledge,
  policy, and practice / edited by Susanna M. Hoffman and Roberto E.
  Barrios.
Description: First edition. | New York : Berghahn Books, 2020. | Series:
  Catastrophes in context ; Volume 2 | Includes bibliographical references
  and index.
Identifiers: LCCN 2019030424 (print) | LCCN 2019030425 (ebook) | ISBN
  9781789203455 (hardback) | ISBN 9781789206487 (paperback) | ISBN
  9781789203462 (ebook)
Subjects: LCSH: Disasters--Social aspects. | Disaster relief--Social
  aspects. | Emergency management--Social aspects.
Classification: LCC HV553 .D569 2020 (print) | LCC HV553 (ebook) | DDC
  363.34--dc23
LC record available at https://lccn.loc.gov/2019030424
LC ebook record available at https://lccn.loc.gov/2019030425

British Library Cataloguing in Publication Data
A catalogue record for this book is available from the British Library

ISBN 978-1-78920-345-5 hardback
ISBN 978-1-78920-648-7 paperback
ISBN 978-1-78920-346-2 ebook

In loving memory of our friend and colleague
Terry Jeggle

Contents

Illustrations

Acknowledgments

A volume that proposes to offer a set of chapters intended to unveil the many vectors of a complex problem takes the concerted effort of many people. We would first like to offer our heartfelt gratitude to our many contributors. Because they come from a variety of backgrounds, they were able to approach the topic from a wide spectrum of standpoints, and, in so doing, were able to produce a compendium of remarkable relevance and caliber. We would also like to thank the trio of peers who reviewed our manuscript. Their insightful comments both affirmed the importance of the work and helped improve it. We are indebted to our executive editors, Mark Schuller, Anthony Oliver-Smith, and Gregory Button, for adding our volume to the outstanding series on the topic of risk and disaster they are compiling for Berghahn Books, *Catastrophes in Context*. It is a crucial subject, becoming more so all the time, thus making the series invaluable. We owe a debt of gratitude to our publisher and editor-in-chief, Marion Berghahn, for her steady encouragement; to Tom Bonnington, assistant editor, who guided the book to completion; Caroline Kuhtz, the production editor; and to Elizabeth Berg, who designed the terrific cover. We would finally like to state our deep appreciation to our very tolerant partners, Ron Doades and Kelly McGuire.

Defining Disaster upon Disaster

Why Risk Prevention and Disaster Response So Often Fail

SUSANNA M. HOFFMAN

> The world of policy requires both formal reasoning for much of the machinery of governance but also a more organic connection to, and recognition of, human beings.
>
> —David Haines as quoted in J. Pajo and T. Powers, "The Anthropology of Policy Emerges"[1]

The chapters that make up the following volume attempt something curious. They propose to describe and advance the rectification of something that is essentially a cipher—the gap that exists between what experts know about risk and disaster and what of that knowledge makes its way into the directives of establishments presiding over the problem and the operations of people on the ground dealing with it.

The issue is not unique. The same sort of gap exists in almost every domain that deals with any and every human predicament. The same sort of disjunction prevails between what researchers unearth in the field of health and what enters canons of medical management and actual practice of medical practice. A similar disjunction occurs in the field of food and nutrition. Up-to-date erudition on what constitutes well-rounded and wholesome nourishment only sluggishly creeps into dietary protocols and onto plates. An identical chasm transpires between what is known, what is endorsed, and what is enacted in the realms of environment, hydrology, education, transportation, and more. All such fields display a rift, and a pernicious one, between what experts recognize and what gets into the governing mandates and executed undertakings.

In the sphere discussed in the volume, that of the perils that menace human communities and the catastrophes that befall them, the gap between knowledge, policy, and practice has led, and is increasingly leading to, dire consequences. That gap has resulted in advancing people's vulnerability rather than diminishing it, it has engendered a furtherance in the number and sorts of hazards that human communities face rather than dispatching them, and it has augmented the miasma surrounding a people's recovery from calamity rather than alleviating it. In some instances, the gap has been responsible for creating endless hardship: it has magnified poverty, enabled disenfranchisement, and led to the founding of enduring recovery ghettos. Indeed, it has engendered what those in the risk and disaster field commonly voice: "First there is the disaster; then comes the real disaster."

Addressing the gap between what is known about hazards and disasters and what enters policy and programs would be important enough considering the circumstances that have prevailed more or less consistently around the world until recently. However, the chasm bears particular and snowballing importance now. The sorts and the scope of both hazards and naturally triggered and technological calamities that people face today have proliferated in the past few decades and with harrowing impact. Because there is little sign that this newfangled exigency will lessen rather than further advance, attending to the breach between what experts comprehend about risks and catastrophes and what gets put into guidelines and operations at this time carries great relevance. It not only bears on present alarming happenstance, but also bears on the imminent future (Hoffman 2016a, 8–9; 2016b).

The corpus of erudition about hazards and calamities is not small. Over the past sixty years, scholars have acquired a great deal of knowledge about every kind of cataclysm and the risks leading to them through systematic research across multiple disciplines. Illuminated have been the causes of mishaps, the quagmires of recovery in the short and long runs, and the increasingly frequent displacement of affected people along with their necessitated resettlement. Yet little has been accomplished in terms of risk reduction; the problem has instead turned into risk creation. Nor has much been accomplished to lessen the brunt of disasters when they happen; indeed, their calamitous clout has amplified. In the meantime, on top of the previously garnered realizations, three new and critical understandings concerning risk and disaster have emerged, each with considerable study behind it. One is that the ever-more-frequent disasters of both geophysical and technological origin across the planet along with increasing conditions of vulnerability are being driven by disturbing contemporary economic, political, and social forces. The second is that both the old and the new sets of disaster-driving factors are merging with further aberrant and exacerbating components, including global warming, coastward migration, and urban densification (Hoffman 2016a, 2017). The third is the now almost totally accepted comprehension that there is no such thing as a natural disaster. All catastrophes are human caused at one level or another. Even seismologists, climatologists, and engineers have come to accept this realization. There may be natural triggers to disasters, but it is what humans choose, do, make, alter, or ignore that results in a calamity's occurrence, including those erroneously called "natural" as well as those deemed "technological"—that is, disasters that are the consequence of flawed human manufacture. No matter if the happening is an earthquake, flood, volcanic eruption, cyclone, wildfire, drought, nuclear meltdown, oil spill, or other pollutant, the underlying determinant is social. We ourselves are creating the hazards and the calamities.

Yet, time and again, little cognizance amplifying risk exposure, disaster causation, or how disasters unfold to favorable or unfavorable circumstance seems to get through to today's burgeoning governmental and nongovernmental operatives or doers on the ground. And the word "burgeoning" is hardly accurate. The numbers of governmental agencies from international to national to regional and especially the nongovernmental organizations (NGOs) dealing with hazards and calamities everywhere have mushroomed almost beyond count. The disaster industry, if we can call it such, today ranks among the fastest growing in the world. Although without question many individuals in positions of authority or engaged in on-site operations in dire situations are well-meaning and well-versed, far too many lack familiarity with the basic components and erudition dealing with the problem. Many are political or career-ladder appointees who have little background in hazards and calamities. Some are well-intended neophytes. Others are benign or not-so-benign opportunists: scores of compelling needy people and heaps of alluring money are involved. Admittedly, the informative literature is extensive. It is also somewhat scattered. Still, comprehensive texts and pertinent journals abound, as do many knowledgeable consultants. In consequence, for example, despite the well-known fact that very few aspects concerning one disaster are transferrable to another, what often emerges are ill-suited cookie-cutter approaches and unfitting stratagems derived from such mis-garnered dockets known as "best practices" and "lessons learned." All told, the upshot has been, as White et al. note, we are at a state of "knowing better and losing even more" (2001, 81).

A Growing Alarm

We are not the first to address the dismaying and increasingly dangerous rift between what is known about risk and disaster and what gets into policy and practice. Warnings about the problem have long existed, but have recently spread. In their article "The Disaster-Knowledge Matrix—Reframing and Evaluating the Knowledge Challenges in Disaster Risk Reduction," Spiekermann et al. (2015) also address the difficulties in integrating research-based knowledge into the policies and practices surrounding disaster scenarios. They access what they perceive as the core barriers in the exchange and implementation of knowledge concerning risk and catastrophe and introduce a means to identify factors hindering the conveyance of information. Their inventory somewhat parallels ours, although theirs is more focused on mechanism. Correspondingly, in his chapter in the book *Disaster Research and the Second Environmental Crisis* (2019, 161), James K. Mitchell directly asks, "Why can't we do better?" He notes that most efforts to contravene risk and disaster up to present have focused on four diverse, but not well-amalgamated themes: (1) improving scientific knowledge and technical intervention, (2) instituting legal restraints on unwanted actions, (3) buttressing existing societal arrangements for reducing vulnerabilities, and (4) developing incentives to accomplish mitigation in combination with bringing in underrepresented groups. He proposes a strategy of empowerment based on uniting all participants in a collective endeavor. Wilson (2006) in his article, "Beyond the Technocrat? The Professional Exert in Development Practice," details how the common understandings of professional and governmental roles lead to missing the crucial point of engagements with other actors. He calls for open spaces and a community of practice. David Mosse, a professor of anthropology at the University of London who has also worked as a social development advisor for several NGOs asks in his chapter in *Development and Change* (2004, 639–671) if good policy is impossible to implement. He observes that most agencies are shaped by the exigencies of their organizations and the need for joint associations rather than by enacting efficacious policy. Strategies are further formed to solidify political support, therefore, making the link between research, policy, and action problematic. White and Haughton (2017, 412–419) note how decision makers in both process and practices of hazard management skew their protocols toward current concerns and, in so doing, shape future guidelines in conformity to current circumstance, thereby impeding new and changing input. Long-term considerations accordingly become located in so-called hazard-scapes, in which risks are fixed and difficult for future generations to reverse. They refer to the practice as the "tyranny of the present." Serafini (2017), speaking of the failures of recovery after the Italian earthquakes of 2016 and 2017, reaffirmed that a common cause of the rift lies with many layers of bureaucracy, each charged with a different aspect of risk and disaster. All progress in consequence duly ends up in a complete stalemate. Noting similar paroxysms in what he calls adversarial countries, Kelman (2012) points the blame of not integrating programs to action on the manifold failures of disaster diplomacy. He cites the double edge of dealing on the one hand with power brokers and on the other with the fear of scrutiny, all combined with internal prejudice, misgivings, and mistrust in governing institutions. Finally, Brondo (2015) calls for the much overdue amalgamation of practitioners within germane academic departments as the means of improving understandings and efficacious initiatives.

What Is Addressed in the Following Chapters

By pulling together contributions from individuals who have been deeply involved in policymaking with those who have extensive experience as practitioners as well as with academic researchers, we attempt in this volume to formulate a comprehensive examination of the chasm that exists between what is known about hazards and events, what gets into dictums, and what gets enacted. Our intent is to offer a triptych that reflects all three aspects in relation to the others. Introduced in the chapters are many concerns that are well established and many that are not, among them the factors that drive vulnerability and disaster construction; the frequent efforts of global and national forums to establish guidelines along with their constant revisitation; the tribulations faced by field personnel confronted with critical needs versus roadblocks; the effects of global warming; the complexity of resettlement; gender; the importance of local people's perceptions and ideology, including their chimeras and delusions. A summary of the book's chapters follows. Further description appears in introductions to the three parts.

Part I of the book, entitled "Illuminating the Fissures: Suppositions, Execution, Agendas, Realities, and Execution," is directed toward an exposition of the problematic fissures between knowledge, policy, and practice from the point of view of people who have both shaped programs and tried to enact them.

This part begins with a chapter outlining the scope of the disjunction and stating many of its manifold, and often covet, facets. Although he is an academic, Roberto E. Barrios has nonetheless directly worked in a number of risk, disaster, and recovery situations. His chapter "Unwieldy Disaster: Engaging the Multiple Gaps and Connections That Make Catastrophes" sets forward many of the basic causal components that debar the integration and realization of information disaster experts have amassed. The chapter further shows how anthropological research methods and theories, a theme that we return to in the book's conclusion, helps explain why disasters not only often persist but also end up distending human and economic costs. Barrios stresses that what is required to bridge the chasm is a combination of understanding human behavior as expressed in each particular cultural circumstance, amalgamated to expert knowledge, material agency, governance practice, and very importantly, a measure of imagination. Barrios also presents the philosophical background that separates the various players involved in the disjunction.

The second chapter, "Advocacy and Accomplishment: Contrasting Challenges to Successful Disaster Risk Management," written by Terry Jeggle, a highly experienced and informed international practitioner, takes the examination of the rift directly to the practitioner's dilemma. Despite the many guiding international mandates and conferences that have asked the many factions immersed in the risk and disaster conglomerate to share information, few have executed efforts to minimize the quandary or share their ken. He points out that there exists no acknowledgment within any international charter that indicates a condition or undertaking in one place transfers, or is applicable, to another. Jeggle asserts that competent disaster risk management advances effectively only when both effort and leadership are localized. He further discusses the advocacy aspect of international, national, and local institutions and outlines the history of directives guiding them, including the United Nations' (UN's) International Decade for Natural Disaster Reduction (1990 to 1999), the Yokohama Plan, and the UN's International Strategy for Disaster Reduction (2000 to 2015) and outlines how terminology shapes perceptions, roles, contexts, and agendas.

In "Natural Hazard Events into Disasters: The Gap between Knowledge, Policy, and Practice as It Affects the Built Environment," Stephen Bender, from his original background in architecture to his many years as an international hazard and catastrophe executor, takes a sweeping look at sovereign states, multilateral development banks (MDBs), NGOs, and the international community. He examines how each defines, shapes, and operates within various vectors of the disaster field: risk reduction, risk management, climate change adaptation, emergency assistance, and post-disaster relief, recovery, and reconstruction. The various agencies involved, he maintains, know very well who is vulnerable and why, what can be done, and who will benefit from their policies and practices. What he discloses about them, however, are the covert issues of power, prestige, and funding among and within these organizations. Bender chronicles how certain concerns arise to claim dominance, which they are, and how they eclipse others. As a result, discontinuities, often deliberate, not only result in total downfall but also lead to it.

Adam Koons is an anthropologist who has spent his entire career as an in-the-field relief agent. His work has taken him to countless countries on almost every continent and almost every sort of fateful situation. He has worked in the context of environmental to technological catastrophes, conflict arenas, and in refugee and resettlement camps. Key to his direct and immediate endeavor has been the rights-based approach derived from the Sphere Minimum Standards directives, a protocol shared among such agencies as the UN-led Inter-Agency Standing Committee, the 190-member consortium of United States–based NGOs, and many other international groups. Although axioms exist, in "Humanitarian Response: Ideals Meet Reality," he finds there almost always remains a disparity between what should happen and what does happen. The challenge lies in the interstice between the ideal and the real-time decision making that by exigency ensues in crisis situations. Both axioms and actions bear implications in terms of ethics, politics, sociocultural desires, and ongoing relief operations.

In "Disaster Theory Versus Practice? It Is a Long Rocky Road: A Practitioner's View from the Ground," Jane Murphy Thomas takes the investigation of the knowledge, policy, practice chasm into a detailed description of several actual recovery projects. Within the portrayals, she deciphers why some of the programs succeeded and others did not. Thomas illuminates barriers, describes the many actors involved along with their positions and roles, then tells when and how the players nurture the project or constrain it. The projects take place in Bangladesh, Afghanistan, and Pakistan-administered Kashmir. Each project has a differing overseeing agency. In her exposé Thomas earmarks cultural issues, organizational behavior, matters of expedience, the muddled meaning of the term "expert," and, as with other chapters, power and politics.

Part II of the book, "Situations and Expositions: Plights, Problems, and Quandaries," moves the discussion of the rift away from agent and operative to an exposition of outstanding plights, problems, and quandaries that vex the disaster scenario. Some of particulars (e.g., gender) that this part addresses have long been known, and some are newly compounding (e.g., climate change and the increasing predicament of displacement). The types of calamities cited are both of quick-onset, albeit that is a misnomer—all have long developing chronologies—and those slow in arrival and recognition. Some are unexpected, and some predictable and chronic. Different aspects of the gap are unmasked again in each chapter in this part of the book.

To begin, Shirley J. Fiske and Elizabeth Marino take on the mounting disaster imbroglio of climate change." They point out that enmeshed within climate change are both slow- and quick-onset occurrences and that both sorts of events contribute to the expansion of devastation. Climate change, in contract to other risks and occurrences, brings up distinct, and often political, chasms between erudition, policy, and action because the scholarship itself may be endorsed or denied. The fundamental dilemma, as the authors point out, is that the acuteness of the catastrophe is largely invisible. The onset of the alteration is by and large incremental. Sometimes it is marked by punctuated events, and other times it creeps up in a continuous way. The fracture between knowledge, policy, and any sort of mitigation, therefore, comes down to local definition and acceptance of the situation. Acceptance depends on several sources: insiders, outsiders, region, state, country, and globe. The authors set forth the social construction of climate events such as floods, hurricanes, wildfires, and rising sea levels, even though to the communities they seem to be forces of nature. They detail how climate change calamities in actuality occur, as with other disasters, in historical and socioeconomic contexts of power, social stratification, income, resource, and social network disparities, although once again outsider agencies often little heed the genuine causations.

Brenda D. Phillips addresses the perennial disaster quandary of gender and its role in the disparity between exhortation, instruction, and implementation in her chapter, "Disrupting Gendered Outcomes: Addressing Disaster Vulnerability through Stakeholder Participation." The matter of gender has no boundary in the risk and disaster amalgam. It crops up at every level and in every facet from original hazard to final recovery, if there is such. To say that the elements embroiled within the gender conundrum are myriad, complex, and clamorous is understating the subject. As Phillips indicates, the neglectful and reprehensible treatment of women both leading to and subsequent to a catastrophe appears an intractable scourge. Despite years of recognition and concerted effort, the mistreatment and disparities of gender within the material, legal, economic, political, and ideological realms of disaster continue. In her all-encompassing survey, Phillips illuminates the totality of predicaments and, in so doing, unveils the pervasive schisms that remain largely unabated between what is known about gender within risk and catastrophe and what does, and mostly does not, happen. The chapter brings up a number of global situations and sets forth what achievements have taken place. She includes a number of new contributions to the topic, including the recent inclusion of men and the predicaments they endure.

As risk situations and disaster impacts burgeon across an increasingly populated globe, the displacement of people and the need for the resettlement has escalated. The question is, Where can people go as land disappears and perils loom? What happens when whole communities or ethnic groups want to move together as one, and not as individuals? Anthony Oliver-Smith who has worked more than forty years in the two arenas implicated in the mushrooming predicament, disaster and development and forced displacement, addresses the growing quandary in his chapter, "Resettlement for Disaster Risk Reduction: Global Knowledge, Local Application." As the quagmire widens, so does the resistance on the part of those who must move and those who must accept newcomers. The gap between what is, in essence, an unchartered situation and old policies and solutions looms especially large. Oliver-Smith's chapter reviews the history of resettlement, then examines the contradictory confluence of environmental disruption and Disaster Risk Reduction (DRR). He lays out the construction of global opinion about the matter, reports on progress and problematic outcomes, and illustrates with germane case studies.

Ryo Morimoto's chapter, "From Nuclear Things to Things Nuclear: Minding the Gap at the Knowledge-Policy-Practice Nexus in Post-Fallout Fukushima," zeros in on another contemporary development, in this case a peril that has sprung up only in recent decades but carries with it annihilating ramifications. His focus is specifically on the latest incident in what has, unfortunately, become a litany of happenings. In his chapter, Morimoto resurrects two older phrases that he finds more applicable than ever. One is "missing expertise" (Rajan 2002), and the other is "a new species of trouble" (Erikson 1994). Morimoto's particular concern is the contamination of and, especially, the vexation of decontamination in a land where the event of contamination itself violated cultural code. He argues that the gap surrounding this peril does not arise from unsuccessful coordination of knowledge, policy, and practice nor from lack of common language or clear communication. The rift lies, he claims, beyond the reach of the simple knowledge-policy-practice collaborations. It is ethnography, the key method of anthropology, he argues, that reveals what decontamination consists of for the locals undergoing the interminable catastrophe of the Fukushima meltdown. He reminds us again of another implicit understanding about disasters: The distribution of risk and vulnerability in society is uneven.

Mark Schuller's chapter, "'Haitians Need to Be Patient': Notes on Policy Advocacy in Washington Following Haiti's Earthquake," sheds light on yet another modern occurrence in what Morimoto in his chapter calls the nexus of disaster and its growing place in the disparity. That occurrence is the rise of today's clamorous advocacy. Schuller's discourse centers on Haiti, a country on the western part of the island of Hispaniola. As a nation, Haiti has become almost eponymous for every predicament noted in this book. The event Schuller details is the devastating Haitian earthquake of 2010 and its grievous continuing post-disaster recovery.

Through his personal participation in a solidarity effort championed by a number of local NGOs directly dealing with the plight of the islanders, he enumerates the actions, obstacles, and frustrations of the advocacy effort as it progresses all the way from its home site to Washington, DC. He describes the mandates of the nonexpert, non-knowledgeable politicians who nevertheless hold sway over programs and funding and tells how the money goes to military and for-profit firms he decries as Beltway bandits for their great influence over protocols, contracts, and aid distribution. He recounts the process and players, the official representatives, the lobbying, and tells of the formation of helpful support groups. His revelatory chronicle discloses the roles played by language difficulties, socioeconomic status, and trust.

Part III of the book, "Revamping Apparatus and Outcome," turns to whether solutions or perspectives exist that might offer an integrative bridge between accumulated knowledge about risk reduction and calamity and the tangled web that has so often led to their suppression.

Susanna M. Hoffman's chapter, "The Scope and Importance of Anthropology and Its Core Concept of Culture in Closing the Disaster Knowledge to Policy and Practice Gap," circles back to Barrios's initial presentation. She advances that anthropology's deep cultural perspective, and with it the inclusion of local, indigenous proficiency, can operate as a mechanism for consolidating scholarly information, externally imposed guidelines, and, ultimately, the production of effective aid. She proposes, as increasingly have others, that integrating a people's own mastery and appreciations along with the other contributing vectors achieves better outcomes in reducing vulnerability and accomplishing restoration. Anthropology's frame of reference incorporates a people's long-garnered understanding of their surroundings along with the ways they have traditionally managed upheavals. It incorporates what they perceive as dangers, and what they want outcomes to be. Their perceptions and desires are not necessarily the same as outsiders'. More importantly, it includes understandings of complex, many-layered, guiding customs a people may share, the disregard of which has caused many risk reduction and disaster recovery programs to flounder. How is space perceived? Who has the actual prescriptive authority to stay or go? How is property inherited and what does its legacy mean? Hoffman recounts a host of details that impact efforts but often are neglected. It is the lack of fusing deep culture, indeed allowing such customs to dictate, that has commonly led to not only to one disaster but also to disaster upon disaster.

After first noting how anthropologists today routinely chronicle the human impact of disasters, the chapter coauthored by Katherine E. Browne, Elizabeth Marino, Heather Lazrus, and Keely Maxwell directly specifies the misalignments between institutions offering aid and communities receiving it and makes precisely and squarely explicit "what is known" about disasters from anthropology's most critical insights. In their chapter, "Engaged: Applying the Anthropology of Disaster to Practitioner Settings and Policy Creation," they earmark the obstacles faced in enabling practitioners and policies to recognize gained knowledge and offer, as no other presentation in this book does, a point-by-point set of recommendations for integrating risk and disaster knowledge into policy and practice. To do this, they ask three questions: (1) What is expressly and currently known about the causes of disaster, reducing impact, and managing impacted communities when calamities occur? (2) What in detail are departures that separate academic work on disasters and practitioner work? (3) And finally, what suggestions can be offered to span and eliminate the chasm? The authors come from a broad spectrum of practicing and academic anthropologists. They cover their topics in bullet point clarity, discussing such points as convergence of outsiders, shunning self-help, the drawbacks of privatization, the flaws of an extraordinary versus normal perspective, divergent measures of success, language and framing use in order to convey advice and more, and conclude with citing instances where academic and practitioner approaches are united.

The final chapter of the volume turns the discourse toward the future and reminds the reader that the time ahead matters. In her chapter "Future Matter Matters: Disasters as a (Potential) Vehicle for Social Change—It Is About Time," Ann Bergman asks if risk and disasters themselves can provide the vehicle for social change. In giving her chapter the subtitle "It Is About Time," she presents a double entendre. The problems of risk and disaster have gone on too long and their impacts even now influence what is coming. Bergman delves into a discussion of utopia versus dystopia in the context of past and looming calamity. She muses about whether dangers and disasters are the new normal, and if so, whether they have agency and provide opportunity. In all these questions, she directs the reader toward an understanding of sustainability.

Further Factors

While together the chapters in *Disaster Upon Disaster* cover a far-reaching panoply of factors involved in the chasm between risk and disaster knowledge, policy, and practice, still more factors exist. Some of them are crucial enough to bear mention.

The first among them straddles a double line, sometimes overt but mostly covert, often said as a facetious characterization but largely deeply believed. It is the seemingly implacable judgement among many policymakers and practitioners that any insight emanating from the academic or scholarly community is the product of someone "airy-fairy" or "ivory tower" and is, therefore, without merit. One colleague practitioner deemed the it "the wall of scorn." Those adhering to the attitude treat academics as if they have never faced "reality" or, at least, the reality that policymakers and practitioners must deal with. Therefore, they know nothing about the nitty-gritty of hazard mitigation or disaster entanglement. Such a posture ignores that, in fact, most academics in the field of risk and disaster engage in situ in assessing actual hazardous situations and have participated in post-event tumult. Some have even been victims (Hoffman 1999). Their research by its very nature takes them to such settings. Those who study risk and disaster have as well typically looked at countless cases of vulnerability and calamity, enough to see the existence of overarching patterns that augment understanding and pinpoint unique distinctions. They can see beyond a singular crisis to the whole collective. Many consult with a wide variety of diverse organizations and speak at conferences attended by all sorts of personages, including governing officials, heads of agencies, and other experts, all of which gives them particular ability in potentially closing the gap between various factions.

There further abides a widespread and unfortunately persistent assumption among many engaged in the risk and disaster enterprise that the only solution to hazards and calamity lies with physical solutions, as in building levees, heightening tsunami walls, and thinning forests. Unfortunately, the public has been long inculcated in the same belief. As a consequence, most funding and authorized programs go toward tangible fixes, not social ones. That the fundamental cause of risk and disaster is exposure and requires social remedies, goes ignored. As a consequence, physical scientists and engineers, and not social scientists, are given primary, and frequently singular, consideration in addressing any malady. As the first several chapters in the book makes clear, authorities also tend to favor economic interests, such as development, tourism, and industry, over matters of mitigation, although these same priorities are themselves leading causes of the burgeoning disaster expansion.

To cap off the conundrum, policymakers and agencies also often do not see the entire discipline of hazard and disaster research as being all that credible. In this, the fault also lies with the scholars of the field. Hazard and disaster knowledge is scattered among a number of disciplines and researchers from the various bailiwicks have yet to coalesce their topics into a single specialty. While some experts attempt to integrate their subject matter with others, some do not. Similarly, the few universities that house disaster centers and give degrees in the subject again by and large advance only the one vector, generally the social one, and neglect the others. Within the fostering universities, as well, the subject of catastrophe is still considered to be marginal and is treated as such. It is also true that the study of risk and disaster itself has yet to develop the three criteria that would establish it as a recognized and accredited field. The subject lacks a unifying set of theories. Although it rightly embraces a diverse set of approaches, each is rather territorially espoused and advanced by a particular discipline, with little, albeit growing, crossover. In addition, while the body of literature necessary to give credence to a field is rapidly growing, it, too, lacks integration. Lastly, over the unfolding of the field, the academics involved have switched the focus of their study in a seemingly erratic manner. Concern has peregrinated from events to recoveries, victims to survivors, extent risks to risk reduction, risk reduction to risk construction, vulnerability to exposure, mitigation to resilience. Some scholars currently even eschew the word "disaster" and admonish others not to use it. The wavering theme and parade of mutating vocabulary have rendered scholars flighty to policymakers and aid establishments and have implied a feckless nature to their knowledge.

Rarely has yet a further contributor to the rift between knowledge, policy, and practice caught much attention, that is, the taking into account certain veiled aspects within the realm of policy and practice. Without a doubt true kindness exists among many disaster management and relief administrations and organizations, and certainly among their staff, but what is often not acknowledged is that the aura of solicitude they frequently procure and the actions they take under that posture involve hidden considerations that work to obstruct outside input. In almost every hazard or disaster situation, one or two organizations and/or practitioners emerge to attain, and then continue to bear, the designation of holding particular sympathy toward the suffering. What is often overlooked is that frequently when such an organization acquires veneration, at the same time, it garners power. Once achieving the esteem of exemplary compassion, the organization then readily gains determination of programs and protocols (Barrios 2017). That, in turn, creates the phenomena of drawing sympathy back to itself, redoubling its cull of rewards, notoriety, money, continuity, and, again, power. The establishment that gains the mantel of sympathy, and with the funds and repute achieved, is often able to create monuments to itself, construct and name buildings bearing its name, and propagate legends lauding its magnanimity. Frequently, it further acquires official, or semi-official, status as chief among organizations, thereby diminishing the import and thrust of other entities. When a particular agency gains the sway of sympathy, the disjunction between expert knowledge and the agency's practices rigidifies. Agencies with the power of sympathy tend to discount expert opinion as being extraneous to their proven success. They also more readily adopt cookie-cutter and best-practices platforms despite their having little or no relevance. Since their prestige as well often allows them access to many locales and situations, they become overextended and find easy answers more adroit. Dominating governmental or private establishments also secure influence over the framing of disaster: what happened, how long it lasts, and what constitutes actual injury. Often they promote dogmas of progress and betterment to justify their actions. It should be noted, however, that in many cases the victims also cultivate and wield the dynamics of sympathy.

In addition, it warrants mentioning that most authorities and agencies direct their policies and programs to what they deem as "communities." Yet the very use of the term "community" can worsen the gap in integrating scholarly expertise and especially in seeing that expert insight reaches an entire populace. "Community" is a word that has sliding definitions. In many cases community is more concept than fact. The word may or may not apply to a composite of survivors or vulnerable and, even if to some degree applicable, it may or may not include all relevant persons. Unquestioned, as it usually is, the term, and the assumption it implies, can connote both a broad sweep of inclusiveness and all-round effectiveness. Most often, in truth, it means the program reached some of the germane persons, such as its minor functionaries or ad hoc groups, but in truth did not make inroads to reach the entire body. In many instances, the use of the term "community" may only provide the persuasive wording of a funding proposal. Most anthropologists as a rule have dealt with authentic communities, where interaction among those involved takes place on a constant and continuous basis. But in most situations today, certainly those of major risk and calamity, community operates perhaps more like a metaphor borrowed from physics; the potential of collectivity into an actual community among a group of people may be there but is largely underlying. The fact of community becomes kinetic only when it is triggered by something such as a catastrophe, and probably last but for a short while. The thought that people compose a community may, thus, be a false instrument of policy and programs and leave the meaningful spread of knowledge and information sidelined in its wake.

Dismayingly, most programs, be they by government or other agency, still also revolve around relief as opposed to risk reduction or creation. The concern with relief, as opposed to prevention, is deeply rooted in history (Dauber 2012) and, despite knowledge particularly concerning risk and exposure, continues to affect the focus and ideology of most authorities and agencies. The notion that the vulnerable are responsible for their own condition is also deeply rooted and rebuffs considerable wisdom to the contrary. Both antiquated notions are thankfully undergoing a slow but steady reassessment via the current popularity of resiliency studies.

Policymakers and practitioners, and for that matter scholars, may also experience program amnesia, but it may not be deliberate. Over time most agencies, officials, and practitioners undergo a great deal of turn-over, taking along with it the chronicle of dangers and dilemmas. The forgetting, however, may also be purposeful, such that faulty programs are used again to promote hidden agendas or because they reflect inertia. Between policymakers, researchers, and locals, there also occurs issues of translation along with other failures of communication. Each separate entity in the complex of risk and disaster, might use different terminology or interpret words or intent differently. In fact, failures of communica-tion are endemic to the entire conundrum. Again, these may be guileless snags, but they may also occur from calculated mishap.

One final point: Disaster agencies in their policies, along with practi-tioners and their actions and academics in their analyses, have tended to be decidedly Western-centric in all their considerations concerning risk and disaster. The orientation is somewhat ingrained because, at the bottom line, the very ideas of risk, recovery, and resilience are themselves Western notions. In every situation concerning surrounding hazards, pro-grams to be implemented, and on-the-ground practice, whether these take place among a Western society people or non-Western, all persons involved need to assess the circumstance in a culturally relative and local manner. That includes what the people assess as a risk, define as a disas-ter, and how they calculate what constitutes recovery. Still more crucial is the inclusion of what knowledge the people themselves have about their surroundings and their perils before assaying, enacting, or installing any judgments, programs, or protocols.

Currently, the gap between risk and disaster knowledge and what enters the directives of policy and the actions of practitioners pervasively continues. With the explorations as to why detailed in this volume, along with the increasing insights of others addressing the issue, the hope of an integrated endeavor moves nearer.

Susanna M. Hoffman (Ph.D., Anthropology, UC Berkeley) is author, coauthor, and editor of twelve books, including *The Angry Earth*, first and second edition, and *Catastrophe and Culture*, two ethnographic films, and more than forty articles. She initiated the Risk and Disaster Thematic Interest Group for the Society for Applied Anthropology and is founder and chair of the Risk and Disaster Commission for the International Union of Anthropology and Ethnographic Sciences. She was the first recipient of the Aegean Initiative Fulbright concerning the Greek and Turkish earthquakes, and helped write the UN Statement on Women and Disasters.

Note

1. Epigraph: Pajo and Powers 2017, 11.

References

Barrios, R. 2017. *Governing Affect: Neoliberalism and Disaster Reconstruction.* Lincoln: University of Nebraska Press.

Brondo, K. 2015. "Understanding Practitioner-Academic Department Relationships." *Anthropology News* 56, no. 5–6, 29.

Dauber, M. 2012. *The Sympathetic State: Disaster Relief and the Origins of the American Welfare State.* Chicago: University of Chicago Press.

Erikson, K. 1994. *A New Species of Trouble: The Human Experience of Modern Disasters.* New York: Norton.

Hoffman, S. 1999. "The Worst of Times, the Best of Times: Toward a Model of Cultural Response to Disaster," *The Angry Earth: Disaster in Anthropological Perspective,* ed. A. Oliver-Smith and S. Hoffman, 134–155. New York: Routledge.

———. 2016a. "Continuity and Change in the Applied Anthropology of Risk, Hazard, and Disaster." *Annals of Anthropological Practice,* special issue edited by A. J. Fass, 40, no. 1, 39–51.

———. 2016b. "The Clash Within: When Gaps and Conflicts Occur Within the Greater Culture." Paper delivered at Society for Applied Anthropology, Vancouver, 30 March–3 April.

———. 2017. "Disasters and Their Impact: A Fundamental Feature of Environment." In *Handbook of Environmental Anthropology,* edited by H. Kopnina and E. Shoreman-Ouimet, 193–205. London: Routledge.

Kelman, I. 2012. "The Many Failures of Disaster Diplomacy." *Natural Hazards Observer* 37, no. 1, 1–15.

Mitchell, J. 2019. "Growing the Constituency: A 21st Century Challenge." In *Disaster Research and the Second Environmental Crisis: Hazard, Disaster, and the Challenges Ahead,* edited by J. Kendra, S. Gabriel Knowles, and T. Wachtendorf, 161–188. New York: Springer.

Mosse, D. 2004. "Is Good Policy Unimplementable? Reflections on the Ethnography of Air Policy and Practice." *Development and Change* 35, no. 4, 639–671.

Pajo, J., and T. Powers. 2017. "The Anthropology of Policy Emerges." *Anthropology News* 58, no. 4, 11.

Rajan, S. 2001. "Missing Expertise, Categorical Politics, and Chronic Disasters: The Case of Bhopal." In *Catastrophe and Culture: The Anthropology of Disaster,* edited by S. Hoffman and A. Oliver-Smith, 237–260. Santa Fe: School of American Research.

Serafini, M. 2017. "No Time for Rules: Anarchy and Entrepreneurship in Post-Earthquake Italy." Presented at the American Anthropology Association Meeting, Washington, DC, 29 November–3 December.

Spiekermann, R., S. Kienberger, J. Norton, F. Friones, and J. Weichselgartner. 2015. "The Disaster-Knowledge Matrix—Reframing and Evaluating the Knowledge Challenges in Disaster Risk Reduction." *International Journal of Disaster Risk Reduction* 13, 96–108.

White, G., R. W. Kates, and I. Burton. 2001. "Knowing Better and Losing Even More: The Use of knowledge in Hazards Management." *Global Environmental Change Part B: Environmental Hazards* 3, no. 3/4, 81–92.

White, I., and G. Haughton. 2017. "Risky Times: Hazard Management and the Tyranny of the Present." *International Journal of Disaster Risk Reduction* 22, no. 412–19.

Wilson, G. 2006. "Beyond the Technocrat? The Professional Expert in Developmental Practice." *Development and Change* 37, no. 3, 501–523.

Part I

ILLUMINATING THE FISSURES

Suppositions, Realities, Agendas, and Execution

The chapters in part I provide a broad overview of the field of disaster risk reduction (DRR) and disaster risk management (DRM) at a variety of scales, from the local to the global, and in a variety of organizational units, from state governments to intergovernmental organizations and non-governmental organizations (NGOs). Terry Jeggle (chapter 2) and Steve Bender (chapter 3) bring a combined eight decades of experience to bear on their description of the recent history of DRR and DRM. Readers looking for a primer on the institutional landscape of DRM will find these two chapters particularly illuminating. Jeggle focuses on the intergovernmental sector and the ways its involvement in DRM has evolved during the past two and a half decades. In a complementary fashion, Bender provides an overview of the ways state governments, multilateral development banks, NGOs, and development and disaster reduction organizations contribute (or do not contribute) to disaster reduction. The core messages of Jeggle's and Bender's chapters can be summarized thus: First, development organizations that transform environments in such a way that they enhance disaster risk are not being held accountable or responsible for the conditions of vulnerability they create. Second, the existence of the knowledge, policy, and practice necessary to reduce disaster risk is not sufficient to actually reduce vulnerability or disaster incidence. What matters is how these three domains of DRM are connected and integrated by human actors—especially those in positions of sociopolitical and institutional influence. Third, despite much lip service given to the cause of DRM, actual commitment of resources to agencies directly involved in DRR remains to be seen in many parts of the world. DRM concerns remain a reactionary trend—that is, DRM concerns only come to the foreground after catastrophic events and are not therefore a preoccupation of everyday governance.

Complementing Jeggle and Bender, Roberto E. Barrios provides an ethnographically detailed account of why and how disasters take shape and magnitude despite the presence of governmental investments in DRM. Using the case of the 2013 disaster in Mexico triggered by Hurricanes Ingrid and Manuel, Barrios gives us a detailed illustration of the themes discussed more abstractly in chapters 2 and 3. Barrios's piece also functions as a link between the three introductory chapters and the section's final contributions by Adam Koons and Jane Murphy Thomas. Barrios's main point is that disasters are engendered as much by the disconnections between DRR resources, political officials, governmental agencies, development organizations and contractors, and disaster-vulnerable populations as they are by the particular ways these various actors are connected in a catastrophe-affected society. The analytical task of the researcher and policymaker interested in disaster reduction is to identify and describe both these disconnections and connections in order to focus on the most effective sites of intervention. In the case of Mexico, we see a country where resources that are meant for DRM are diverted by political officials, and where relationships between government agents and development contractors prioritize short-term financial gain over disaster prevention. The case of Mexico clearly illustrates Bender's and Jeggle's observation that having all the pieces necessary for effective DRM does not guarantee DRR. What matters is the ways human actors connect, or refuse to connect, these various elements of disaster mitigation.

While Bender's and Jeggle's chapters focus on issues of management and prevention, chapters 4 and 5 complete our review of the challenges of integrating disaster knowledge and practice by looking at the emergency response and long-term reconstruction phases of disasters. The core message communicated by both Koons and Thomas is that disaster response and reconstruction experts—who are often not a part of disaster-affected populations or intimately acculturated into local mores—must remain flexible and capable of negotiating standards, procedures, and project objectives with the culturally complex populations they set out to assist. Failure to negotiate and adapt recovery aid packages—whether they be immediate life-sustaining assistance or long-term aid like the construction of a hospital—with disaster-affected people (who have priorities that are nuanced by locality-contingent livelihood pressures) guarantees the failure of relief efforts. In the case of relief organizations charged with the task of delivering life-sustaining aid in the immediate aftermath of catastrophes, Koons explains how the knowledge of disaster relief experts is summarized in a collection of humanitarian aid standards that have been compiled and revised over the course of decades of work by countless committed agency workers. This knowledge, Koons warns us, cannot be rigidly applied but must be adapted and negotiated with local priorities, a negotiation that requires perspicacity and flexibility on the part of aid workers. In a similar fashion, Thomas presents us with a number of cases in which multi-million-dollar disaster mitigation projects either failed due to lack of negotiation between NGO experts and local leaders or became successful when social scientists acted as intermediaries between external international institutions and disaster-affected populations.

CHAPTER 1

Unwieldy Disasters

Engaging the Multiple Gaps and Connections That Make Catastrophes

ROBERTO E. BARRIOS

Introduction

During the first half of the twentieth century, emergency management authorities, governmental officials, and the public predominantly viewed disasters as unavoidable effects of natural hazards and technological malfunctions (Hoffman and Oliver-Smith 1999). Since the 1970s, a number of critical geographers, historians and anthropologists have proposed a different view of catastrophes (O'Keefe, Westgate, and Wisner 1976). These scholars have demonstrated that catastrophes are by no means natural events but are, instead, long-unfolding processes engendered by human practices that enhance the destructive and socially disruptive capacities of geophysical phenomena and technological malfunctions (Bankoff and Hilhorst 2004; Hoffman and Oliver-Smith 1999; Maskrey 1993). The recognition of disasters as socioenvironmental and socio-technological processes enabled a dramatic rethinking of the concept of "natural disasters." Use of the term "natural" suggested that the mitigation of calamity was fully out of people's control because "nature" was considered to stand in polar opposition to, and outside of, "society" and "culture" in established ways of thinking about disasters. The vulnerability approach, as this new perspective came to be known, presented a great challenge to policymakers and disaster risk managers: If disasters were not solely natural, and if they took form and magnitude in relation to human practices, then they could be prevented.

During the first two decades of the twenty-first century, national governments, intergovernmental organizations, and nongovernmental organizations (NGOs) seem to have made great strides in their capacity to respond to and manage catastrophes, but recent reports suggest these institutions have not been as successful at reducing the human practices that create disaster vulnerability. The Hyogo Framework for Action 2005–2015 (HFA), for example, was an international agreement signed by 168 of 191 member states of the United Nations (UN) designed to reduce disaster losses, particularly in terms of human lives and the destruction of socioeconomic and environmental resources at a global scale (*Global Assessment Report on Disaster Risk Reduction* [GAR] 2015). The creation of the Hyogo Framework was accompanied by the establishment of a series of reports drafted by the United Nations International Office for Disaster Risk Reduction (UNISDR) known as the GAR, which were meant to track the progress achieved under the international agreement. In 2015, the milestone year that marked the completion of the Hyogo Framework's timeframe, the GAR reported,

> The expected outcome of the HFA has only been partially achieved. Twenty-five years after UN Member States adopted the International Decade for Natural Disaster Reduction (IDNDR) and ten years after the adoption of the HFA, global disaster risk has not been reduced significantly. While improvements in disaster management have led to dramatic reductions in mortality in some countries, economic losses are now reaching an average of U.S.$250 billion to U.S.$300 billion each year. . . . More critically, both the mortality and economic loss associated with extensive risks in low and middle-income countries are trending up. (UNISDR 2015, xiv)

In this chapter I explore how anthropological research methods and theories help us explain why, given that social scientists have made great strides in identifying how disasters take shape and magnitude, disasters seem still to not only be persisting but also actually increasing in their human and economic cost in some parts of the world. More specifically, I want to show how explaining the persistence of disasters requires a combination of anthropological interests in human behavior, the imagination, expert knowledge, material agency, and governance practice. A critical reader may object that this is an overly ambitious objective, and that such a broad focus runs the risk of leading to platitudes that explain very little. Nevertheless, I will argue that the nature of disasters requires just such a combination of analytical interests, and that such is the challenge of understanding why disasters persist and why gaps between knowledge, policy, and practice exist.

Quite a few years ago, Anthony Oliver-Smith (1999) commented that disasters challenge a neat separation between nature and culture because they take form and magnitude at the nexus of dialectical relationships between human practice and the materiality of the world (which often reacts to people's actions in unexpected ways; see Pickering 2008). I have always found Oliver-Smith's words to resonate with the ideas of another renowned scholar of human-material relations, Bruno Latour (1993), who has argued that modern ways of thinking are those that claim to separate what is natural, material, objective, and biological from what is cultural, social, political, subjective, and meaning-laden. Hence, modernist ways of engaging, governing, making knowledge, and thinking about the world are those that create divisions between the physical or hard sciences and the social sciences and humanities. They are also those that separate knowledge into neatly and narrowly focused disciplines like physics, engineering, biology, political science, sociology, and anthropology.

At the same time, Latour points out, the world in action is not one that observes those differences, and it is one where what is meaningful and cultural and what is material and objective are intimately intertwined and co-constituted. Analyses that explain disasters as effects of a root cause that lies strictly in the objective material (i.e., where disaster is reduced to a natural hazard) or in the sociocultural worlds (i.e., social constructivist approaches) are flawed because they miss the inseparability of things like meaning and matter, society and environment, and politics and scientific knowledge. The relevance of Latour's observations toward analyzing and narrowing the gap between academic knowledge on disasters and policy and practice is that disasters take form and magnitude through inseparable co-constitutive relationships between subjects we are accustomed to separating into different fields of inquiry (e.g., semiology studied by humanists, politics studied by political scientists, environment studied by biologists and ecologists, the material world studied by geophysicists), when what we need is a transdisciplinary approach that cuts across these artificial differences. It is also worth noting that the institutions designed to reduce disaster incidence are designed in ways that sustain the above-mentioned separations between fields of inquiry and practice, creating gaps between policy, various forms of knowledge (e.g., landscape engineering, geology, anthropology, political science), and political culture and jurisdiction.

To illustrate these points, I use several examples from my own research on hydro-meteorologically triggered disasters in Mexico in the context of the global capitalist political economy—sometimes referred to as globalization, sometimes as neoliberalism—and climate change. Furthermore, I also detail how the persistence of disasters is not only the effect of gaps between knowledge, policy, and practice, but also the result of connections and intimacies (some unspoken, some tacitly accepted) between the factors and agents that shape disasters. Disaster mitigation then becomes a matter not only of identifying gaps, but also of recognizing and naming ties between those elements that are assumed to remain separate in modern forms of governance and thought, yet remain closely related in human-environment relations.

Finally, I conclude by reviewing how addressing the gaps and connections between materiality, politics, and disaster risk knowledge requires us to reconsider the priority given toward fiscal and capitalist indicators as measures of social well-being and development. Disaster mitigation requires us to critically examine the environmental impact of a global political economy driven by commodity production and consumption. In addition, it requires us to foreground a concern with environmental justice and land use planning that earnestly addresses the livelihood and housing necessities of those most impacted by disasters on locally relevant terms. Taking on these issues, in turn, demands a reimagining of governance and development in a way that challenges socio-evolutionary teleologies rooted in capitalist or neoliberal visions of so-called progress and addresses the postcolonial politics that enhance disaster risk in Mexico, among other places.

Gaps between Scientific Knowledge and Governance

On 15 September 2013, I found myself in El Zócalo, the central plaza of Mexico City, with a population of 35 million. El Zócalo is a 700-year-old space of statecraft that precedes the colonization of Mesoamerica, and is where, every year during El Grito, the country's state-sponsored celebration of national independence, Mexicans celebrate Mexican statehood. In 2013 President Enrique Peña Nieto of the PRI (Institutional Revolutionary Party) found himself fending off accusations of electoral fraud and decreasing support in public opinion polls. To counteract these trends and informal indictments, the city government invested close to a million Mexican pesos to hire musical ensembles for the entertainment of crowds, while the PRI party spent an undetermined amount to bus in spectators from rural areas to fill in for those city residents who chose to boycott the celebrations downtown.

At El Zócalo, police clad in riot gear controlled the bused-in attendees who had been promised 150 pesos, free transportation, and a spectacle of popular music, military parades, and fireworks. Over my head, the huge national flag that regularly adorns the central plaza waved under a strong southeasterly wind, its colossal undulations beneath rapidly moving clouds creating an impressive effect. The winds that animated the massive flag were the remnants of Hurricane Manuel, a category one storm making its way across the country's Pacific Coast. Within a twenty-four-hour period, Manuel would coincide with rains and winds precipitated by another similar phenomenon, Hurricane Ingrid, which was saturating the country at the same time from its Gulf of Mexico coast.

The storms interacted with centuries-old land-use patterns, settlement, political structures, and social inequity-making practices, with the effect of causing landslides and floods in rural and urban areas in the states of Guerrero, Veracruz, and Oaxaca. At least two hundred people were killed, an estimated sixty thousand were displaced, and 50,120 homes were destroyed; all told the storms caused 58.5 billion Mexican pesos ($4.64 billion) in damage (Beven 2014; Pasch and Zelinsky 2014). The 2013 disaster affected the affluent, middle class, tourists, the urban poor, and marginalized rural communities in varying ways. The highway connecting Mexico City and the Pacific Coast resort of Acapulco washed away at various locations, cutting off communication between the two areas. In Acapulco itself middle-class housing developments sponsored by state agencies slipped under water, demonstrating a failure at the intersection of urban planning, development, and governmental flood risk assessment programs. Torrential rains also caused flash floods in the city, trapping up to forty thousand tourists in their hotel rooms, requiring extraordinary efforts on the part of state authorities to coordinate the airborne evacuation of vacationers via the local airport, and bringing the tourism industry to a standstill. In the rural areas of the State of Guerrero, economically marginalized and predominantly indigenous communities had a dramatically different experience. In the town of La Pintada, a landslide claimed sixty lives and buried a substantial part of the settlement.

In the immediate aftermath of the storm, several government officials were quick to invoke climate change as a primary culprit behind the disaster. The National Secretariat of Environment and Natural Resources issued a press release confirming that Hurricanes Ingrid and Manuel were phenomena associated with global warming, which they said promised to generate more extreme events like severe storms and droughts. The press release referenced the fifth report of the Intergovernmental Panel on Climate Change (IPCC), claiming that oceanic waters surrounding Mexico have experienced a rise of 0.44°C in their first seventy-five meters of depth, which the agency pointed to as a primary driver in the formation of tropical storms and hurricanes (Ponce 2013). On a similar note, Maria Amparo Martinez, the director general of the National Institute of Ecology and Climate Change (INECC) noted that the seawaters surrounding Mexico are now referred to as *la alberca caliente* (the warm pool) because of its temperatures that average 26–27°C (Ponce 2013).

The attribution of the Ingrid-Manuel disaster to climate change, however, met with criticisms on the part of the Mexican social scientists who study disasters. They viewed the statement as a rhetorical strategy to represent the catastrophe as an event whose mitigation is out of the capacity of Mexican state agencies. As buttress, press releases pronounced climate change as an anthropogenic phenomenon that was engendered by development practices that occurred beyond the country's borders and was, therefore, out of state agencies' control. Furthermore, the attribution of Manuel and Ingrid's catastrophic effects to climate change obscured several gaps and connections between scientific knowledge, governance, development practices, and material agency that were also implicated, and perhaps even more directly, in the 2013 catastrophe.

First among the gaps was a lack of coordination between existing risk maps used by Mexico's National Center for Disaster Prevention (CENAPRED) and urban development initiatives. In 2001, for example, Acapulco's municipal government purposely modified the city's Urban Development Plans to authorize construction in areas identified by CENAPRED as high flood risk and not suitable for settlement. Such was also the case for the housing developments in Acapulco's peri-urban area known as Laguna de Tres Palos. This part of the urban landscape was, in fact, heavily affected during the 2013 floods, which destroyed government-subsidized housing created for middle-class families (Nájar 2013).

According to Mexico's federal secretary of agrarian, territorial, and urban development, Acapulco city officials had the legal and geographic instruments necessary to regulate urbanization and prevent much of the flooding disaster. Among these instruments were documents like the Urban Development Plan and CENAPRED's National Risk Atlas, but these items alone did not impede the construction of either irregular or government-sanctioned settlements in flood-prone zones. Political and economic interests, it turned out, could override existing knowledge about disaster risk in the shaping of Acapulco's urbanization. In the words of Secretary Jorge Carlos Ramirez Marin: "It's a story of legalized disaster, a story that legalizes a lack of regulation, for Guerrero has, since 1998, a Plan of Urban Development, which they modified in 2001. What was first an obligation became optional. They used their authority, and a law that prohibited construction was changed because it would only be accepted when it was of obvious [financial] benefit to the metropolis" (Muñoz 2013, 1).

The invocation of climate change by some government officials as the cause of the Ingrid and Manuel catastrophe, then, created a discursive gap that isolated climate as the key element in creating the 2013 disaster and downplayed the role of land use, development, and political practices at the local level. This is what anthropologists of disaster often call the representation of disaster as "natural" hazard—that is, the objective material world as the maker of disaster. However, in Mexico disasters also feature a key political-economic dimension that exceeds the mitigating efficacy of technological and scientific knowledge alone.

It is noteworthy that the gap between disaster risk knowledge and land use in Acapulco featured a flipside of connections and intimacies between political culture, capitalist tourism development, and the local environment's material agency that shaped the 2013 disaster. CENAPRED is charged with the task of drafting the risk maps and issuing warnings when geophysical hazards present themselves, but the making of development decisions that impact land use and engender vulnerability lie outside of its jurisdiction. This power, instead, is granted to local municipal presidents and local governments, where elected officials, as in the case of Acapulco, are left to choose freely between short-term—and often personal—economic gain and long-term disaster prevention. This is a classic example of those global development trends identified in the GAR. In the making of decisions like those made by Acapulco city officials, political culture, development initiatives, and material agency connect and become mutually intertwined in such a way that they engender a disaster.

What is also of interest here is that there is nothing natural or inevitable about such gaps and connections; instead, people create them in the act of parceling out jurisdiction to different governmental agencies at varying levels of national, state, and municipal government in a manner that produces vulnerability through development projects. Meanwhile, disasters take form and magnitude across these boundaries as a result of the ways people engage local environments through development practices. This is what Latour (1993) would call hybrid networks or collectives that implode separations between nature and culture, objects and subjects, meaning and matter, and science and politics.

Gaps in Disaster Risk Knowledge Making

The risk maps commissioned by CENAPRED and compiled in the Mexican National Risk Atlas are credited with the capacity to identify areas susceptible to catastrophes when hydro-meteorological hazards present themselves. But here lies another gap in disaster reduction: Although funding for creating risk maps is ample, ethnographic interviews suggest that the accuracy of information within the maps is often questionable. Funding is provided to municipal presidents who, disaster researchers suspect, pocket the better part of donated funds. The remaining funds are used to contract freelance geographers who do minimal work to meet CENAPRED's guidelines. As a geographer working in the high-landslide risk city of Teziutlán, in the state of Puebla, commented over the course of my ethnographic research: "The plans are not trustworthy. If CENAPRED gives 100,000 pesos to the municipal government to draft a risk map, the municipal president will give a geography graduate student 10,000 pesos to do the job and keeps the rest. The student then does a quick cut-and-paste job, and the map is not accurate" (author's fieldnotes 2013).

At the CENAPRED offices, on the other hand, the National Risk Atlas is fetishized as an objective representation of Mexico's risks as they really exist. During an ethnographic interview with the staff of the hydro-meteorological risk division, officials proudly displayed the maps on flat-screen monitors as the organization's primary accomplishment in DRR. The risk map then becomes a form of representation that claims to represent the real, and becomes more real than what it represents while at the same time omitting the messy connections and intimacies between post-colonial political culture and techno-scientific risk map–making practices, such as corruption, siphoning of risk map funds, or subcontracting of risk maps to exploitable graduate students. I use the term "postcolonial" to express the idea that the colonial order of things did not vanish with nine-teenth-century independence movements in Latin America, but continued to unfold and transform in the postindependence period and as an effect of subsequent imperial, capitalist, and neoliberal global relationships between Mexico and world powers, including the United States. What is particularly relevant here in Mexico and other Latin American countries is the establishment of government as a mechanism for personal gain during the colonial period and the perpetuation of this political culture until the present.

It is also true that even when CENAPRED maps are accurate, hazards such as landslides, which occur at highly localized levels, manage to elude the scale of those risk maps that are drafted to the agency's standards. So there are two gaps here: the first are gaps in disaster reduction between CENAPRED and local governments that commission the production of risk maps, and the second is the gap of representation between existing tech-nologies for representing risk and vulnerability and the actual behavior of disaster-triggering geophysical phenomena on the ground.

Gaps in Disaster Response

In the absence of effective creation and application of disaster risk knowledge in some parts of Mexico, hydro-meteorologically triggered disasters remain events that government can respond to only through the institution of Civil Protection and Emergencies (Protección Civil y Emergencias), a state agency developed in the aftermath of Mexico City's 1985 earthquake and charged with the task of disaster response and management.

The efficacy of Civil Protection—which, in collaboration with CENAPRED, is responsible for disseminating and enforcing evacuation orders and assisting populations during catastrophes—is also contingent on Mexico's unique postcolonial political practices, bringing about another set of gaps and connections that shape disasters. Take what government officials refer to as the Year of Hidalgo (El Año de Hidalgo), a name given to periods when newly elected political administrations take charge. The Year of Hidalgo refers to the popular saying, "En año de Hidalgo, que chingue su madre el que deje algo" (In the year of Hidalgo, may he who leaves anything behind fornicate with his mother). The proverb references Miguel Hidalgo, a Mexican Revolutionary War hero whose insurgent campaign became renowned for subaltern looting of elite estates.

Today, the year of Hidalgo refers to the practice in which outgoing political appointees pillage state property, regarding computers, vehicles, and any portable state resources as loot to be taken at the end of their tenure. In landslide-prone cities like Teziutlán, the Year of Hidalgo can have a significant effect on the response capacities of Civil Protection officials. First, as in many other nations including the United States, directors of Civil Protection are themselves political appointees who are given the positions as rewards for their assistance during incoming municipal presidents' campaigns. These officials, who are expected to become a key linkage between university disaster researchers and CENAPRED technicians in mitigation efforts, often lack proper training as disaster and risk managers, and their leadership styles can vary from disengaged to authoritarian.

In 2014, during Teziutlán's Year of Hidalgo, the incoming director of Civil Protection, a cell-phone vendor, obtained the position as a reward for providing communications services for the new president during the election year. Because of his limited background in DRM, researchers from the National Autonomous University of Mexico (UNAM), CENAPRED technicians, and local political officials committed to landslide mitigation had much work to do to bring the new director up to speed. But mitigation researchers and technicians committed to disaster reduction in Mexico are not passive in the face of such gaps. During one of my ethnographic experiences in Teziutlán, CENAPRED technicians and UNAM researchers and graduate students organized a tour of a landslide hazard monitoring station in a vulnerable neighborhood. Their intention, in this case, was to create alternative connections that would buttress their DRR efforts and mitigate the effects of political culture on the operations of Civil Protection. Furthermore, the work of connection-making among these technicians and researchers did not simply apply to political appointees alone but to themselves as well. A good share of my ethnographic experiences involved moments of gendered bonding among predominantly male scientists whose social bonds are key to effectively disseminating and circulating disaster risk knowledge.

In 2014, however, the incoming director was facing his own challenges of the Year of Hidalgo: all the spare tires of Civil Protection's vehicles had suspiciously disappeared the day before he took office, and the vehicle he was assigned to drive appeared to have been sabotaged by his predecessor, who was a member of the political opposition. The new director of Civil Protection was then left to drive his own car in his duties as first responder, an aging and unreliable 1968 Dodge Falcon. The Year of Hidalgo, one could say, is a postcolonial specter that haunts any attempts at disaster reduction in this part of Mexico.

The Gap of Representation

Teziutlán represents not the exception in Mexico, but rather the norm. DRR researchers whom I have interviewed over the course of my ethnographic work have, time and time again, bemoaned the state of Civil Protection offices that are often poorly equipped—either because of theft or underfunding—and inadequately staffed. The actual conditions of Civil Protection, however, do not match the ways the agency is represented in the form of official reports.

While disaster mitigation and response on the ground is a messy entanglement of political culture, technoscientific knowledge-making, and land use practice and regulation, at the level of international public policy national-level officials represent Mexico as a state at the forefront of DRR. This disjuncture became the subject of conversation when one of Mexico's leading experts in landslides shared an anecdote over the course of an ethnographic interview:

> I was at a UN disaster risk reduction meeting in Switzerland once, and the National Director of Civil Protection was there. He was listing the investments by the Mexican State in disaster response, claiming every office was equipped with computers, every unit with ambulances and four-wheel drive vehicles for evacuation. The Director of the United Nations International Strategy for Disaster Reduction (UNISDR) was there, and when he heard all of this, he congratulated Mexico for its advances. Later on I told the Director, how can you congratulate Mexico? How can you believe the things the director of Civil Protection told you? That is not how things are. (author's fieldnotes 2014)

Like CENAPRED's risk maps, institutional reports used for the purposes of global risk assessment enact what Jean Baudrillard (1995) refers to as simulacra and Michael Tausig (1993) has called the mimetic power of the representation over that which it claims to represent. In this form of power, the representation becomes more real to its observers than that which it represents and simultaneously obscures the telling details of Mexico's disaster mitigation shortcomings.

The Gap among Academics and Researchers

A few weeks after the 2013 catastrophe triggered by Hurricanes Ingrid and Manuel, I walked down a cobblestone street in Mexico City's Tlalpan neighborhood where the Center for Research and Post-Graduate Studies in Social Anthropology (CIESAS) is located. I was accompanied by one of the center's most prominent disaster researchers, a geographer who is also the founding member of La Red, a network of Latin American disaster researchers (or *desastrólogos*, as they like to call themselves) that is credited with the creation of the social vulnerability approach to catastrophes. As we walked, we discussed the recent national debacle over the hurricanes, and I brought up the issue of climate change and its possible role in the combined event. My senior colleague turned to me and, with a stern look, said, "You are an anthropologist. How can you seriously say this was caused by climate change?" Just a couple of weeks later, I experienced a similar exchange. While spending a leisurely Sunday with two graduate students working on disaster issues, one declared, "I think climate change is a discourse."

I found both my junior and senior colleagues' comments worthy of reflection. In the United States circles of disaster research in which I move, greenhouse gas–driven climate change is widely accepted by my colleagues as a tangible material fact. One could say that declaring the certainty that the climate change that has occurred since the mid-nineteenth century is the effect of human actions that can be evinced through atmospheric science is an unofficial requisite among disaster anthropologists. These same colleagues find themselves in opposition to political officials such as Senator Jim Inhofe (R-OK), who has called the National Academy of Science's claims on anthropogenic climate change "the greatest hoax ever perpetrated on the American people," and who insists that only God, not people, can control climate (Inhofe quoted in Lapidos 2014, 1).

In the case of the United States, political culture also creates a gap between academic knowledge about hazards and disasters and public policy, but the properties of this gap are almost the inverse of the gap in Mexico. In the United States, campaign contributions and political pressure on the part of energy companies—especially those involved in fossil fuel extraction—and their lobbyists align with political leaders like Senator Inhofe and many others (including newly elected president Donald Trump) on both sides of the American political spectrum, Democrat and Republican, to the effect of limiting the creation of carbon dioxide emission controls necessary to actually counteract climate change trends. Granted, much publicity has been given recently to the agreement signed by 195 nations at the 2015 Paris climate conference, but that accord, if honored, promises to cut only about half of the emissions that need to be reduced to prevent catastrophic sea level rise, severe droughts and flooding, widespread food and water shortages, and more-destructive storms (Davenport 2015).

In Mexico, however, discussions of climate change do not evoke public denial on the part of leading political figures. Instead, anthropogenic climate change is widely accepted in political circles as a material fact. President Felipe Calderón (2006–12), for example, made climate change one of his administration's focal concerns, leading to the drafting of the National General Law of 2012 and the creation of the National Institute of Ecology and Climate Change. Compared to the United States, Mexico appears to be leading the charge in international climate change mitigation efforts, therefore narrowing the gap between policy and scientific knowledge. So why are my fellow social scientists skeptical?

Part of the challenge confronted by the anthropology of disasters is that climate change is not merely a tangible effect of human development practices, but is also a discursive-political phenomenon that is mobilized by social actors in different contexts and in various ways to accomplish political and economic objectives. Invoking Arjun Appadurai (1996), we could say that, like modernity, climate change is now at large. As climate change becomes an object of governmental concern in various localities across the globe, it is bound to become entangled with location-specific histories of political culture and postcolonial social orders, leading to mutations in its discursive manifestations.

Despite Mexico's advances, some disaster researchers who live and work in the country remain skeptical as to the motivations behind this trend in public policy. Fernando Aragón, a regional expert on climate change and urbanization, for example, has suggested that the Calderon administration's focus on climate change was not so much a genuine concern with the anthropogenic phenomenon but was instead a diversionary tactic to draw attention away from other political agendas that had devastating repercussions for the Mexican populace, such as initiating a drug war that has cost tens of thousands of lives in the past decade alone (Aragón 2015).

In Mexico and the United States, the positions social scientists take concerning climate change are manifesting in relation to regional political culture. The varying views of political officials and economic interests—for example, nonrenewable energy companies and drug cartels—concerning climate change, have effects on academic perspectives, creating differences among researchers and unexpected relationalities between scholars and prominent government agents and agencies.

The Gap between Development and Sustainability

While anthropogenic climate change is certainly a slow-onset disaster that threatens to dramatically affect human life as we know it, the case of the 2013 disaster triggered in Mexico by Hurricanes Ingrid and Manuel illustrates the kind of political and development practices that enhance the rising risk identified in the GAR (UNISDIR 2015). While climate change mitigation is important, many disaster researchers in Mexico are concerned that national-level politicians are using the phenomenon as a scapegoat, with the effect of obscuring development practices that also play a key role in giving catastrophes form and magnitude.

The city of Teziutlán presents another case in point. The site where the city is located was chosen by Spanish colonial settlers for mining purposes because of its richness in minerals. The locality was suitable for the size of the original colonial settlement, but the town eventually outgrew the original location during the course of mining and maquiladora development booms of the twentieth century. The settlement lies on one of the major ridges of the Sierra de Puebla and extends from lower elevations in the north to higher elevations in the south. The city's colonial center sits on top of the ridge, where there is no landslide risk, but the greater area where Teziutlán has spread is characterized by several ravines that have been settled in a piecemeal fashion during the past eighty-five years.

One of the key figures in the history of Teziutlán is Manuel Avila Camacho, national president of Mexico from 1940 to 1946. Following political convention of favoring one's region of provenance, Avila Camacho ensured the execution of a number of mining development projects in and near Teziutlán during the 1940s. The project initiated a rise in population that pushed working-class settlements into hillsides. Additional development-related urban growth occurred between 1980 and 1990, when clothing manufacturing operations were expanded in the area. This latter regional development strategy lacked urban planning initiatives that ensured the settlement of working-class neighborhoods in areas with limited disaster hazards, adequate social services, and affordable means of transportation. Consequently, since the 1930s Teziutlán has seen a process of exponential population growth that has not been followed up by urban development practices to mitigate landslide risk. Working-class families have populated landslide-prone ravines that are close to the city's center and clothing factories, because these sites provide proximity to worksites as well as the city's central market areas, where basic commodities are cheaper than in outlying areas.

Resettlement projects like that of Ayotzingo on the outskirts of Teziutlán, which was created as a response to landslides that occurred in 1999 that killed at least two hundred people, have proven unpopular among working-class Teziutecos. The traits that make the site unattractive include undependable provision of public services, intermittent access to potable water, higher transportation costs to and from places of work, higher cost of basic commodities including food, and diminished size of housing structures and land parcels. Anecdotal accounts report that up to 50 percent of originally resettled Teziutecos have abandoned the resettlement site in search of more convenient and less costly places to live, even if these places are prone to landslides. Landslide risk, then, is a product of Teziutlán's own success within a model of development that prioritizes capital gains and people's participation in transnational commodity production processes as cheap labor.

Beyond Mexico's national borders, DRR must also go beyond climate change mitigation. The Paris Agreement, while important and necessary, does not address other forms of environmental destruction that play a role in the kind of disaster risk identified in the GAR. Mining operations, hydroelectric dams, deforestation induced by large-scale commercial agriculture and cattle ranching, sprawling and improperly planned urbanization and tourism development, the transformation of river deltas and hydrological systems for the benefit of commercial maritime operations, are just a few examples of the development practices that enhance disaster risk. When combined with stark socioeconomic inequities that limit socio-spatial mobility and housing options, these development practices lead to enhanced disaster vulnerability.

Part of the challenge of disaster mitigation in the context of an increasingly globally interconnected political economy structured largely on tenets of capital accumulation is that the well-being of nations is primarily assessed in terms of economic growth. Despite much current lip service given to human development and quality of life indicators, in general people are living in a time when capital gains and gross domestic products (GDPs) are given precedence as the means of assessing nations' accomplishments. These indicators, in turn, are dependent on development practices that replicate capital through commodity consumption and a global consumer society that requires the destruction of natural resources for the production of consumer goods, be they electronics, fuels, or touristic experiences. The focus of Teziutlán's development on mining and textile manufacturing to fulfill manufacturing and consumption demands elsewhere is a case in point. Addressing the environmental destruction and disaster vulnerability of a global capitalist consumption society will require reconsidering the fundamental ways we currently define what well-being is and what kind of human-material relations are involved in it.

Recognizing Connections and Narrowing Gaps

The anthropology of disasters demonstrates that catastrophes are multidimensional phenomena that are as much physical, material, and technological as they are sociopolitical. The latter, perhaps, is the most challenging dimension in which to intervene. Current compartmentalization of political jurisdiction in Mexico, for example, relegates agencies such as CENAPRED (that produce technoscientific knowledge about geophysical hazards) to a monitoring role, with no real power when it comes to regulating land use and development. Development decisions, in turn, remain the privilege of municipal presidents and state governors, many of whom favor their own financial bottom lines over long-term sustainability, socioeconomic equity, and DRR; this is another legacy of Mexico's colonial past. Urban development that focuses on the financial gains of real estate investors, hotel companies, and maquiladora manufacturers, in turn, repeatedly fails Mexico's urban poor, who continue to populate flood- and landslide-prone zones in search of employment opportunities and public services, which are concentrated in urban areas.

What the case of Mexico illustrates is that, while disaster mitigation is hindered by socially created divisions of political jurisdiction that keep scientific researchers and their risk maps from the decision-making process in urban development, disasters themselves are phenomena that are characterized by intimacies and connections between political and academic cultures, such as varying positions toward climate change among disaster researchers in the United States and Mexico, control of funding for risk map development, and human values and materiality (i.e., the ways environments react to development projects). Addressing the question why disasters persist, then, requires us pay attention not only to the ways knowledge, policy, and practice are divided by gaps but also to how they are intimately interconnected through people's actions and socio-material relationships, because it is these connections that engender catastrophes.

Addressing the identified challenges of disaster mitigation is no easy feat. As Anthony Oliver-Smith (1999) once noted, disasters are not external to societies, but are generated within them; mitigating catastrophes requires nothing short of a transformation of the human environment relations that characterize human communities at various scales, from the local to the global. Narrowing the gaps and counteracting the connections between political culture, development policy, material agency, and expert knowledge that shape disasters involves addressing the major sociopolitical challenges of nations such as Mexico.

Conclusions

Certainly, one could imagine a utopian state of affairs where dispassionate and un-self-interested disaster mitigation researchers and scientists hold veto power over development projects and land use plans. These experts could ensure that regional economic initiatives observe identified risk zones, avoid the destruction of sensitive environmental zones, and include housing programs that integrate subaltern and working classes into urban fabrics with limited exposure to geophysical hazards and ample access to public services. Such transformations, however, are easier imagined than executed. The anthropological record is filled with examples of failed attempts by centralized states or "rules of experts" to enact modernist societal transformations (Ferguson 1999; Holston 1989; Rabinow 1995; Scott 1998).

But perhaps the analysis of the gaps and connections that shape disasters presents us with an alternative for action. As we see in the case of Teziutlán, disaster researchers and technicians who work for CENAPRED and UNAM recognize the need to create alternative connections between themselves, political appointees, hazard monitoring technologies, and the environment's material agency. Rather than imagining a radical top-down societal transformation that may come someday in an indeterminate future, they manifest their agency through social practices intended to impart their knowledge among political appointees, hoping to raise awareness about disaster risks. These practices can serve as a point of departure for a diffused movement in disaster reduction that empowers mitigation specialists from disciplinary backgrounds that range from the social sciences to engineering, and encourages more intimate and disseminated connections between the creators of disaster knowledge and the societies they work in.

Roberto E. Barrios is professor of anthropology at Southern Illinois University–Carbondale. During the past twenty years, he has conducted ethnographies of disaster recovery in Honduras and Mexico, and in New Orleans, Houston, and Southern Illinois. His work focuses on the inherent assumptions about the nature of communities and people embedded in disaster recovery policy, and how disaster survivors interpret, reconfigure, and sometimes resist these assumptions. He is author of *Governing Affect: Neoliberalism and Disaster Reconstruction* and has published various articles in the journals *Disasters, Annual Review of Anthropology, Identities,* and *Human Organization.*

References

Appadurai, Arjun. 1996. *Modernity at Large: Cultural Dimensions of Globalization.* Minneapolis: University of Minnesota Press.

Aragón, Fernando. 2015. "La Ecología Política del Cambio Climático en México." Seminario Permanente: Vulnerabilidad Social a Desastres. 9 July. Center for Research and Post-Graduate Studies in Social Anthropology, Tlalpan, Mexico.

Bankoff, Greg, Georg Frerks, and Dorothea Hilhorst, eds. 2004. *Mapping Vulnerability: Disasters, Development, and People.* London: Earthscan Press.

Bankoff, Greg, and Dorothea Hilhorst. 2004. "Introduction: Mapping Vulnerability." In *Mapping Vulnerability: Disasters, Development, and People,* edited by G. Bankoff, G. Frerks, and D. Hilhorst, 1–9. London: Earthscan Press.

Baudrillard, Jean. 1995. *Simulacra and Simulation.* Ann Arbor: The University of Michigan Press.

Beven, John L. 2014. *National Hurricane Center Tropical Cyclone Report: Hurricane Ingrid.* Miami: National Hurricane Center.

Davenport, Coral. 2015. "Nations Approve Landmark Climate Accord in Paris." *New York Times,* 12 December. Retrieved 8 August 2016 from http://www.nytimes.com/2015/12/13/world/europe/climate-change-accord-paris.html.

Ferguson, James. 1999. *Expectations of Modernity: Myths and Meanings of Urban Life on the Zambian Copperbelt.* Berkeley: University of Califronia Press.

Global Assessment Report on Disaster Risk Reduction (GAR). 2015. "Making Development Sustainable: The Future of Disaster Risk Management." Geneva: United Nations Office for Disaster Risk Reduction (UNISDR). Retrieved 18 May 2016 from https://www.unisdr.org/we/inform/gar.

Hoffman, Susanna, and Anthony Oliver-Smith. 1999. "Anthropology and the Angry Earth: An Overview." In *The Angry Earth: Disaster in Anthropological Perspective,* edited by A. Oliver-Smith and S. Hoffman, 1–16. New York: Routledge.

Holston, James. 1989. *The Modernist City: An Anthropological Critique of Brasilia.* Chicago: University of Chicago Press.

Lapidos, Juliet. 2014. "Meet the Republicans' Top Guy on the Environment, James Inhofe." *New York Times.* 12 November. Retrieved 10 August 2016 from http://taking-note.blogs.nytimes.com/2014/11/12/meet-the-republicans-top-guy-on-the-environ ment-james-inhofe/?_r=0.

Latour, Bruno. 1993. *We Have Never Been Modern*. Cambridge, MA: Harvard University Press.

Maskrey, Andrew, ed. 1993. *Los Desastres No Son Naturales*. Bogotá, Colombia: La RED and Intermediate Technology Development Group.

Muñoz, Alma. 2013. "Dañaron casi 17 mil casas las tormentas de septiembre pasado, según la Sedatu." *La Jornada*. 17 October. Retrieved 12 August 2016 from http://www.jornada.unam.mx/2013/10/17/sociedad/036n2soc.

Nájar, Alberto. 2013. "El Día que Acapulco Quedo Bajo el Agua." *BBC*. 19 September. Retrieved 14 August 2016 from http://www.bbc.com/mundo/noticias/2013/09/130918_acapulco_guerrero_mexico_dia_quedo_bajo_agua_ingrid-manuel_huracan_lluvias_an.

Oliver-Smith. 1999. "What Is a Disaster? Anthropological Perspectives on a Persistent Question." In *The Angry Earth: Disaster in Anthropological Perspective*, edited by A. Oliver-Smith and S. Hoffman, 18–34. New York: Routledge.

Oliver-Smith, Anthony, and Susanna Hoffman, eds. 1999. *The Angry Earth: Disaster in Anthropological Perspective*. New York: Routledge.

O'Keefe, Phil, Ken Westgate, and Ben Wisner. 1976. "Taking the Naturalness out of Natural Disasters." *Nature* 260, 566–67.

Pasch, Richard, and David Zelinsky. 2014. *National Hurricane Center Tropical Cyclone Report: Hurricane Manuel*. Miami: National Hurricane Center.

Pickering, Andrew. 2008. *The Mangle in Practice: Science, Society, and Becoming*. Durham, NC: Duke University Press.

Ponce, Norma. 2013. "Ingrid y Manuel Asociados con el Cambio Climático: Semarnat." *Milenio*. 30 September. Retrieved 15 August 2014 from http://www.milenio.com/tendencias/Ingrid-Manuel-asociados-cambio-climaticoSemarnat_0_163184144.html.

Rabinow, Paul. 1995. *French Modern: Norms and Forms of the Social Environment*. Chicago: University of Chicago Press.

Scott, James C. 1998. *Seeing Like a State: How Certain Schemes to Improve the Human Condition Have Failed*. New Haven, CT: Yale University Press.

Taussig, Michael. 1993. *Shamanism, Colonialism, and the Wild Man: A Study in Terror and Healing*. Chicago: The University of Chicago Press.

United Nations International Strategy for Disaster Reduction (UNISDR). 2015. *Making Development Sustainable: The Future of Disaster Risk Management. Global Assessment Report on Disaster Risk Reduction*. Geneva: United Nations Office for Disaster Risk Reduction.

CHAPTER 2

Advocacy and Accomplishment

Contrasting Challenges to Successful Disaster Risk Management

TERRY JEGGLE

Introduction

Historically, disasters are recognized as destructive events that have already occurred or are feared to be imminent, and are a looming threat to the public. This is particularly the case when casual discussion refers to events associated with the onset or occurrence of severe natural phenomena. The popular conception of so-called natural disasters misrepresents the underlying roles that human decisions, actions, and socioeconomic conditions have in creating disasters. The extent of damages and losses from hazards such as severe storms, floods, droughts, land degradation, earthquakes, landslides, wildfire, environmental degradation, failure of physical infrastructure, or disease and other public health concerns are never caused entirely by either natural or geophysical circumstances.

Unlike the readily understood needs for emergency assistance provided by relief services at the time of a disaster, many different perceptions and professional activities are required for mitigating disaster risks to minimize public vulnerability to harm beforehand. Yet these anticipated and peremptory abilities are seldom associated with the more commonly acknowledged specialized skills required for disaster management services that exist in all countries.

Public apprehension is conveyed with consequences uppermost in mind when people use the common figure of speech, "It will be a disaster," yet they give less thought to their own roles in the anticipation or possible outcomes of unmitigated risks. The social values of education, health, safe public infrastructure, environmental stewardship, and contingent issues of government administration and budgetary allocations are intrinsic to the well-being of all societies, but their crucial roles in maintaining a safe population have historically been largely absent from disaster discourse except following a disaster.

Governments may strive to minimize damages and protect people from disaster effects, but they are more often compelled to act by the political consequences of human fatalities, physical destruction, and economic losses of an imminent or recent disaster event. Only infrequently does one encounter sustained public interest or official government efforts to identify explicitly vulnerable people within a society who are exposed to hazardous conditions. A widespread commitment to disaster risk management (DRM) has not yet become obligatory in most countries, even though there are examples of protective infrastructure such as flood barriers, or the legal adoption of land-use regulations and construction standards that may or may not be enforced.

This chapter examines some of the issues that either contribute to or impede the successful accomplishment of DRM through institutional advocacy and global professional experience in various countries. Effective DRM depends on the availability and use of knowledge grounded in citizens' understanding of public exposure to disaster risk, and the subsequent incorporation of that insight into official policies and established operational arrangements of governments. Numerous gaps and limitations remain in most countries to stimulate the sustained commitment of institutional resources specifically dedicated to DRM endeavors that exist throughout a society.

By referring to examples of institutional advocacy and international DRM policy agendas or frameworks, this chapter will illustrate the relative influence that external promotional efforts have sought to demonstrate since 1990. Measures of success need to be grounded in countries' own cultural, geographical, or political contexts related to local needs or prior experience rather than necessarily being motivated by comprehensive or idealized theoretical frameworks. Often a studied and collective reflection of a previous devastating disaster within a country has proven to be a galvanizing experience that exceeds other calculated promotional efforts to institute enhanced national awareness through individual policies or a sustained integrated national DRM strategy. This was the case in Indonesia after that country completely revised its national DRM strategy following the Indian Ocean tsunami of 24 December 2004. It is also a policy that Bangladesh has pursued since its founding following the unprecedented losses of the 12 November 1970 cyclone, given the country's geographical characteristics, extreme socioeconomic vulnerability of its population, and an official policy of incorporating disaster risk awareness within its national development planning processes.

These examples illustrate that effective DRM can be only a relative concept based on the acceptance of public values that are matched by the sustained use of cumulative knowledge, technical abilities, and institutional resources within a society that can reduce public vulnerability to disaster risks. Dedication to DRM must remain a continuous process developed through generations of people and remembered experience as both demographic characteristics and the nature of likely threats to collective well-being continue to change. It is folly for either a country or a local community to respond only to the immediate consequences of a severe or threatening event, or to remain oblivious to lessons provided by disasters elsewhere.

The chapter will examine the features of institutional advocacy, policy considerations, and related aspects of realizing DRM accomplishments, or not, in a global context. Many gaps exist, in part because of different understandings of concepts among essential actors. Limitations can also result from the different circumstances or inconsistent applications of national experiences that do not easily respond to prescribed frameworks or externally motivated encouragement. The discussion will address three challenging areas that affect successful DRM: the definition and understanding of DRM, the roles and limitations of international and institutional advocacy, and critical features for accomplishing future DRM in practice.

Definition and Understanding of Disaster Risk Management

There are several reasons why gaps exist among the use of established knowledge, policy formulation and related organizational arrangements, and the realization of lasting DRM accomplishments. The lack of a common understanding about basic concepts and an inconsistent use of language among the different people who need to be involved in effective DRM is an important factor. It has now been more than twenty-five years since national governments began to recognize DRM as a distinctive set of political perspectives and operational abilities than those previously associated with the conventional expressions of emergency relief or humanitarian action used at the time of a disaster. Different public perceptions, institutional responsibilities, technical abilities, and specific professional interests that address current socioeconomic conditions and consider future probabilistic hazard consequences characterize DRM. Rather than responding to immediate damaging events as disaster management typically does, the emergent outlooks of DRM depend on evolving forms of risk governance. A lack of anticipating disaster risks, efforts to minimize people's exposure to future hazards, and systematic failures to respond to threats before they become a disaster distort a common understanding of the purpose, needs, and resulting values of DRM. An acute sensitivity to these needs or the long-term perspectives required seldom exist within the expected capabilities of disaster response or humanitarian assistance agencies.

Understanding Disaster Terminology

Since the early 1990s international institutions have sought to define and distinguish the features of DRM and disaster risk reduction (DRR). Beyond a generalized understanding of efforts to try to prevent disasters, DRR has been understood more specifically from at least the early 2000s until 2015 to refer to the conceptual framework and elements involved and practiced to minimize vulnerability and disaster risks throughout a society in order to avoid or reduce the adverse impacts of hazards and related disasters (United Nations International Strategy for Disaster Reduction [UNISDR] 2004). UNISDR (the secretariat of ISDR) elaborated on this description in 2009 by referring to DRR as "the concept and practice of reducing disaster risks through systematic efforts to analyse and manage the causal factors of disasters, including through reduced exposure to hazards, lessened vulnerability of people and property, wise management of land and the environment, and improved preparedness for future adverse events" (UNISDR 2009b, 10–11).

A revised definition for DRR was finalized during 2016 by an intergovernmental expert working group of the United Nations (UN) to accord with the Sendai Framework for Disaster Risk Reduction adopted at the third UN World Conference on Disaster Risk Reduction convened in Japan in March 2015. The technically formulated but politically negotiated definition of DRR was adopted as being "aimed at preventing new and reducing existing disaster risk and managing residual risk, all of which contribute to strengthening resilience and therefore to the achievement of sustainable development" (UNISDR 2016, np). The inclusion of both strengthening resilience and achieving sustainable development through DRR greatly expands the subject's implied expectations.

The two terms, DRM and DRR, have often been used indiscriminately or incorrectly as synonyms in casual discussion rather than by recognizing their respective differences of systematic activities (DRM) and the conceptually expressed purpose or set of principles to produce DRR. An annotation of the revised 2016 DRR definition clarifies the terms' distinctions by stating, "DRR is the policy objective of DRM and its goals and objectives are defined in disaster risk reduction strategies and plans" (UNISDR 2016,

16). It remains to be seen to what extent such fine programmatic considerations will be realized in practice.

Since 2000, there has been a growing audience of international interest and institutional efforts to build a larger and more diverse global community to realize DRM activities. These efforts to clarify intentions increasingly took the form of citing examples drawn from individual countries or practice in particular circumstances. These formative national efforts and initial technical activities evolved through various motivations and are elaborated in *Living with Risk: A Global Review of Disaster Reduction Initiatives* (UNISDR 2004). In what has since evolved into a recognized sourcebook for the expression of DRR objectives and DRM activities, the volume provided an initial comprehensive description of policies and accomplished DRM practices undertaken by more than two hundred practitioners working in more than sixty countries. Since 2009 the *Global Assessment Report on Disaster Risk Reduction* (GAR) has become a biennial review and analysis of DRR strategies and multidisciplinary analysis from around the world. Each edition highlights a principal subject area of activity and charts progress against current international disaster and development agendas (UNISDR 2009a, 2011, 2013a, 2015a).

Despite the growing amount of information being disseminated, the language used to convey the sometimes-malleable concepts of disaster-related interests held by different actors can limit one's understanding of the range of technical, academic, operational, and public concerns associated with disasters. The growing information also reflects variations on whose interests are being served by resulting arrangements. These different perspectives complicate the communication required to translate specific knowledge into mutually agreed and beneficial policies. Divergent interests can potentially obscure the most critical resources required by the people who are either facing disaster risks or those who are later directly affected by disasters. The lack of clear or consistent uses of language creates difficulties for different parties to establish common ground or to pursue their respective priority roles in the middle of often-contentious economic, social, or political circumstances present in disaster-affected environments. The challenge grows when there are competing external or intervening views of different courses of action seeking to influence already overwhelmed local authorities during a disaster. Deliberate, informed, and participative local decision making that is universally advocated in DRM theory is routinely disregarded by governmental, intergovernmental, and relief agency staffs due to the demands of urgent or efficient disaster management practice.

The absence of common terminology or the presence of an institutional bias can compound confusion arising from contrasting initiatives in crises. The different meanings and professional implications of hazards, emergencies, crises, disasters, catastrophes, threats, security, risk, preparedness, prevention, recovery, rehabilitation, and reconstruction can vary depending on who is using the words and for what purpose. The media often further interprets or confuses institutional interests and professional contexts by its reliance on generic characterizations used in reporting with less attention given to careful analysis. These multiple expressions determine whether journalists are using the language and concepts in a precise technical sense expected by academic inquiry or as understood by technical professionals, rather than more simply conveying casual public narratives instead. Much public commentary about disasters concentrates on the immediate emergency actions, rescue, and relief requirements of survivors who are struggling in the middle of dramatic images of destruction and loss. Such emphasis given to the consequences of a disaster obscures a population's prior recognition about its exposure to disaster risks.

Disaster managers and people with other professional roles have different understandings of words for some disaster-related functions. Disaster managers understand mitigation to be the lessening of the potential adverse impacts of physical hazards (including those that are human-induced) through actions that reduce hazard, exposure, and vulnerability. When climate scientists use the same word, their accepted meaning is "human intervention to reduce the sources or enhance the sinks of greenhouse gases" (Intergovernmental Panel on Climate Change [IPCC] 2012b, 561).

Terms commonly used by people working in the same language or with shared cultural references can convey different expectations among others working elsewhere or in different governing or social contexts. The word "territorial" is frequently used in Central and Latin America to refer to interstate or adjacent geographically described areas that transcend political boundaries in both development and DRM policies and programs. The term is not as commonly used elsewhere as a measure of transnational scale or adjacent operational scope suited to DRM practice. Other common expressions of DRM and development principles such as local community participation, free and open access to information, transparency, accountability, stakeholders, and gender equity can provoke different perceptions depending on cultural or political contexts.

Another example of professionally confounding description can result in contexts when external parties are working closely with local authorities to support disaster assistance. Concerns expressed about public needs and local security by different actors after a disaster can be ambiguous depending on an individual's perceived operational responsibilities.

A local municipality's mayor responding to residents' immediate require-ments is likely to have a different view of their priorities from the view of a foreign security officer evaluating the personal safety of staff working for an intervening foreign nongovernmental organization (NGO) at the same location.

As crucial as operational and cultural contexts are in disrupted disaster circumstances, such differing perspectives of affected populations and disaster-related interveners can easily be overlooked. An international policy expectation for countries to undertake early preparedness for recovery can be confusing and express an uncertain operational role in countries that have neither understood nor fully instituted functional dis-tinctions between emergency assistance, DRM practices, or the respon-sibilities assigned to technical agencies that would be involved in the specific functions concerned. Many such conveniently expressed planning modalities are considerably more complicated and demanding to imple-ment in practice, often with unexpected local ramifications.

Understanding Disaster Risk Management Practice

When disasters are considered in precise terms for study, practice or pro-fessional commentary by people who have not themselves been affected, there is a tendency to place actor's roles either within the specific domains of emergency or disaster management, or else with a primary empha-sis given to their involvement in the preexisting circumstances of DRM. These latter functions are typically associated with risk assessment, ame-lioration, and risk management with significant attention given to the additional social dimensions of vulnerable populations. These aspects par-ticularly relate to the socioeconomic and physical aspects of exposure to risk that underlie the physical effects of specific hazards. Such distinctions in practice and the resulting divisions seen in institutional arrangements demonstrate a crucial feature in understanding DRM in practice. The views and concerns of people directly affected by potential or actual disaster circumstances are very different from those of external commentators.

The legislated or assigned responsibilities of most NDMAs (national disaster management agencies) ordinarily reflect conventional disaster management responsibilities (sometimes inaccurately identified as phases) of preparedness, response, and recovery, occasionally with a ten-tative acknowledgement of the importance of prevention. Government agencies identified with disaster responsibilities seldom include compara-tive demographic, environmental, or socioeconomic analytical competen-cies that are essential for addressing prevalent disaster risk issues. These incomplete perceptions about emergency or disaster management and

DRM roles or misplaced assumptions about NDMA competencies reflect additional difficulties in defining the essence and recognizing the capacities required for assured risk management practice.

Fundamentally, DRM effectiveness depends on a society's anticipated and analytical considerations of social conditions and designated populations' relative existential exposure to harm or loss, and the resulting actions taken to minimize potential consequences. By contrast, emergency or disaster management practice consists of institutional arrangements and technical means focused on physical readiness and skilled tasks to protect people's well-being and property at the time or immediately following possible loss or damage from a severe event.

Despite the very different abilities required to identify and manage risks in contrast to responding to impending emergency or disaster conditions, there is often an additional expectation among government authorities that NDMAs have the technical and managerial abilities to coordinate the recovery and reconstruction of devastated communities after disaster events. In practical terms, search and rescue teams, emergency medical staff, or firefighters are not proficient in abilities required to rebuild schools, reestablish local commerce, plan and undertake the construction of permanent shelter, or "build back better" to reduce future disaster risks. In circumstances following a disaster where there have been no prior institutional arrangements for managing recovery, physical reconstruction and socioeconomic rehabilitation activities, we too easily assume that NDMAs have the necessary managerial capabilities and technical resources to coordinate the range of required social, economic, and physical infrastructure duties.

Without the intervention of very experienced or highly placed directing government officials, or the ad hoc creation of an alternative high-level coordinating authority, few NDMAs have the planning and management skills or the technical resources to implement DRM practices within their existing mandates or competencies. Even fewer governments have prior contingent arrangements either to anticipate the consequences of identified disaster risks, or to plan and implement recovery strategies following a disaster. Where such analytical abilities or planning capacities might exist in academic institutions or commercial enterprises, only infrequently are they closely associated with or routinely engaged by NDMAs other than on a periodic consulting basis. The Gujarat State Disaster Management Authority in India demonstrated a noteworthy example to bridge this gap by creating a comprehensive reconstruction and rehabilitation program immediately following the 26 January 2001 Bhuj earthquake. It engaged both well-qualified managers and a diversified complement of international and domestic private, public, and NGO resources to implement the recovery process in a rare success. The rapid installation of a systematic and inclusive system that embodied local practices demonstrated what could be accomplished through informed, focused, and well-resourced contingent DRM capabilities (Thiruppugazh n.d.).

Habitual terminological references and their organizational appropriation for specific operational contexts frame people's identities and suggest their respective roles in disaster situations. Subject interests and abilities, such as those associated with water, health, gender, shelter, safety, or even family needs and livelihoods, convey distinct cultural characteristics that are pertinent to understanding DRM in practice. Defining features, viewpoints, and consistently expressed terminology will be different among people who have themselves been exposed to unmitigated disaster risks, who have survived a disaster event, or who have indelibly suffered personal losses from one. These issues need to be addressed through professional activities that extend considerably beyond conventionally understood disaster management because the crisis invokes widely varying definitions, expectations, or requirements in the eyes and experiences of others.

This divergent set of perceptions and a variable set of values are particularly acute among local survivors, yet such insight seldom emerges from media reporting at the time of crisis. Disaster-affected populations may be consumed by basic survival needs and trauma or inhabit inequitable social structures; these populations invariably have very different personal and cultural contexts than the people who identify themselves as being disaster managers, public administrators, or external commentators. Meanwhile, disaster management experts or institutional recovery planners speak of "creating resilience through partnership," the expectation of unspecified "accountability," and to "build back better" without creating future disaster risks in a seeming, or actual, foreign language. Representative beneficiary participation and decision making is often an invitation for the emergence of dominant power brokers charitably referred to as "local community leaders."

Beyond terminology, successful DRM depends on intended audiences and practitioners being motivated by a common understanding of the principles involved. The public expression of efforts to motivate public authorities and engage practitioners in a country must be clear, unambiguous, and—most importantly—must relate to circumstances through a shared terminology. This obligation needs to be consistent with the cultural contexts and the physical conditions where people live and work in their society, with evident relevance to their well-being. Common awareness and concerns need to be cast in terms consistent with the affected people's own interests, rather than in theoretical frameworks or generic statements that represent distant ideals driven by a threat of harrowing loss. The issues that can either encourage or impede the effective transition from expressed concepts to motivating sustained actions determine the efficacy of institutional advocacy. The next section considers the progress of principles, efforts, and results from international DRM advocacy since 1990.

The Roles and Limitations of International and Institutional Advocacy

Since 1990 a number of practitioners have pursued institutional advocacy internationally to stimulate a more explicit understanding and resulting DRM action in countries. They have sought to do this by creating an identifiable community of professional interests, official government commitments, and public engagement to advance the wider realization of DRM in principle, policies, and accomplishments. Since the inception of a DRM global agenda expressed in terms of disaster reduction, there have been challenges to attract the interests and to combine the resources of existing humanitarian and disaster management practice, the technical abilities of risk assessment and mitigation activity, and the aspirational role of governing interests in national development. The comprehension of DRM encompasses an even wider range of subject matter and professional disciplines that combine multiple geophysical features and cultural and demographic dynamics, as well as socioeconomic, political, and commercial motivations. The role of DRM has been configured to recognize the importance of creating a common understanding and complementary actions to reduce people's vulnerability and public exposure to identified disaster risks and potentially hazardous conditions, but the complexity of the task has proven to be easier articulated than accomplished. This underlying and layered rationale contrasts with the established immediacy of disaster management responsibilities and the familiar specialized emergency services that respond to disaster events when they occur.

As DRM advocacy has evolved, it has been stimulated and further guided, primarily by international institutions, to advance a global political agenda that could create first an outlook, and then a movement, to engage and link many actors involved in their individual professional pursuits for a common purpose in different countries. In both political and professional terms, the expression of DRM has signaled what is essentially a new professional discipline that can easily assume different dimensions depending on one's own perceived role in the subject. International advocacy for DRM policy engagement has foremost been directed toward the necessary involvement of government responsibilities at all scales of administrative and financial interests. The realization of DRM activities depends on the knowledge, expertise, and resources of technical and scientific abilities, academic research and learning, commercial enterprise, financial allocations, and the supporting public interests of civil society. Actors in all these domains engage in promotional activities related to their respective interests, which certainly extend beyond the parameters of DRM. The advocacy referred to in this discussion, however, is understood to comprise the promotion of specific DRM principles and common objectives that need to be embraced collectively if they are to be successful in a country.

Participants acknowledge the values of advocacy as a strategic marketing process that works through the dissemination of information and experience, the stimulation and organization of interdisciplinary contacts, and the intended expansion of professional collaboration. There are many ways to pursue these functions, and many successes have materialized to varying degrees as products of advocacy during the past twenty-five years. However, we need to be cautious so we do not misidentify intended methods or expected outcomes as accomplished DRM practices. The continued relevance of advocacy can be seen only if it is able to progress through increased DRM understanding, resulting policy acceptance, and an evident expansion of mitigating activities. To be of lasting value, accomplishments need to be suited to the requirements of individual countries, but they also need the ability to meet commonly accepted standards of effectiveness and durability.

The extent to which we can realize such ambitious expectations remains to be seen. I discuss the major global advocacy initiatives for DRM since 1990 below. A crucial matter to note is the distinction between the tendency of advocacy to rely on rhetorical or repetitive statements of prescriptive actions that other parties should pursue and the tangible investments and substantive actions that the implementing parties need to adopt if DRM is to be realized. The limited availability of technical expertise and the absence of material resources among the most vulnerable populations are only two examples of practical restrictions that can limit accomplishments regardless of what may have been astute analyses of the risks and the expression of well-intentioned official policies.

The Conceptual Foundations and Institutional Influences on Advocacy

Disaster mitigation has been a professional concept in disaster management thinking for many years although it has received less systematic international attention and limited public recognition compared to the political imperatives of humanitarian assistance and emergency prepared-ness and response activities. Initially, mitigation concepts and the techni-cal subjects involved were predominantly rooted in academic disciplines of natural science, hazards studies, engineering, and public health (Jeggle 2001). Pioneering social science research related to vulnerability and per-ceptions of risk, hazards, and disasters have been pursued in faculties of geography, sociology, psychology, and anthropology rather than being incorporated as analytical functions of NDMAs.

Considering the consequences and civil responsibilities associated with disasters, and the crucial roles that civil servants must perform in any successful DRM strategy, it is surprising that DRM has historically only seldom been a common subject in higher education programs of public administration. This gap has prevailed even though other aspects of actuarial, political, financial, and state security risks are integral to other corresponding professional studies.

Scientists originally proposed a global expression for countries to ele-vate the political attention and resource requirements needed to reduce the rising cost and consequences of so-called "natural" disasters (Jeggle 2001). Such an international focus on the distinctive subject of DRM advo-cacy, policy, and accomplishments dates from the designation by the United Nations General Assembly of the International Decade for Natural Disaster Reduction (IDNDR) from 1990 to 1999 (UN 1989). This political declaration was the first of several global advocacy efforts to encourage a systematic political approach of countries and activities supported by international organizations. From its beginning, the very concept of the IDNDR has included numerous expectations ranging from declarative encouragement to the much more demanding commitments of devel-oping technical capacities, public acceptance, and resource allocations if DRM were to become a political reality leading to safer populations and fewer disaster consequences.

The International Decade for Natural Disaster Reduction, 1990–1999

The primary goals of the IDNDR were to improve the capacities of coun-tries to mitigate the effects of natural disasters, particularly in the areas of assessing the potential for disaster losses, establishing early warning systems, and encouraging the creation of disaster resistant infrastructure (UN 1989). The goals of IDNDR were stimulated by an interest to increase the effective application of science and technology globally as a basis for preventing the occurrence and limiting the rising costs of disasters. This rationale was intended to encourage governments and international organizations to allocate more resources to advance scientific research and to share advanced technical assistance among national multidisciplinary technical committees.

The declared intention of the IDNDR for increasing commitments to reduce disaster risks reflected the negotiated interests and concurrence of countries, but the UN did not budget official funds for realizing the agreed objectives. The resulting IDNDR program of promotional activities was funded by a succession of ad hoc or voluntary donor contributions from countries. The funds that governments provided typically reflected miscellaneous foreign assistance grants or other national, economic, or technological interests. By bringing a diverse group of professional interests and government officials together to address disaster risks, the UN hoped that the IDNDR would stimulate the creation of strategic national disaster prevention policies and resulting actions in countries. The promotional mechanism of IDNDR provided initial definition of fundamental DRM concepts and introduced a coherent platform for countries to exhibit existing scientific knowledge and technical means of accomplishment for practitioners. Early DRM proponents anticipated that resulting DRM policy guidance would engage governments and involve supporting organizations. They focused emphasis on low- and middle-income developing countries that the technologically advanced donor countries considered to be more exposed to serious disaster consequences and to have fewer technical capabilities or only limited institutional arrangements to address disasters.

The scope and resulting work program of the IDNDR activities were agreed by member states of the UN at the First World Conference on Natural Disaster Reduction, convened in Yokohama, Japan, from 23 to 27 May 1994. The principles, a strategy, and a plan of action to realize IDNDR's objectives from 1994 through 1999 were expressed in the *Yokohama Strategy and Plan of Action for a Safer World: Guidelines for Natural Disaster Prevention, Preparedness and Mitigation* (hereafter Yokohama Strategy; UN 1995). This technically astute but politically negotiated foundation document included broad policy guidelines for the prevention, preparedness, and mitigation of the then-designated natural disasters; these guidelines could apply to community, national, and regional levels of official responsibility with the support of the international community. It was anticipated that the Yokohama Strategy would provide a basis for government officials, scientists and academics, and local practitioners to relate more closely by jointly promoting disaster reduction. The resulting IDNDR Plan of Action for the period from 1996 through 1999 included activities for advocacy, policy promotion, and international coordination, and was budgeted at $6.53 million. The 1999 budget of $3.27 million provided for one director, thirteen professional officers, and seven administrative staff to administer the program (UN 1999b, 109–14).

The principles outlined in the Yokohama Strategy have remained a valid expression of relevance and professional involvement essential for effective DRM practice. Others consistently restated the essence of these principles in slightly different ways over the next twenty years, but the emphasis has provided a solid foundation for the subject. The following principles have maintained a consistent focus for engaging professional interests and abilities:

1. *Risk assessment* is a required step for the adoption of successful disaster reduction policies and measures.
2. *Disaster prevention and preparedness* are of primary importance in reducing the need for disaster relief.
3. *Disaster reduction policies and measures* are integral elements of national development planning and investment strategies.
4. *Strengthening national and institutional capacities* are top priorities, and are essential as a basis for follow-on activities in the twenty-first century.
5. *Early warnings of impending disasters,* and their effective dissemination, are key factors in successful disaster reduction.
6. *Preventive measures* are most successful when they involve participation of the people affected at all levels of administrative or operational responsibility.
7. *Vulnerability* can be reduced by the application of appropriate development activities focused on targeted beneficiary groups and by the education and training of the communities concerned.
8. *The necessary technology to reduce disasters* needs to be shared and to be made freely available as integral parts of technical cooperation among all members of the international community.
9. *Environmental protection,* as a critical component of sustainable development, is essential for disaster prevention and mitigation.
10. *Each country bears the primary responsibility for protecting its people, infrastructure and national assets from the impact of natural disasters.* The international community needs to demonstrate strong political determination to mobilize and to use existing resources efficiently, including financial, scientific, and technological means, bearing in mind particularly the needs of developing, and least developed countries.

Although the tangible accomplishments of IDNDR were modest at the time, the global initiative was significant in defining and projecting

DRR concepts that would evolve in many countries in the following years (UNISDR 2004). Some countries, including Australia, Chile, China, Colombia, Germany, Japan, South Africa, Vietnam, and others were motivated by the global IDNDR initiative to embrace strategic DRM policies that could project a wider global influence in a nascent field. The global promotion of DRR principles also spurred shared interests and stimulated collective DRM policies, newly formulated legislation, and regional cohesion around countries' shared disaster interests. Periodic technical meetings and joint political discussions encompassing shared economic, social policy, and environmental interests in regional geographical settings nurtured collective DRR relationships, particularly in Central America and the Caribbean, among the Andean states in South America, and throughout the South Pacific.

There was considerably less embodiment of DRM in key development sectors even as IDNDR acknowledged the feasibility and means of doing so. Few countries adopted comprehensive early warning systems, although some, like Bangladesh, expanded previous efforts. Most significantly, IDNDR succeeded in planting the seeds of and stimulating an extended dialogue about the feasibility of reducing disaster risks through anticipation and tentative risk assessment or mitigation activities in those countries where government authorities were prepared to commit to them (UN 1999a).

The essential features of DRM have remained largely as they were first expressed in the IDNDR principles in 1994, although additional subjects would demand greater attention as they arose in later international development discourse. The growing demands in particularly vulnerable countries for adapting to climate variation and change, the expanded policy reliance on sustainable development attributes that could build resilience, and more-assertive advocacy for gender equity would become explicit in future DRM advocacy agendas.

The International Strategy for Disaster Reduction, 2000–2015

At the conclusion of the IDNDR in 1999, member states of the UN adopted continuing institutional arrangements that were shortly institutionalized as a global political agenda from 2000 to 2015. When the preparatory arrangements were confirmed and designated as the UNISDR (United Nations International Strategy for Disaster Reduction) from 2002, it was significant that the defining adjective "natural" was dropped in the transition from IDNDR to UNISDR. The altered expression signaled a growing professional recognition of the human actions required to mitigate disaster risks by minimizing public exposure or reducing human vulnerability.

While recognizing that humans cannot alter threatening forces of severe natural phenomena, the broad approach adopted by ISDR to extend its appeals beyond the prior scientific rationale and technical mitigation techniques encompassed multiple dimensions of disaster management, risk governance, and sustainable development objectives in countries. As the ISDR proceeded, the emergence of climate change as a potent global concern further extended the linkages between disaster consequences and severe weather events, with an anticipated increase in financial resources to pursue climate change adaptation as an instrument of advancing DRR (IPCC 2012a).

The expanded outlook of corresponding DRR component interests also corresponded with the first decade of the twenty-first century and a reemphasized global agenda to mobilize national commitments to development issues. The UN designated prioritized, targeted, and measurable Millennium Development Goals (MDGs) to focus countries' attentions and stimulate their own resource allocations toward the achievement of ten fundamental development objectives. The MDGs did not specifically address the consequences of disasters, but ISDR accommodated this parallel international millennial program to concentrate global efforts that could drive coordinated national policies to accomplish a set of common objectives to provide a renewed impetus for DRR.

Although disasters remained the defining feature of both DRM and DRR, the development rationale was expanded considerably among principal promoters after 2000. The strategy became a movement to attract additional subject interests and a corresponding anticipation of possibly more-abundant resources for implementing risk reduction concepts. The UNISDR (the secretariat of the ISDR) sought to marry the positive endeavors and potential resource commitments of national development to the tentative efforts to reduce the consequences of disasters, while continuing to embrace the operational relevance of disaster preparedness manifested through existing NDMAs. The complementary interests of both national development objectives and DRR became a cornerstone of advocacy during ISDR, even though the respective implementing actors remained hesitant to demonstrate common cause in practice among their distinct operational counterparts.

The ISDR focus was considerably elaborated by its defining *Hyogo Framework for Action 2005–2015: Building the Resilience of Nations and Communities to Disasters* (HFA; UN 2005). This document was timely, and 168 countries at the World Disaster Reduction Conference in Kobe, Japan, adopted it from 18–22 January 2015, less than a month after the devastating Indian Ocean tsunami of 26 December 2014. Although the conference had been planned for more than two years, the 230,000 human deaths and unprecedented losses of $15 billion from the tsunami galvanized international DRR attention at the time.

The HFA reiterated earlier DRR concepts and principles, while advocating for the means and activities to be pursued by so-called partners in national governments, international institutions, and many other potential actors to implement ISDR's five priorities for action:

1. *Ensure* that the DRR is a national and local priority, with a strong institutional basis for implementation.
2. *Identify, assess, and monitor* disaster risks and enhance early warning.
3. *Use* knowledge, innovation, and education to build a culture of safety and resilience at all levels.
4. *Reduce* the underlying risk factors.
5. *Strengthen* disaster preparedness for effective response at all levels.

Beyond understanding the interpretation of the diplomatically stated fourth priority of action to mean lessening risk wherever people may be exposed to it, or, more simply, mitigation practice, these objectives spanned well-expressed policy expectations of long standing. The need for public policy and legislation to protect citizens from harm, for risk assessment and effective early warning, for public understanding and knowledge about safety and dangers, and for the necessity of timely preparedness for effective crisis response were are all expressed as basic principles of sound governance. In what should have been the most revealing priority for benefits in accomplishing DRM in practice and the intended engagement of development interests through MDG subject areas, the fourth priority of mitigating risks remained obscure and less accomplished (UNISDR 2011, 2015a).

Challenges in translating a clear understanding of DRM and policy intentions continue in the development of managerial, resource, and technical capacities through professional engagement in many domains of society. These include professional activities in such sectors as food and livelihood security, public health, social welfare, shelter and other critical lifeline services, public infrastructure, spatial planning, and natural resources and environmental management, as well as in the expanded use of insurance, commercial, or financial measures. As crucial as these subjects are for accomplishing DRM, they remain distant from the discourse, training, or capabilities that continue to define the traditional functions of disaster management as they are practiced.

Does More Advocacy Produce Better Disaster Risk Management?

With its appeals and calls to action at all levels of international, regional, national, and local levels of involvement, the HFA simultaneously extended a universal appeal for everyone to engage in a systematic work plan that was both comprehensive and prescriptive in nature. Secretary General Ban Ki-moon of the UN declared in 2011, "Disaster risk reduction is everyone's business," as he encouraged the institutional support from a coalition of interests spanning disaster preparedness, climate change, and various UN development endeavors around the world (UN 2011, np).

This considerably expanded mission progressively assumed multiple forms and became a distinguishing feature of UNISDR compared to the more reserved scientific presence of IDNDR. In the earlier formative decade IDNDR had focused primarily on the technical and academic research communities, scientific meetings, and specially designated demonstration projects, while displaying less-robust interactions with government policies or stimulating widespread public interest. The ISDR sought to exceed this modest prior impact with a more assertive market-ing strategy targeting global and national proponents of both disaster management and international socioeconomic development practice. The growth of ISDR interests was manifested in its expanded association within the UN system and among other international development organi-zations, although in most countries the designated national authority for DRR promotion remained with the NDMAs. Because disasters invariably defined them, in many countries the NDMAs remained under-resourced or politically marginalized and dedicated to routine disaster management activities with slight involvement in countries' national development or parallel climate change adaptation strategies.

The UNISDR, ISDR's secretariat placed within the UN in Geneva, was mandated to coordinate a coherent professional global identity for the considerably expanded professional interests being included within the growing domain of DRR relevance. By embracing marketing techniques to build an audience beyond international institutions, the HFA effectively became a branded framework. The resulting activities of UNISDR have continuously depended on obtaining more resources to maintain its insti-tutional viability for the cause of promoting DRR interests and adoption, rather than implementing DRM in practice. This latter tangible activity has been left to individual countries and academic and technical institutions, or to be pursued within the routine activities of professional disciplines that are not necessarily identified explicitly with disaster management roles.

In addition to individual UN offices and organizations, other interna-tional organizations, such as the International Red Cross and Red Crescent

Movement, international development financial institutions like the World Bank, the Organisation for Economic Co-operation and Development (OECD), or bilateral national technical assistance organizations also encouraged the adoption of DRR concepts or supported DRM activities within their own program activities. The actual value of countries' investments in DRM practice globally remains difficult to ascertain, in part because of the absence of accounting criteria and few disaggregated data related to risk reduction in either disaster management or development expenditures. However, with the expanded involvement of different organizations, a diffusion of subject interests that have adopted DRR terminology, and the diverse audiences being addressed, by 2015 UNISDR's promotional activities were being administered by nearly 120 international staff working in eight regional offices. More than 99 percent of UNISDR's 2014–15 biennium activities were funded by voluntary contributions. UNISDR projects the program's 2016–17 work program financial requirements to be $75 million (UNISDR 2017).

Critical Features for Accomplishing Future Disaster Risk Management in Practice

The functions of institutional advocacy and sustained professional practice both have their purposes, but their respective roles and audiences should not be confused. The relative emphasis of organizational advocacy about what should be done to reduce disaster risk conceptually and the combined efforts required for actually doing DRM remain very different propositions. The time required to realize these different actions varies greatly: advocacy is episodic whereas successful DRM commitments need to be continuous, transcending individual governments' terms of office. Since advocacy is largely pursued by external bodies marketing a cause, there is a risk that its effectiveness will decline over time if collective understanding of disaster risk and explicit investments are not transformed into practice by responsible implementing parties. As of this writing in 2019, after more than twenty-five years of promotional activity and growing disaster losses, there is abundant institutional documentation that advances the combined political, social, and economic rationales for countries' governments and local communities to turn DRM advocacy into practice (Asian Development Bank [ADB] 2013; Hallegatte et al. 2017; Jha and Stanton-Geddes 2013; UNISDR 2013). The strategic institutionalized commitments and assured allocation of existing human, material, and technical resources to do so remain glaringly inadequate to the task in nearly all countries.

Without official adoption of DRR concepts, or the limited application of often existing technical capabilities that are not routinely aligned with DRM practice, there is the potential that theoretical views or analytical concepts that external commentators formulated will be qualitatively different from the accumulated experiences of countries. Idealized motivating concepts such as building resilience conflict with practical organizational and resource realities. Local authorities and practitioners routinely struggle with numerous implementation requirements, limited budgets, and often-competing interests in their areas of responsibility. To simply exclaim, "Build back better!" is very different from actually rebuilding damaged communities and restoring productive livelihoods after a devastating disaster. It is even more challenging to do so with the foresight that understands prevailing risk and reduces people's vulnerabilities before a future threat occurs.

Addressing such constraints or revising locally held perceptions of intended beneficiaries in disaster-prone countries or already marginalized populations can be tedious, lengthy, and difficult to sustain. Actual conditions may not conform so easily to the expressions of prescriptive guidance or the encouragement of intermittent external support. This persistent divide between the expression of political solidarity and the more tangible challenges of providing the necessary technical abilities and managerial capacities with only limited material support limits ISDR's ability to translate advocacy into DRM practice.

After more than twenty-five years of concerted promotional international efforts and expanding political agendas to encourage—if not materially induce—lasting accomplishments in DRM practice, it is time to reassess the means that DRM has used. Four key motivations must be reasserted in coming years if DRM practice is to become a lasting contribution to countries' vitality and the enabling of the equitable well-being of their populations.

Emphasize Public Exposure to Prevailing Risk

The salient context of DRM is one of dynamic risk in a society that various segments of the population experience and tolerate to different degrees. Each affected group of people is located in different political, social, economic, and geophysical circumstances. They pursue different livelihoods and are affected by different types of risk. The social sciences demonstrate that there are relative and differential perceptions of risk both within and among societies. These variations are routinely obscured by any discussion of disasters determined solely by the forces of an individual type of hazard. Frequently, an official overconfidence about authorities being prepared and able to respond with emergency services to a severe natural phenomenon is confused with an often-absent responsibility to identify or minimize risks before imminent disaster threats materialize.

The concepts and language of disaster management or humanitarian assistance represented by the specialized abilities of NDMAs are crucial for responding to disasters as they occur, but they are not generally suited to fulfil the expectations of DRM. The specific consequences of a disaster are determined by the complex, often long-standing and unique human contexts in which the disasters occur. While the minimization of adverse human consequences, losses, and destruction from potential disasters may motivate DRM interests, those interests' wider requirements for anticipating, identifying, and mitigating risk factors have failed to captivate official policy commitments or to obtain sufficient institutional and material resources. NDMAs have too easily assumed that DRM is a precursor or extension of responding to imminent crises, even though DRM roles are very different from disaster-driven effects.

The primary attention accorded to NDMAs as the locus of DRM responsibilities and wider subject oversight in most countries is inconsistent with the agencies' origins, training, and technical abilities. NDMAs seldom possess the anticipatory outlooks and analytical requirements to assess and alleviate human vulnerability and public exposure to risk. Few disaster-defined service organizations possess sufficient political authority to command a national commitment to invoke combined interministerial actions to address broader environmental, climate, or emergent risks. Disaster management services are frequently concentrated in primary population centers, with outlying populations or distant communities served poorly, if at all. Socially motivated abilities such as the anticipation of special needs or requirements to compensate for deficient community social services in marginalized communities are uncommon in most NDMAs.

It is telling that nearly all governments have some crisis or emergency management institutions, but very few maintain facilities other than hazard-specific technical institutions to monitor comprehensive conditions of public risk. This dereliction of public responsibility is in stark contrast to the thorough official investigation of all critical aircraft or other transportation accidents that occur in any country. Technical and often public inquiries are routinely conducted to determine the detailed causes of these incidents, and particularly to identify the human or systemic circumstances that demand remedial action in the public interest. Similar comprehensive forensic analyses of disaster causes within a society are rare, with slight effort dedicated to the remedial correction of the even more costly continuation of public exposure or social vulnerability to risk.

The political, socioeconomic, cultural, demographic, and geophysical conditions of risk and the technical abilities to address them are found throughout a society rather than within the confined abilities of disaster management departments. The overarching political authority for successful DRM needs to be situated at the highest levels of government, with pertinent planning and evaluation processes able to be coordinated through linked public administration responsibilities. The implementation of risk management needs to be vested in many technical departments of government and additionally addressed through civil society entities, institutions, and commercial enterprises. These issues are the same issues that drive much of the academic research that explains why people inhabit inherently hazardous places, or that document the limited force of public policies to minimize people's vulnerability to potential disaster consequences. Despite existing knowledge, DRM effectiveness continues to be constrained by official reliance on specialized disaster services invoked because of emergent crises, rather than on the means to engage other integral professional functions that are responsible for the assured vitality of the society.

Resolve Inadequate Resource Commitments

There is little doubt that DRM advocacy has advanced officials' ability to recognize and discuss the subject, but as countries have reported their activities in the HFA monitor or during their biennial statements to global platforms, complaints about the lack of human, technical, and material resources are a common refrain. Rhetorical intentions and prescriptive advice have been more-frequent components of international declarations than have accompanying resource commitments. While some countries have increased the budgetary resources or technical programs dedicated specifically to national DRR programs, the collective international disaster and development assistance communities have been much less inclined to do so.

Ultimately, individual national governments and subordinate authorities need to determine their own rationale and provide the necessary means to realize DRR in their own cultural, political, and socioeconomic contexts. It is essential for national leadership to provide or develop domestic resources to implement risk management strategies consistent with public needs and interests.

A comparison of how international organizations and governments allocate financial resources for responding to disasters and managing their antecedent risks provides insight into predominant global emphasis. Countries routinely record budgetary allocations and annual expenditures for domestic emergency activities, but there are few examples of budgetary obligations to plan for or to allocate resources to conduct explicit risk management activities. When pressed to indicate financial commitments for DRM practices, government authorities are prone to highlight aggregated expenditures for disaster management or otherwise suggest that disaster reduction costs are reflected elsewhere within the internal accounting of, as they put it, "various other departments."

It is difficult to determine the amount of funding allocated to DRM, although a few references suggest that it has never been much. Even among major donor countries, amounts committed to disaster reduction compared to emergency assistance have been minuscule. Despite spending $3.05 billion on disaster response between 1985 and 2004, the United States allocated only $195 million specifically for disaster prevention activities during the same period (Jha and Stanton-Geddes 2013).

There are considerable disparities between the resources committed by the external assistance community and international organizations to development investment when compared to disaster relief assistance and DRR (Kellet and Caravani 2013). One estimate places the losses from disasters in developing countries between 1992 and 2012 at $862 billion. This amount is equal to 28 percent of the $3.03 trillion provided to the same countries through external development investment during the same period. All the disaster-related aid of that international assistance amounted to $106.7 billion (35.2 percent of the total), but only $13.5 billion was allocated to reduce disaster risks. This expenditure for future disaster prevention or reduction amounted to 12.7 percent of all disaster-related assistance and only a minuscule 0.45 percent of all the development aid of $3 trillion during the period. The $13.5 billion provided by the international community for DRR is a meager commitment when compared to the actual disaster losses of $862 billion incurred globally during the same period. With such a poor value proposition, the dominant donor nations are unconvincing when they exhort financially strapped and highly indebted developing countries to invest their own limited resources in DRR.

Some international organizations have pursued initiatives to support DRR institutional capacities in disaster-prone countries with the anticipation that the recipients will eventually be able to assume full supporting obligations. In recent years some UN organizations have revised international program strategies by committing more resources to support national DRM capacities. However, since the budgetary allocations are often included in broader program strategies and sometimes within the context of existing emergency management commitments, it remains difficult to determine precise expenditures for individual DRM activities. However, the following examples suggest some cogent approaches for tangible commitments that enable countries to develop their own DRM resources.

Since 2005 the United Nations Development Programme (UNDP) has supported DRM policies, explicit risk analysis, and DRR institutional capacities within national development strategies. The organization has allocated funds to sixty countries, averaging $200 million annually. The *UNDP Strategic Plan, 2014–2017: Changing with the World* combines future organizational commitments to climate change and disaster risk assessment, climate and disaster risk governance, preparedness and early warning, in addition to resilient recovery and crisis response within national development contexts (UNDP 2013).

The World Bank established the Global Facility for Disaster Reduction and Recovery (GFDRR) in 2008 to support disaster-prone developing countries in their efforts to pursue DRM policies and to develop their institutional capacities. In contrast to only advocating for the wider use of DRM policies, the GFDRR has provided financial resources and technical support for countries to make national investments and to design risk-financing programs with investment partners. It has pursued these activities in the subject areas of risk identification, risk reduction, preparedness, financial protection, and resilient recovery in the expanding global contexts of exposure to natural hazards and the effects of climate change. The program began with an initial annual global investment of $35 million and fulfilled its subsequent investment program of $68 million in fiscal year 2015.

Because global public health concerns have become more readily included in DRR discourse as an aspect of national development needs, the World Health Organization (WHO) has similarly sought to develop and implement comprehensive health programs that encompass contingent emergency risk management operational frameworks as intrinsic features of the various community and national health programs that it supports. The recognized inadequacy of the organization's global threat monitoring and emergency preparedness arrangements regarding the seriousness of the Ebola outbreak in West Africa in 2014–15 stimulated further organization-wide reforms to improve its organizational technical, budgetary, workforce, and operational capabilities. Nearly $120 million was budgeted for WHO's core emergency operations for the 2016 and 2017 fiscal years, demonstrating WHO's integration of extended risk management outlooks and expanded emergency operational capabilities to complement the organization's traditional technical and normative roles in supporting national development programs in recipient countries' heath sectors.

Pursue Realistic and Viable Risk Management Practice

Diplomatic interests and international advocacy over twenty-five years have developed a widespread adherence to DRR concepts by associating the subject with complementary interests and additional development agendas. What has become an unbounded breadth and diversity of risk in society has clouded the definitive roles of DRM and the daunting possibilities of realizing DRR concepts. In contrast to the excesses of the Sendai Framework for Disaster Risk Reduction, serious proponents of DRM need to acknowledge a deliberate long-term realism to develop and sustain an incremental adoption of the subject.

An updated outlook will require a collective sensitivity and the appreciation of technical experience within individual societies in what is essentially a social process. The accomplishment of DRM cannot be reduced to comprehensive frameworks made up of time-bound targets, dozens of indicators, cross-referenced goals, and similar instruments of project management. Such universally prescribed aims overstate simplified causes and echo conventional development slogans at the expense of deliberate understanding, durable professional substance, locally and culturally validated experiences, and inherent public interests. Many different actors certainly need to be involved, but crucially their contributions will depend on adhering to their respective work disciplines shaped to minimize public risk.

DRM needs to address in specific terms applicable to locally relevant circumstances the many aspects of social, political, economic, commercial, environmental, technical, and human concerns involved in the range of disaster risk-related matters. Rhetoric does not travel well beyond conference venues. Politically couched progress reports or periodic program documents do not reflect the realistic compromises that must be forged in local conditions. To speak simply of "all stakeholders" or the "need for active participation at all levels to achieve concrete results" in diplomatically negotiated declarations diminishes the likelihood of substantive and relevant actions addressing real needs in specific locations. In some cases DRM states encouragement for local translation into practice as desirable outcomes, but seldom are necessary efforts so readily financed or undertaken simply because DRM has expressed them.

Sustained DRM practice proceeds less through the adoption of DRR policies implemented by a disaster management agency than through the complementary efforts of many other professional relationships and activities. They exist among technical disciplines, academic faculties, government authorities, and their interactions with local people's concerns or initiatives. Some of these planning and implementation relationships may be strong, explicit, and formalized, but many are more likely to be fluid, casual, and often motivated by a dynamic individual. The routine work of many people continuously involved in their familiar contexts avoids the language of crisis while their activities are informed by the practical requirements of identifying local so-called problems. Ideally, they work to minimize people's exposure and vulnerability by addressing lasting human and societal needs rather than by episodic interventions.

Conclusions

DRM practitioners pursue the practical means, resources, and mechanisms with their understanding of locally perceived threats to public well-being, physical features of place and habitation, social elements, and existing cultural characteristics in a country. Many of these resources may already exist in societies even if they are not routinely identified with disasters or specific risk management roles. DRM practitioners might not use the word "disaster" to describe these responsibilities, nor will that word regularly appear in the name of the organizations involved. Similarly, lasting community values are not regularly associated with NDMAs where the institutional culture is frequently derived from uniformed forces focused on responding to crises or disruptive events.

Crucial DRM contributors are more likely to be identified with study, governance, and business, or to be known by other professional designations. Those contributors include people engaged in education, public health, social services, human and physical geography, sociology, anthropology, engineering, geology, meteorology, hydrology, environmental management, conservation of natural resources, food production, forestry resources, marine and coastal habitats, urban and regional planning, construction, public administration, economics, energy, public utilities, transportation, public works, information and communication technology, and media, among others. This is where DRM happens. These professional responsibilities should be acknowledged and their needs supported as being essential to DRM practice.

It is equally important that more attention be given to crucial circumstances in which policymakers or commercial interests are creating or worsening risks by their daily work, actions, or decisions. In such cases, few of these risk creators recognize that they are contributing to disaster risks because they focus on other perceived attributes such as economic opportunities, returns on investment, competitive advantages, or shareholder value. Risk reduction will become an accepted public value and a serious political obligation only when the most crucial operational roles are expressed as community expectations and not defined only in terms of unforeseen disasters. The exercise and fulfilment of risk management needs to comprise essential public, professional, and commercial responsibilities for people's well-being and public safety.

Terry Jeggle worked internationally in disaster and risk management assignments for more than forty years in seventy countries. He spent sixteen years managing emergency operations and refugee or crisis situations for CARE, an international NGO. He was the program and training coordinator, and later the executive director, of the Asian Disaster Preparedness Center (ADPC) in Bangkok, Thailand, from 1990 to 1995. From 1995 to 2009 he was the senior policy officer for the secretariats of the United Nations International Decade for Natural Disaster Reduction (IDNDR) and the United Nations International Strategy for Disaster Reduction (UNISDR) based in Geneva.

References

Asian Development Bank (ADB). 2013. *Investing in Resilience: Ensuring a Disaster-Resistant Future.* Mandaluyong City, Philippines: Asian Development Bank.

Global Facility for Disaster Reduction and Recovery (GFDRR). 2016. *Managing Disaster Risks for a Resilient Future: A Work Plan for the Global Facility for Disaster Reduction and Recovery 2016–2018.* Washington DC: Global Facility for Disaster Reduction and Recovery. Retrieved 24 June 2016 from https://www.gfdrr.org/sites/default/files/publication/GFDRR_Work_Plan_2016-18.pdf.

Hallegatte, Stephane, Adrien Vogt-Schilb, Mook Bangalore, and Julie Rozenberg. 2017. *Unbreakable: Building the Resilience of the Poor in the Face of Natural Disasters.* Climate Change and Development Series. Washington, DC: World Bank. doi:10.1596/978-1-4648-1003-9. Retrieved 15 February 2017 from https://www.gfdrr.org/sites/default/files/publication/Unbreakable_FullBook_Web-3.pdf.

Intergovernmental Panel on Climate Change (IPCC). 2012a. *Managing the Risks of Extreme Events and Disasters to Advance Climate Change Adaptation.* Edited by C. B. Field, V. Barros, T. F. Stocker, D. Qin, D. J. Dokken, K. L. Ebi, M. D. Mastrandrea, K. J. Mach, G. -K. Plattner, S. K. Allen, M. Tignor, and P. M. Midgley. A Special Report of Working Groups I and II of the Intergovernmental Panel on Climate Change (IPCC). Cambridge, UK: Cambridge University Press.

———. 2012b. "Glossary of Terms." In *Managing the Risks of Extreme Events and Disasters,* IPCC, 555–64. Cambridge, UK: Cambridge University Press.

Jeggle, T. 2001. "The Evolution of Disaster Reduction as an International Strategy: Policy Implications for the Future." In *Managing Crises: Threats, Dilemmas, Opportunities,* edited by U. Rosenthal, R. A. Boin, and L. K. Comfort, 316–41. Springfield, IL: Charles C. Thomas.

Jha, A. K., and Stanton-Geddes, Z., eds. 2013. *Strong, Safe, and Resilient: A Strategic Policy Guide for Disaster Risk Management in East Asia and the Pacific. Directions in Development; Environment and Sustainable Development.* Washington, DC: The World Bank. Retrieved 11 February 2017 from http://documents.worldbank.org/curated/en/2013/02/17423304/strong-safe-resilient-s-strategic-policy-guide-disaster-risk-management-east-asia-pacific.

Kellet, J., and A. Caravani. 2013. *Financing Disaster Risk Reduction: A 20 Year Story of International Aid.* London: Global Facility for Disaster Risk Reduction and Overseas Development Institute. Retrieved 11 February 2017 from htpp://www.odi.org.uk/sites/odi.org.uk/files/odi-assets/publications-opinion-files/8574.pdf.

Thiruppugazh, V. n.d. "What Has Changed after Gujarat Earthquake 2001." Asia Science and Technology Foundation. Retrieved 2 February 2017 from http://www.jst.go.jp/astf/document/43abst.pdf.

United Nations (UN). 1989. *The International Framework of Action for the IDNDR. Annex to United Nations General Assembly Resolution A/44/236 of 22 December 1989.* New York: United Nations.

———. 1995. *Yokohama Strategy and Plan of Action for a Safer World: Guidelines for Natural Disaster Prevention, Preparedness and Mitigation* (Yokohama Strategy), World Conference on Natural Disaster Reduction, Yokohama, Japan, 23–27 May 1994. New York: United Nations.

———. 1999a. *Activities of the International Decade for Natural Disaster Reduction Report of the Secretary-General.* A/54/132—E/1999/80 of 21 July 1999. Paragraphs 64–81. Retrieved 30 November 2016 from http://www.un.org/esa/documents/ecosoc/docs/1999/e1999-80.htm.

———. 1999b. *International Decade for Natural Disaster Reduction Report of the Secretary General Addendum, Final Report of the Scientific and Technical Committee of the International Decade for Natural Disaster Reduction.* A/54/132 Add. 1 of 18 June 1999. Retrieved 30 November 2016 from http://www.un.org/documents/ecosoc/docs/1999/e1999-80add1.htm.

———. 2005. *Hyogo Framework for Action 2005–2015: Building the Resilience of Nations and Communities to Disasters* (HFA). Extract from the final report of the World Conference on Disaster Reduction (Kobe, Hyogo, Japan, 18–22 January 2005). (A/CONF. 206/6). Retrieved 22 April 2016 from http://www.unisdr.org/wcdr.

———. 2011. "As Risks from Disasters Rise, UN Chief Calls for Better Preparedness." United Nations News Center. Retrieved 2 December 2016 from http://www.un.org/apps/news/story.asp?NewsID=38327#.WEIba7U8LYU.

United Nations Development Programme (UNDP). 2013. *UNDP Strategic Plan, 2014–2017: Changing with the World,* Executive Board of UNDP, Second regular session 2013, 9–13 September, New York, Document DP/2013/40. Retrieved 30 April 2016 from http://www.undp.org/content/dam/undp/library/corporate/Executive%20Board/2013/Second-regular-session/English/dp2013-40.doc.

United Nations International Strategy for Disaster Reduction (UNISDR). 2004. *Living with Risk: A Global Review of Disaster Reduction Initiatives.* New York: United Nations. Retrieved 2 December 2016 from http://www.preventionweb.net/publications/view/657.

———. 2009a. *Global Assessment Report on Disaster Risk Reduction 2009 (GAR 09): Risk and Poverty in a Changing Climate.* Geneva: United Nations.

———. 2009b. "2009 UNISDR Terminology on Disaster Risk Reduction." Retrieved 3 April 2016 from http://www.unisdr.org/we/inform/publications/7817.

————. 2011. *Global Assessment Report on Disaster Risk Reduction 2011 (GAR 11): Revealing Risk, Redefining Development*. Geneva: United Nations. Retrieved 24 April 2016 from https://www.unisdr.org/we/inform/gar.

————. 2013a. *Global Assessment Report on Disaster Risk Reduction 2013 (GAR 13): From Shared Risk to Shared Value*. Geneva: United Nations. Retrieved 24 April 2016 from https://www.unisdr.org/we/inform/gar.

————. 2013b. Preventionweb, International Day for Disaster Reduction 2013. Retrieved 5 December 2016 from http://www.preventionweb.net/english/professional/publications/tags/index.

————. 2015a. *Global Assessment Report on Disaster Risk Reduction 2015 (GAR-15): Making Development Sustainable. The Future of Disaster Risk Management*. Geneva: United Nations Office for Disaster Risk Reduction (UNISDR). Retrieved 24 April 2016 from https://www.unisdr.org/we/inform/gar.

————. 2015b.Working Background Text on Terminology for Disaster Risk Reduction 2 October 2015, reissued on 23 October 2015 with technical corrections. Retrieved 28 February 2016 from http://www.preventionweb.net/drr-framework/open-ended-working-group/.

————. 2016. "Terminology." Working text on terminology for the Second Session of the Open-ended Inter-governmental Expert Working Group on Indicators and Terminology relating to Disaster Risk Reduction held in Geneva from 10–11 February 2016. Issued on 3 March 2016. Reissued with factual corrections on 24 March 2016. Retrieved 15 April 2016 from http://www.preventionweb.net/events/view/47136.

————. 2017. *UNISDR Work Programme 2016–2019*. Geneva: United Nations Office for Disaster Risk Reduction (UNISDR). Retrieved 14 February 2017 from http://www.unisdr.org/files/51558-workprogramme20162021pri.pdf.

CHAPTER 3

Natural Hazard Events into Disasters

The Gaps between Knowledge, Policy, and Practice as It Affects the Built Environment through Development

STEPHEN BENDER

Introduction

The discussion at hand is about the risk to the built environment from natural hazard events. The built environment comprises all social and economic infrastructure—all components such as housing, schools, medical facilities, water tanks and sewers, bridges and highways, banks and shops, and places of public assembly and worship—placed on the natural landscape whether through private or public, planned or spontaneous, regulated or self-built structures of all types that surround human populations. The built environment for the purposes of this discussion comes about through development, which we can think of as betterment. Whatever the built environment component, it is conceived and placed on the landscape for a purpose by and for someone's betterment as someone sees it. But the concern here is why so much of the built environment, no matter how it came about, no matter who owns or operates it and for whatever purpose, is demonstrably vulnerable to natural hazard events such as earthquakes, volcanic eruptions, floods, hurricanes, landslides, and the like. The discussion is also why the impact of these events often ends in a disaster—the overwhelming of a societal unit to the point that it must call for and depend on outside assistance.

This discussion introduces efforts to understand more specifically why on occasion certain groupings of institutions that work towards or take responsibility for the creation of the built environment to improve the lives of certain populations ultimately end up putting those populations at risk. The risk manifests itself in the vulnerability to natural hazard events of the economic and social infrastructure whereby the infrastructure components and their associated populations are subject to destruction and harm. It will identify some challenges of those groupings or stakeholders to reduce vulnerability to the built environment and its associated populations before the triggering natural hazard event occurs. In doing so the discussion presents risk to natural hazard events from the perspective of the built environment in the context of development. The emphasis is on the role of the stakeholders' reigning development policies and practices including globalization, trickle-down economics, and community-based capitalism and collective efforts. Through mistakes but mostly through misconception, deception, and depredation, development facilitates the creation and continuance of vulnerability but without accounting, in a visible and transparent manner, for who pays, who benefits, and why. Beyond a critique of development per se, the task is to look at how those who carry out development efforts can take on responsibility and accountability for risk, and how such development efforts can avoid creating the exposure of populations, including migrants and the settled poor, to life-threatening risk, decrease degradation of ecosystems, and increase resilience of basic social and economic infrastructure.

At issue is the need to look at specific areas—gaps—between knowledge, policy, and practice for restructuring the development agenda (purpose, resources, beneficiaries, and operators) toward risk reduction coupled with responsibility and accountability. These identifiable gaps are important in order to understand a society's relationship to risk from natural hazard events, how it comes about and what can be done about it. Specific discussion is needed about qualitative and quantitative goals to avoid putting the built environment at risk that compromises ecosystem structure and function, and that places human populations, particularly the poor, in harm's way. The discussion will lay open obfuscating assumptions and dated paradigms in order to increase an understanding of disasters in development rhetoric: "Who is vulnerable and why?" "Where and when are they vulnerable?" and "Why are they vulnerable?" It proposes a more-holistic, more-integrated, and more-egalitarian framework to reduce the risk of catastrophic loss in the context of both the development community and the disaster management community.

This discussion has been informed by a previous period of evolution of thought as to the origins and causes of catastrophic loss of economic and social infrastructure and death and injury of associated populations. This period begins with the Peruvian earthquake of 1970; countries from around the world responded to the disaster and local religious figures preached as in the past that natural hazard events and their destruction are punishment for the sins of the people; their words can still be heard today. The period's end is marked by the publication in 1984 of "Prevention Better Than Cure" by Hagmann (1984) and "Natural Disasters: Acts of God or Acts of Man" by Wijkman and Timberlake (1984) that directly link disaster losses with development decisions, and by the aftermath of the Mexican earthquake of 1985 when the government's immediate concern was accessing the counsel of international experts as to how groups of untrained and unauthorized students could set up relief operations before trained official forces could begin similar actions. The period informs by examining the reasons why one village of adobe structures in the mountains of Peru was 80 percent destroyed and not ten miles away another village was 80 percent standing following the earthquake. This evolution of thought influenced the response in the late 1980s from the military authority in a Central American country responsible for civil defense to a question concerning the occupation of ten thousand inhabitants in the capital city's highly hazardous landslide zone. The authority stated there were no other settlement options for those refugees from the region's civil war. The reality of situations made visible by the period mentioned above sheds light on national and local governments failing to act to reduce increasingly recognized population vulnerability through collective policy at the international, national, and regional levels.

Sovereign states, multilateral development banks (MDBs), nongovernmental organizations (NGOs) and specialized development and disaster agencies (DDAs) at all geopolitical levels are stakeholders that directly promote development and their efforts are directly affected by disasters. They are composed of people and organizations acting on behalf of the recipient populations through their efforts to carry out development and disaster mitigation in the context of those populations' expressed or understood wants and needs. Through knowledge, policy, and practice, these stakeholders deal with natural phenomena as they relate to cultural, economic, social, political, and religious actions and to the built environment as expressed through societal units and their culture. The term "knowledge" as used here is meant to encompass cultural facts or conditions of knowing which constitutes familiar information gained in part through policy and practice (or experience). The term "policy" is meant to include all forms of prudence and wisdom including scientific knowledge whether codified or not in social relations and culture. Policy constitutes high-level overall planning and guidance for embracing the general goals and acceptable procedures especially of a governmental body or other structured societal organization. The term "practice" is meant to include direct observation of or participation in events as a basis of knowledge, including scientific research and traditional building practices.

These three elements—knowledge, policy, and practice—are not linear in their application no matter what order we give them. However, they influence one another, overlap, and coexist even when societal actions produce visible contradictions in their content. They are descriptors of societal expression.

Knowledge, policy, and practice may provide a society with the wherewithal to take action to manage risk. But there are no assurances that in any given instance any or all three of these elements have contributed to or are positioned to effectively deal with disaster risk reduction (DRR). Nor is there any assurance that the contributions of the three will act in unison or that they will not be at odds with one another. This is increasingly visible in societies distinguished by their evolving participatory democracies and free market economies. Whether or not these two are desirable, enduring, or endearing forms of government and economic organization, they are for the most part the contexts of what gives shape to knowledge, policy, and practice regarding disasters, risk, and development to the most people around the globe. In fact, such evolving societal settings may actually foster a lack of action toward managing risk and reducing the vulnerability of the built environment. In the past and at present in many countries the rate of economic and social infrastructure growth—particularly housing, potable water systems, transport lines, energy and industrial production, and commercial centers—turns out to reflect an ever-greater share of economic and social infrastructure at risk. This matter is directly related to the global economy and its markets, the concentration of population centers in coastal areas, import and export activities, and degradation of the structure and function of ecosystems both urban and rural by the related natural resource exploitation for economic gain not only in the developing world, but also in the industrialized world.

There is an underlying supposition that knowledge, policy, and practice will lead to disaster reduction through risk management. Unfortunately, there are too few examples of a societal unit proclaiming and acting on these three elements individually or in a coordinated priority fashion to achieve such a goal. On balance, it appears less an issue that societies lack access to the knowledge, policy, and practice to reduce disasters and more an issue that societies do not pursue by choice risk reduction on a priority basis. Such choices flow out of a consideration of for whom development takes place, why, and who pays, as well as from how development is monitored, measured, and reported beginning with monetary units per capita and associated calculations.

The conundrum is straightforward. Calls for societies to use development as a tool—as the principal tool—for managing risk to natural hazard events are not new. Unfortunately, practice, knowledge, and policy—in that order—have in some instances not made contributions to reducing risk depending on the hazard type, population and built environment components, and acceptable levels of risk. There exist few examples where a societal unit has explicitly stated that disaster risk reduction—lessening of dependency on outside intervention in the aftermath of a natural hazard event—is a priority of development as manifest in the society's own guidance and actions. Such guidance and actions include the national constitution, executive orders, laws and ordinances, budgets, judicial actions, regulation, and enforcement.

The following examination is aimed at illuminating those situations where knowledge, policy, and practice have made efforts in reducing disaster risk. Also illuminated are those situations where these three elements still fall short of effectively lessening the impact of natural hazard events and in strengthening resiliency.

Gaps as Discontinuities, and What Else?

Gaps between knowledge, policy, and practice are not necessarily manifestations of arrogance, ignorance, disregard, misunderstanding, avoidance, malfeasance, errors, and omissions. Gaps can occur naturally in a society, particularly in those societies moving simultaneously toward stronger participatory democracies and free market economies. Gaps also can be deliberate manifestations of competing claims by groups in society using government and economic development to achieve parochial ends.

Imagine a Venn diagram with three equal intersecting circles—knowledge, policy, and practice—each overlapping with the other two. And they are necessarily inside a larger circle or bubble that can be labeled "development" with a subtitle "with its built environment." Each circle individually as well as collectively presents the opportunity for both ex-ante and ex-post disaster reduction actions. In the past, concepts such as the disaster cycle and the window of opportunity following a disaster have framed prevention and mitigation actions, but outside of or parallel to development. Most recently and forcefully the concept of resilience has come to the fore to address both ex-ante and ex-post actions, but not necessarily in the context of development.

The issues of power, prestige, funding, independence, and the lack thereof, often cause discontinuities between knowledge to policy to practice, and often cause discontinuities—sometimes deliberately—depending on the circumstances. A feedback loop from practice to policy to knowledge often suffers the same discontinuities.

From a development perspective, apparently the growth of the gross domestic product (GDP) for upwardly mobile national economies with strong and weak participatory democracies alike is a predictor if not an indicator of increasing disaster vulnerability. It is also a predictor of a widening gap between not only the rich and the poor, the vulnerable and the safe, but also between disaster losses and investment in resilience. GDP reflects financial transfers regardless of the cause or their effect. It is becoming more and more visible that the greater the GDP growth, the greater society's built environment exposure to natural hazard risk. This is precisely because economic growth at present has a direct correlation to the creation of risk to natural hazards (and an inverse correlation to risk reduction) given the policies and practice, and use of knowledge under which development takes place. GDP growth year over year does not identify the growth of disaster losses or the expenditures to address loss due to natural hazard events. Nor does GDP growth identify or reflect the paucity of public or private budget allocations to improve resilience of existing economic and social infrastructure. In many regions of the world, the long-term rate of growth of disaster losses is apparently but not surprisingly outpacing the growth of national GDP, making the need for strengthened resilience ever more urgent.

The declaration of and response to a disaster does not necessarily lead to social change that benefits the marginalized, poor, and vulnerable populations. But why has development not been addressing these populations before the hazardous event takes place? More generally, post-disaster efforts can have a positive, enabling economic and social change effect in a society, but what was development enabling before the disastrous events took place? Disaster shocks can open political space for the contesting and concentration of political power controlling development. But what was the development power structure doing before such inevitable events? Without understanding the development context in which disasters take place, it is difficult to understand the potential and shortcomings of knowledge, policy, and practice.

More on Development

"It's the risk, stupid," to paraphrase a political slogan used in the presidential election campaign in the United States in 1992. The question is, "How do societies choose to deal with risk? The key aspects of the type and extent of increasing vulnerability of the built environment present in less-developed and more-developed countries after almost fifty years of "modern" development and disaster management efforts are the following:

• Increasing damage and destruction of social and economic infrastructure in both the public and private sectors.

- Relatively few changes to zoning laws and building codes to increase the resilience of the built environment to natural hazard events. These few changes almost always come about following an event with catastrophic loss, and then are implemented only in the impacted area or in as limited an administrative place as possible.
- Fierce opposition from the private arena of all sectors to increase requirements for resilience through land use planning, zoning and building regulations and enforcement.
- Continued encroachment of the built environment in known hazard-prone areas with well-funded, politically driven support overcoming dedicated, but limited opposition.
- Public- and private-sector failure to provide proper regulation and enforcement of master plans, zoning regulations, and building codes; and failure to adequately fund public-sector implementation of the plans, regulations, and codes that do exist.
- Failure to adequately deal with the impact of not only formal but also, and just as importantly, informal and illegal expansion of the built environment in the name of development—from settlements to monocrop agriculture, sea ports, mining, electrical power generation, manufacturing, and roads.

More on Disasters

The most prominent manifestation of the development-disaster continuum over the past five decades is the creation of the disaster sector alongside the development sector alongside the previously created environment sector with the full participation of sovereign states, but led by the NGOs with support from MDBs and DDAs. Although the disaster sector has evolved under the watchful eye of the development sector, disaster management interests are often tasked with, or at times seek, responsibility for managing emergencies as well as overseeing risk reduction of the built environment by using specialized agencies at all administrative and geographic levels—from national disaster czars to local civil defense colonels. However, there has been a lack of follow-through by governments and a very parochial, reticent view by the private sector to managing risk on a sector-by-sector basis. Moreover, funding shortfalls exist in national and sector budgets, both public and private, as counterpart participation in international conventions and agreements and intra-national programs under legislative authorization dwindle. In addition, there has been relatively little direct investment in DRR. This situation reflects in part the lack of assignment of responsibility and accountability for the losses that ultimately result from unmanaged risk. At the international level the result is the conditioning of sovereign states through development assistance by the DDAs to follow what is, in effect, "If it is worth doing it will be supported by MBDs and NGOs" as well as the broader array of DDAs organizations. At the same time, progress achieved by the disaster sector has occurred even as traditional sectors continue to carry out policies and practices that construct risk in the built environments in the name of development.

Globalization adds to the risk to natural hazard events through development. The proponents driving globalization as they take over communities assume little or no responsibility for risk to loss. They do make manifest their interest in financing and constructing recovery and reconstruction, while often re-creating vulnerability. Globalization also impacts response to declared disasters and risk reduction through enabling opportunities for the growth of donations, particularly attempting to capture remittances, along with lending and borrowing between and among private institutions, MDBs, and sovereign states.

Lack of resilience is not often the product of chance. Often discontinuities come about willfully and deliberately, depending on the circumstances. How else in the world today could there be the lack of disaster resilience at all levels? At present, knowledge to policy to practice does not often lead to disaster reduction. And feedback loops from practice to knowledge to policy often suffer the same discontinuities.

The next sections look at four groups of stakeholders: sovereign states, MDBs, NGOs, and DDAs.

Sovereign States

In recognition of their priority status among societal groups, without a doubt sovereign states have the most influence in shaping risk and development at all levels of society as it pertains to the built environment. They are perhaps the group most capable of developing policy to manage risk. In general, their policy prerogative and want comes not always from knowledge or practice, but from dogma and desire. Sovereign states have become quite adept at managing their parochial interests even in the face of their collective decisions. They are the principal agents for creating agreements, binding or otherwise, that call for risk reduction. These actions often include participating in a reporting system that focuses on goals or outputs (identified by policy objectives) such as the number of lives lost, number of people impacted, and economic, physical, and financial losses by sector. But far too often the sovereign states reporting ends up including only data on inputs—the number of committees created and their meetings, laws and regulations and their dissemination; and information generated and disseminated through public awareness campaigns, education, and training. Practice trumps policy, and the knowledge base is left aside.

Some factors that condition actions by governments follow:

- In the past fifty-odd years of the modern development period, a great deal of the vulnerability of populations and their economic and social infrastructure has been increased, or vulnerability has been created by government and private sector development actions where no vulnerability existed before.
- Actions between governments at whatever level are solely within the prerogative of those sovereign states.
- Governmental actions exist side by side with the private sector's actions, with varying degrees of coordination, collaboration, cooperation, and control.
- The private sector may well depend on public infrastructure for significant if not critical components for its functioning. The private sector generally focuses its development and thus its natural hazard risk management actions on the infrastructure that it owns and operates. More recently, the private sector through globalization has begun to address natural hazard risk reduction concerns of the work force that it employs.
- The DDAs still address natural hazard risk management as a separate rather than an integral part of their mainstream development efforts.
- Governments, no matter their form, as well as the private sector and the DDAs, are at the stage where they always consider their resources as limited and so they continually frame the politically based choices between development options and disaster reduction options. These three stakeholders commonly define and approach risk reduction to natural hazard events as existing outside the development agenda. Their approach in dealing with climate change and its impact on development is a prime example.
- Until recently and largely due to the catastrophic results of natural hazard events that have prompted repeated disaster declarations by the same national, provincial, and municipal governments, there has been some discussion but less action on defining acceptable levels of risk and investing in risk reduction actions in the context of development.

The United States' natural hazard DRR effort is only now beginning to take on the dimensions of broader and more-pervasive risks at the natural environment-society interface. The nation does not yet have a natural hazard risk management approach in any way comparable to the Clean Water Act of 1972 or the Clean Air Act of 1963 with their regulations and enforcement. Great forces are acting through democratic and market processes to thwart if not prohibit federal, state, and local action and to deny the creation of mandated private sector action. Knowledge based on ever-more-abundant experience makes manifest the impact of natural hazard events on the built environment, the potential for such events in the future, and the ways to avoid the disastrous consequences of those events. Yet government and private sector policies are often absent or inappropriate for the present and foreseeable risk management challenges. These policies are a direct result of actions by some to maximize profits and minimize government interference in any and all transactions that impact the creation or transfer of wealth through the built environment. Practice trumps policy in the face of knowledge.

As a society Americans might think that public service infrastructure—education, health, water and sanitation, transport, energy, and telecommunications—is developed to standards that ensure life safety if not continuity of service even if privately owned or operated. This is not the case. The public sector, shaped by conditioning societal forces, is just as apt as individuals and private corporations to relegate risk management to a secondary position behind cost savings, reducing taxes and promoting economic growth. Practice trumps policy and knowledge.

At the international level as related to developing countries, both public and private sector actors alike acknowledge the evolving shift among governments in recognizing the explicit and implicit costs of disasters. These actors also note that these governments are now more attuned to citizens' rising expectations regarding their government's capacity to manage emergencies and respond to disaster losses. They further note the political price to be paid for getting it wrong. The trend for the public is to use insurance and other risk transfer mechanisms to cover losses without investment in mitigating economic and physical losses before the natural hazard events occur. The amount of public sector uninsured losses to natural hazard events continues to grow, as does the extent of government liability. Policy trumps knowledge and practice.

The price associated with political will, which is the willingness to risk taking an unpopular position, often means having faith in what knowledge—including traditional as well as scientific knowledge—can bring to public policy even while acknowledging that some doubt exists. This is often referred to as the "no-regrets approach" to natural hazard risk management, particularly to climate change risk reduction.

Leadership to create policy need not be free of doubt, but it does face the issue of practice as exemplified by the cost of risk reduction. Avoidance of accountability and responsibility for risk as well as investment in risk reduction through risk transfer is currently the favored strategy for dealing with risk to the built environment.

Doubt exists even among some scientists as to the origins of current climate hazard conditions, yet the majority of nations believe that global warming must be addressed now. Those individuals who deny climate change and global warming often do not hold any doubt about the deniability of population-induced climate changes. It is with certainty they hold that changes beyond those demonstrated by linear trends are not believable. Their approach is compounded by urban and rural areas experiencing land use change and natural resource management derivations far exceeding in rapidity and intrusiveness what has occurred in the past. These deniers have no previous experience with these increasing natural hazard event impacts under these human-made conditions, let alone natural hazard events not experienced in recent history. Policy trumps knowledge in the face of a need for practice.

Multilateral Development Banks

The World Bank, African Development Bank, Asian Development Bank, Inter-American Development Bank, Caribbean Development Bank, and their sister institutions operate as multilateral development concerns. They are primarily lenders to sovereign states although now the primary source of borrowed capital for economic and social infrastructure by most developing countries is private capital markets. Nonetheless, much of the lending of MDBs along with bilateral lending is aimed toward poverty alleviation whether through direct investment in social infrastructure and human services, or indirectly through transport, energy, and agriculture. Such programs are defined as development through economic growth, trade, and globalization.

It would appear at present that there is a restrained effort in the MDB community to bring substantive, coordinated contributions of knowledge, policy, and practice in natural hazard vulnerability reduction to borrowers' and lenders' development actions. While MDBs lend and grant resources for rehabilitation and reconstruction from damage and loss of built environment components following a disaster, they recognize that much of the economic and social infrastructure components damaged or destroyed and much of the capital made available for rehabilitation and reconstruction were originally granted by the same institutions for development.

MDBs have a long-standing policy concerning nonintervention in sovereign state affairs except when their own interests such as repayment of loans are threatened. No matter what the origin or causal factor of the risk to the built environment arising from their loans, the MDBs assume no responsibility for the creation or management of risk. MDBs most often do not make public the physical structures developed with proceeds from their loans. While MDBs have long been involved with financial risk management concerns, they have recently supported assessment of economic risk, but still avoid involvement in physical risk management. Financial risk remains the primary concern because it directly addresses the capacity of the borrower—most often a sovereign state—to pay back the loan.

Support for the calculation of economic risk comes with the growing recognition that disaster losses of nations are most definitely decreasing the growth of impacted economies, disaster reconstruction funds notwithstanding. Avoidance of involvement in physical risk management is to avoid any culpability for damage or destruction by leaving such matters to the borrower using best local practices. The DDA stakeholders are talking about the financial risk to borrowers and lenders in harming each other's financial mutually interdependent status. Policy trumps practice.

MDBs have not led the creation of public policy risk management institutional capacity across sectors because this is outside the purview of accepted policy focus even if practice-based knowledge is available. Lack of sector policy is due in part to the confusion over, and political expediency of, attending to natural hazard-induced crisis management on a one-off basis. Risk management is not yet defined as central to development knowledge or practice. Policy trumps knowledge and practice.

Nongovernmental Organizations

At the international and regional levels, NGOs may be further along than MDBs in holding discussions on disaster—development linkages. But NGOs are just as challenged as the development banks to carry out mainstreaming of disaster risk management (DRM) into development. They are leaders among those acknowledging that there are no precedents for the climate variability or sea level rise that many countries are experiencing, where they are actively involved with development and disaster assistance programs. There are divisions and turf areas in both NGO and MDB institutions built around funded and highly visible yet dual programs of development and disaster management. Unfortunately, the risk management of natural hazards has not been identified, made visible, nor acted on in a manner sufficient to rethink both disaster management and development program approaches. This is particularly true in the case of economic and social infrastructure such as schools, health facilities, potable water systems, and rural transportation, energy generation, and agriculture. The exception is those projects developed under the title of climate change adaptation; those are apt to be deliberately not identified as development or disaster management programs. Climate change adaptation projects are increasingly identified as part of environmental programs, which for the past fifty years have maintained a sector identity and presence apart from development and disaster management. Policy trumps practice and knowledge.

Disaster and Development Agencies

The call, the case, and the opportunity for investing in resilience are as much a part of every development action as are the cost and benefit, winners and losers, and beneficiaries and benefactors, whether manifest as such or not. Unfortunately, from the point of view of obscuring risk in development actions the DDAs are at a stage where at best these agencies see investment in resilience—to reducing risk to levels of loss to below disastrous situations—as something to be brought in from the outside to development considerations.

The investment in resilience has reached levels, particularly when dealing with climate change adaptation, whereby project proposals and budgets are presented to reflect "with and without" risk reduction measures. On the one hand, this is mandatory in order to access financial support reserved for climate change adaptation and other environmental objectives. Such support often operates outside mainstream development and risk management processes. On the other hand, such actions are presented with and without risk reduction measures in order to identify the investment in resilience that is taking place alongside investment in development. Policy trumps practice.

Behind the with and without approach is the now three decade's DDA accommodation of the creation of the disaster sector. This sector came about through the accommodation of the even-longer-standing existence of dual development and disaster response and humanitarian assistance programs separate in design and implementation in many of the organizations and institutions of the DDA. Practice trumps policy and knowledge.

International political organizations including groupings of sovereign states have a particular way of dealing with sovereign state reporting of outputs (few and far between) and inputs (often qualitative in expression). When plausible, the DDAs often prefer to scale the qualitative elements of measurement to express them in quantitative terms. Post-disaster project execution evaluations become the end of project progress reports citing means but not verifiable outcomes. On that basis, the DDAs can make subtle yet often very general analysis, comparisons, and categorizations. Practice trumps knowledge and policy.

The DDAs, in the name of sovereign states and in response to political pressure, can engage in counterproductive actions that are in direct contradiction of their policy. DDAs collect inappropriate and potentially harmful goods (clothing and medicines, respectively) for disaster-impacted populations in the face of internationally ascribed donation policies. Practice trumps policy and knowledge.

In summary, risk reduction to natural hazard events is now more broadly discussed in the DDAs but effective policies and intervention schemes are still not consistently part of the development agenda. Presently the DDAs' rewards system—both real and implied—seems to be at odds with efforts to invest in risk reduction. DDA success is highly focused given the mounting human and built environment losses on doing a better job of humanitarian assistance and disaster relief and rehabilitation. In parallel, the transformation of disaster management as risk reduction into a development-centered approach is at best at the beginning stages of discussion at most policy and administrative levels. Specific economic sector entities both public and private are beginning to discuss the creation of their sector-specific emergency management capacity involving specialized agencies. A shift to a development-centered approach in policy calls into play development theory and practice, particularly as it addresses the poor and poverty, gender equality, civil society, and highly vulnerable populations targeted in the MDGs and Sustainable Development Goals (SDGs). It also calls into play administrative decentralization, land use management, planning control/regulation/enforcement, governance, transparency, corruption, and accountability. Practice trumps policy and knowledge.

Closing

Still there are ample reasons to believe that as the focus on DRR grows, the influence of disaster risk finance policy will commensurately overtake the definition of the problems and the solutions that stakeholders seek, create, and use.

The past fifteen to twenty years has witnessed a major shift in DRM thinking. Governments and their development partners have begun to pivot from a focus on merely the consequences of disasters to recognition of their responsibility to reduce natural hazard event impacts through active risk management. The dimension of disastrous natural hazard events is growing but is still dwarfed by the dimension of other challenges such as civil and religious wars, terrorism, ethnic cleansing, and genocide. Efforts and investment to enable proactive risk management investments remain disappointing and uneven. There are many factors responsible for these modest results. Arguably the most important factor lies in the fact that altering the risk management landscape requires systemic change—change affecting fundamental concepts of development, functions of government, and the roles of sovereign states, MDBs, NGOs, and the DDAs. Another factor lies in the changes in the relationships between government and the private sector. These are not small matters, and alteration of the development-disaster management landscape will not happen quickly. These stakeholders must look for DRR actions to lessen the number of disasters even as the location, frequency, and severity of hazard events change. The established physical form-giving forces of development do not cede control easily, nor do their constituents. Nevertheless, these actions are preferred by some as the prime motivational force to reshape development rather than waiting and reacting to catastrophic losses. There are two main paradigms:

- *Disaster losses giving shape to development:* Losses are growing at such a pace that post-disaster needs will lead to the overshadowing of the forms and functions of what is commonly understood as development planning throughout the world.
- *Risk reduction shaping development:* Development that inter alia limits the impact of natural hazard events so that there is a discernible difference between development dealing with betterment that includes less vulnerability to natural hazard events, and development dealing primarily with the aftermath of natural hazard events in the name of recovery and the dictates of the marketplace for returns on investment in the vulnerable society.

Nation-states need to know much more about natural hazard risk and about society as it has evolved over the past six decades in terms of time and place in relation to physical character, built environment-related social relations, and culture. Those nation-states in the throes of a representative democracy governed by laws that protect not only the rights of individual citizens but also society at large must constantly identify, discuss, and decide on risk reduction actions in terms of the gaps between knowledge, policy, and practice. For these societies, choices as to what is known about who is at risk, and why and what can be done about it, especially for the poor, are paramount. Those countries in the throes of a free market economy whose presence and power strives toward maximizing gains with minimal expenditure of one's own capital must constantly identify, discuss, and decide on who will pay, how, when, and to whom for the losses that are the result of constructed risk through economic development. These discussions should be based on who will be responsible and accountable, and how to deal with populations that are knowingly and unknowingly subjected to exposure and who are powerless to reduce their vulnerability to natural hazard events.

The following is a list of issues and related indicators that will contribute to risk reduction on a priority basis. They point to the resilience of the built environment at all geographical scales and places. These issues and indicators make visible the link between history and culture to society's manifestation of the built environment. These issues and indicators should be considered before disasters constitute the societal norm for the outcome of natural hazard events. These issues and related indicators explore the many paths available for effective risk reduction of those most vulnerable and least able to bring political, economic, and institutional weight to bear to address their vulnerabilities and losses. The issues and related indicators needed are as follows:

1. Monitoring water quality and quantity for all populations and their needs. Water is the primary natural resource concern across the globe. It is, in one form or another, an issue in all types of natural hazard events. Water is on a trajectory to become too scarce and the opportunity to solve this concern is narrow and rapidly closing.
2. Monitoring the role and impact of natural hazard event-induced migration on development actions by sectors using visible development indicators. Disaster-induced intra- and international migration ranks with globalization, civil strife, terrorism, and war in terms of how to meet the challenges and opportunities to those on the move.
3. The creation and implementation of risk transfer schemes that include economic and physical as well as financial indicators to support risk reduction as well as emergency response.
4. Monitoring the participation and impact of remittances in developing countries' hard currency inflows and their use particularly after a disaster so as to determine the best use of remittances and why, and by whom.
5. Creating and disseminating on a cost-free basis as public good natural hazard information using qualitative and quantitative indicators covering discrete population groups, geographical places, and hazard types.

6. Monitoring national indices focusing on sovereign state–specific, sector-specific, population group–specific and geographic place–specific vulnerability, risk, repetitive losses, and disaster declarations.
7. Monitoring and reporting on the investment in social and economic infrastructure sector by sector with corresponding estimates of the probable maximum loss due to natural hazard events, expressed in not only financial but also economic and physical dimensions.
8. Monitoring and reporting in a more consistent manner on how national and local governments have influenced natural hazard event consequences and outcomes traced through loss of life, population affected, and physical, economic, and financial loss.
9. Monitoring and reporting on DDAs and governments using their catalytic role to develop actions that deliver real-time rewards for investments while producing benefits tomorrow. Such structures should involve incentives and linked benefits—not to produce long-term dependency or subsidies, but to energize a risk management framework that not only tolerates long-term thinking, but also rewards it.
10. Presenting an explanation of critical moments in time—most appropriately presented as timelines—where societies' approach to creating their built environments have been significantly impacted by natural hazard events; governmental responses through the executive, legislative, and judicial responses to natural hazard risk and disasters; and economic response to post-disaster, reconstruction, and pre-natural hazard events.
11. Creating and reporting on a composite country resilience and built environment vulnerability indicator using existing indices and other measurements outside the purview of sovereign states—failed states, repetitive disaster states, rogue states, and so on—as well as authorized development and disaster-related measurements including the human development index, GDP, probable maximum loss by sector, disaster losses, and population affected by events and at risk.
12. Monitoring and reporting on verifiable and sustained DRR through executive power, legislative and regulatory frameworks, and judicial review in the context of development:
    • the use of disaster reduction actions through development that leads to verifiable social change that benefits marginalized, poor, and vulnerable populations;
    • the use of economic and social infrastructure development to enable risk reduction actions before disastrous natural hazard events; and

- the identification of the opening of political space for the contesting and cooperative use of political power in the wake of disaster shocks.

Conclusions

In the past when faced with the risk of annihilation of societies due to the threat of nuclear war, the risk management response of sovereign states was a strategy called MAD—mutually assured destruction. Now a risk faced around the globe is again generated by human action. It is, put simply, development-induced vulnerability to catastrophic natural hazard events. But this time the risk should be managed by a mutually acceptable collective strategy, mutually assured survival (MAS).

Stephen Bender, registered architect, served the Organization of American States (1979–2005), creating and directing a technical assistance program dealing with natural hazards and disaster risk reduction as part of international development assistance. He earned a bachelor of architecture degree at the University of Notre Dame (1968), served as a Peace Corps volunteer and trainer (1968–71), and served as a university lecturer and researcher at Rice University, where he earned a master of architecture degree in urban design (1974). He has served as a consultant to the Aga Khan Development Network/Focus Humanitarian Assistance, Asian Development Bank, German Technical Cooperation Agency, Inter-American Development Bank, Organization of American States, United Nations International Strategy for Disaster Reduction, and the U.S. Agency for International Development.

References

Hagmann, G. 1984. *Prevention Better Than Cure*. Stockholm: Swedish Red Cross.
Wijkman, A., and L. Timberlake. 1984. *Natural Disasters: Acts of God or Acts of Man? An Earthscan Book*. International Institute for Environment and Development.

CHAPTER 4

Humanitarian Response

Ideals Meet Reality

ADAM KOONS

Introduction

The term "humanitarian response," as used in this chapter, is response to disasters and crises that are severely affecting people in particular defined locations, and that are precipitated by particular episodes of natural or human-made disruptions such as drought, or conflict, or both simultaneously. In the midst or aftermath of a disaster or crisis, humanitarian response actors seek to save lives, alleviate human suffering, provide basic human needs, restore well-being, and maintain human dignity.

A considerable number of formal and informal mandates, policy frameworks, and guidelines addressing humanitarian response prevail. They include such tenets as the Rights-Based Approach, Do No Harm, the Humanitarian Charter, and the Sphere Project's minimum standards in humanitarian response. There are also numerous global collaborative bodies, such the United Nations (UN)–led Inter-Agency Standing Committee and InterAction, the 190-member consortium organization of United States–based nongovernmental organizations (NGOs), working to adapt and operationalize the policies and standards more effectively. They accommodate and anticipate or predict what *should* happen during emergency response. However, disasters are messy. They do not follow the plans or standardized rules. Each is somehow different and unique. There is always a difference between what *should* happen and what *does* happen.

This is the gap. Even as emergency relief workers understand well the disconnect between the ideal (the *should*) and the reality (the *does*), the challenge always lies in making real-time immediate decisions that have implications in ethics, politics, culture, socioeconomics, and relief operations. This chapter will explore some of these dilemmas from the perspective of a full-time practitioner. The discussion is in no way a criticism of the standards, but rather in support of them as vital and tremendously useful components for aid workers, and so is an exploration of how the standards fit into effective humanitarian response.

Background

It is important to set the context to understand the comments that follow in this chapter. How am I qualified to present this discussion? I am a full-time humanitarian response practitioner and have been an international NGO decision maker, with thirty-five years working in this domain. Recent positions have included global director of humanitarian response for a large international NGO and senior vice president of global programs for another humanitarian-focused international NGO. With these and earlier positions, I am very often a member or leader of the first team to hit the ground in an emergency response; other times I am in constant communication with that team. I then normally stay involved in some way throughout the response and into the recovery period. My same standing occurs in longer-term and chronic humanitarian situations, such as complex emergencies, conflict-derived crises, and refugee or displacement circumstances. Though I am a beneficiary of others' research, I do not conduct research, and the realities of the job generally preclude sufficient time for in-depth analysis.

Therefore the following discussion is based on direct personal experience in most of the large-scale disasters of recent times, including, among others, the Darfur (Sudan) crisis starting in 2003; the 2006 Yogyakarta Earthquake in Indonesia; Hurricane Katrina in 2005; the Pakistan floods of 2010 and 2011; the refugee situations in Sudan, Somalia, and Syria over many years; the Haiti earthquake of 2010; the Horn of Africa famine/drought in 2011; the Libyan war in 2011; the Philippines typhoon in 2013; the Ebola crisis in Guinea during 2015; the Nepal earthquake in 2015; and the ongoing crises in Afghanistan, Iraq, and Syria. In order to broaden the ideas represented here and to verify my own ideas, I have interviewed a number of similarly experienced aid colleagues on the topic presented here.

My own role as stated has often been as response team leader, arriving with the first external aid workers to join with local first responders to supplement them with specialized expertise, human resources, and funding. Variously, I have been involved in conducting initial emergency needs assessments, designing response programs, writing funding proposals, coordinating within UN systems with responding agencies, negotiating funding with donors, and implementing programs.

When we arrive, often within twenty-four to seventy-two hours of the precipitating event, we frequently find the local population serving in the immediate first responder role without any particular expertise or resources, but just because they are also victims and are helping each other. In certain settings, local people are frequently quite expert because they have repeatedly been faced with the same sort of disaster in their locations and have developed good skills, although usually insufficient resources. We also sometimes find local authorities trying to undertake their own first steps in response, although they are often unprepared and untrained. Sometimes local authorities do not assist or are completely overwhelmed.

Where the organization already has an existing long-term, well-established development program, and an operational infrastructure, we enlist our own local staff. These offices are usually already taking action according to various levels of existing plans and preparedness, and guidance from headquarters. In such cases, our incoming team comprises emergency specialists with skills not already found in the local programs. In other cases, where we have no preexisting program or office, we literally arrive with backpacks on our shoulders and no infrastructure or support system waiting for us. At the same time, teams from dozens of other similar organizations are going through the same process and the UN is setting up its own structure.

Before moving on, it is appropriate to define the term "aid workers," and to make a distinction between foreign and local aid workers. "Aid worker" is a very broad and ill-defined category of people. Some are seasoned professionals with specialized training in a range of technical and operational topics who have been conducting disaster response for decades in a variety of roles. Others are specialists in logistics, emergency shelter or water systems, nutrition and food delivery, or generalists with a wide and deep knowledge of all of the aspects involved in emergency response, including fund raising, coordination, donor regulations, and more. Aid workers know the processes, standards, protocols, and coordination systems, and generally know one another from different NGOs, UN agencies, and donors institutions from their many years of interaction around the world. They have a wealth of experience adapting to different emergencies and knowing what works and does not. But, at the same time, many emergency staff are young and eager workers with little on-the-ground experience who must depend on strictly following existing guidance and protocols (if they are even familiar with them, and veteran workers often have serious doubts about that), and/or take direction from a veteran team leader. Still others are thrown into a response from the very differently structured and paced development side of international work due to circumstances or country experience. In short, they were at the right place at the right time or else happened to be available to deploy. They, too, are well trained, but not in emergency response. They may be struggling to adapt to the very different lightning pace of the situation: the urgent needs to save lives, chaos, high stress, and the potentially traumatic and unique mechanisms of disaster. Still others with no experience or training even remotely relevant somehow arrive saying they "want to help," often unaffiliated with an official professional agency or organization.

Most importantly, the largest and most critical cadre of relief workers are local professionals (as differentiated from local victims who also participate significantly in relief), working in their own country to aid their own compatriots, but often left following the instructions of outsiders in leadership roles. Whereas international professionals may arrive with a wealth of global experience, specialized skills, the ability to adapt and compare from past lessons learned, and knowledge of international systems and standards, their local counterparts have invaluable local understanding and perceptions. Sometimes their own communities, friends, and families have been victims of the disaster.

I include in this discussion both natural disasters and conflict-driven humanitarian disasters because, in reality, the same aid givers work in both environments and very often are faced with complex emergencies that overlay both categories in the same place simultaneously, such as drought during a civil conflict.

Guidelines

Each disaster (or conflict-driven humanitarian crisis) occurs through a convergence of factors in time and place including local culture, economics, politics, ecology, infrastructure, and more. The factors' convergence makes the result unique, though perhaps similar to other disasters elsewhere in certain patterns and superficial appearance. Nonetheless, with no time for careful research or intensive analysis, a viable, feasible, effective, cost-responsive plan must be developed and implemented quickly in order to save lives and prevent suffering.

Fortunately, aid workers have a large body of guidance literature, standard operating procedures, frameworks, standards, and principles. These have been built over decades of response experience, post-action analysis, and a commitment to continue improving our responses to save more lives, reduce continued suffering, and to help people retain and reestablish dignity faster and better. Most aid workers also focus on the principle of "build back better."

These templates for response have primarily been developed by practitioners themselves, so they are, in fact, realistic and based on experience, applied research, and analysis from within the humanitarian community. The templates are driven and funded variously by UN agencies or initiatives and often by consortia of agencies that include international NGOs, universities, governments, international organizations such as the International Federation of Red Cross and Red Crescent Societies (IFRC), and UN bodies. The templates range from statements about universally agreed humanitarian principles and aid worker behavioral standards, to specifics of how much water to provide a refugee.

The collective body of guidance is meant to systematize disaster relief and humanitarian action and to prevent starting from scratch each time and reinventing the wheel. The guidance gives aid workers a head start to save time, resources, and—most importantly—lives. Its basic goal is to provide the following:

1. An agreed-on set of humanistic and humanitarian principles
2. Common framework of agreed targets
3. Minimum standards
4. Common language to enhance collective and collaborative action and coordination
5. Standard methods for assessment, design, and implementation
6. A way to prevent starting from blank in every new situation
7. A head-start for those without sufficient prior personal experience

Though it is not the purpose here to describe and examine each of the guidance documents and agencies that provide such humanitarian roadmaps, it is useful to present an idea of the range and variety of these tools. Some of the key ones include these:

1. Humanitarian Charter
2. Code of Conduct of the International Committee of the Red Cross and Red Crescent Societies (ICRC)
3. Core Humanitarian Standard
4. Model Act for the Domestic Facilitation and Regulation of International Disaster Relief and Initial Recovery Assistance
5. Collective Responsibility: Perceptions, Expectations, and Realities of NGO Coordination in Humanitarian Leadership
6. International disaster response laws, rules, and principles, developed variously by different agencies (IFRC, International Council for Voluntary Agencies, U.S. Office of Foreign Disaster Assistance, and others)

7. Humanitarian Practice Network
8. Sphere Project minimum standards in humanitarian response
9. Livestock Emergency Guidelines and Standards
10. Field Operations Guide for Disaster Assessment and Response, U.S. Agency for International Development (USAID) / U.S. Office of Foreign Disaster Assistance)
11. Numerous documents on best practices in every sector or setting that are prepared and circulated by NGOs, donor governments, and UN agencies
12. Real-time evaluations undertaken by the UN during emergency responses
13. World Humanitarian Summit, under preparation for years, that occurred in May 2016

In and by itself, the guidance covers a very broad range of protocols from general principles of behavior, ethics, and perspective to very specifically quantified material targets according to sector (e.g., food, water) or topic (e.g., refugees) of response. The directives are sufficiently broad—for example, they urge attention to local culture and indicate the number of liters of water to provide per person—that they might seem like comparing apples and oranges. But since apples and oranges are both fruit, the standards all invoke the same intent: improving the outcomes of disaster and humanitarian response. They are all important and they all connect in their own way. Those such as the humanitarian principles derived from the ICRC, for instance, require aid to promote

1. humanity,
2. neutrality,
3. impartiality, and
4. independence.

Likewise, the *Code of Conduct for the International Red Cross and Red Crescent Movement and Non-Governmental Organisations (NGOs) in Disaster Relief* include the following tenets (IFRC and ICRC n.d.):

1. The humanitarian imperative comes first.
2. Aid is given regardless of the race, creed or nationality of the recipients and without adverse distinction of any kind. Aid priorities are calculated on the basis of need alone.
3. Aid will not be used to further a particular political or religious standpoint.
4. We shall endeavour not to act as instruments of government foreign policy.
5. We shall respect culture and custom.
6. We shall attempt to build disaster response on local capacities.

7. Ways shall be found to involve programme beneficiaries in the management of relief aid.
8. Relief aid must strive to reduce future vulnerabilities to disaster as well as meeting basic needs.
9. We hold ourselves accountable to both those we seek to assist and those from whom we accept resources.
10. In our information, publicity and advertising activities, we shall recognise disaster victims as dignified humans, not hopeless objects.

At the other side of the spectrum are such guidelines as the universally endorsed Sphere minimum standards (Sphere Project 2018), that include, for example, such targets for disaster victims as ensuring access to

1. water: 15 liters per person per day,
2. shelter: 3.5 square meters per person,
3. sanitation: 20 people per toilet, and
4. food: 2,100 kcals per person per day.

It is worth noting that the Sphere manual is impressively comprehensive, far beyond simple metric targets, and immensely valuable for creating common points of reference and ensuring that a wide range of considerations are taken into account within the numerous elements of a humanitarian response. The manual explains each target in detail, and sets each into context, while justifying the targets and discussing possible nuances. For example, in the shelter and settlement section of the Sphere manual, the discussion includes various types of shelters, notes on displaced people being hosted by other households, debris removal, the use of public structures like schools and health facilities, property ownership and land rights issues, site selection and drainage, settlement patterns, and health vectors in settlements. It acknowledges cultural practices, climate, design participation, materials and materials access, and a host of other issues surrounding the topic. None is meant to be prescriptive; all are intended to ensure that aid planners and program designers include such considerations. The other sectors of Sphere, such as water, sanitation, and hygiene; food and nutrition; health care; and others, use the same approach.

A word about the origins of the standards and their alignment with reality is important. All are an outgrowth of practice itself, derived from field practitioners who have collectively over time felt the need to codify the thinking that guides implementers as a result of experience, skills, knowledge, and numerous deployments or assignments. They are the result of a thoughtful, analytical, and consultative processes of developing, to the degree possible, universal statements from a multitude of examples. They are cumulative lessons learned. In this respect, the ideals embodied in the guidance are generalizations of the reality of humanitarian response and represent the ideal composite of what aid workers should strive and hope to do.

Applying generalized principles and standards in specific disaster instances invokes the question of whether there is a disaster universal or whether all disasters are, in fact, locally specific and unique. For aid workers and the victims of disaster there are undeniable universals, such as our collective desire to provide aid equitably, neutrally, and impartially; and to do no harm. Equally, we strive to enable victim dignity, reduce suffering, facilitate a return to normalcy and stability, and provide the essential material components of survival and well-being. The universals for victims include trauma, loss, self-preservation, resourcefulness, self-help, mutual and collective aid, empathy, and community cohesion, among others.

There are also numerous obvious specificities and unique characteristics in every case. Some are based on the fundamental local circumstances of the people impacted, their culture, economics, politics and power, geography, social networks, and structures of interaction. Some are based on the effects of the particular disaster on local physical infrastructure, support systems, physical access to affected areas, the response of existing coping mechanisms, government response, community response, access to resources.

The Dilemma / Challenge

Here is the issue framing this chapter: With all of the excellent and relevant guidance that is at the aid workers' disposal, what happens when aid workers hit the ground for a particular response?

We find a unique situation, but one often framed by known or predicted and expected patterns, behaviors, and needs. Strategies and actions for a particular response—which have to emerge immediately—can be developed only after matching what we should generally do (the existing standards and guidance and protocols) with the specific situation at hand. Decisions that comply with, and use, the standards should be made by informed calculation. However, in each particular emergency we always need to ask, which of the many standards and guidance might work right now, and to what extent?

Simply put, is there a gap between the guidance and what we need to do in a particular response? And does the guidance adequately cover all the bases? Put another way, how prescriptive should, or can, the guidance—the ideals of good response—be? To what degree can aid workers depend on the guidance to steer their on-the-ground decisions and activities? Do the existing policies, practices, and tools fit any given current specific situation? And, a corollary to this is, must we know the entire guidance toolkit to determine what to do, and to know the range of choices at our disposal?

The following discussion is built around a general set of questions:

1. Realities: Are there really gaps between the ideal and reality?
2. Response Program Planning: What are the types of decisions or situations that create the greatest gaps or divergence from the generic guidance?
3. Gaps: If there are gaps, does it matter?
4. Many Gaps: What are the consequences of these gaps on our program effectiveness?
5. Gap Consequences vs. Expediency: Can such gaps be bridged?

Realities

To begin to assess the existence and degree of gap, it is necessary to know what types of factors influence the divergence or alignment of universal guidance from each individual disaster. It is this intersection that enables (i.e., hinders or facilitates) the development of actual strategies put into action in a given response. When a response is planned (at dire speed in most cases), some of the kinds of factors that must be accommodated include these:

1. Universally agreed standards, principles, and guidance known by the responders
2. Guidance and protocols from the responding agency headquarters that aid workers must follow
3. Host (affected) government perceptions and reception of external assistance
4. Host government response capacities, systems, and political will
5. Local government (national, provincial, district, etc.) preparedness for the current crisis
6. Local community preparedness for the current crisis
7. Competing or contradicting or hidden political agendas
8. Geographic setting (e.g., rural or urban), whether accessible (and if so, by what transportation means)
9. Safety and security of aid workers and victims (e.g., natural hazards, conflict, civil turmoil)

10. Local coping strategies (from past similar disasters) or new strategies (based on current knowledge and resources), social safety networks
11. Sociocultural and political context
12. Universal disaster victim characteristics and traits (as defined in research and studies)
13. Locally specific disaster victim responses (linked to coping strategies and preparedness)
14. Aid worker skills and experience
15. Aid worker knowledge of the local context (e.g., resident vs. external)
16. Local structure and capacities of the responding agency (national agency, international agency, development specialized or humanitarian experience, technical skills, staff size, etc.)

Added to these factors are additional elements requiring consideration in mounting every humanitarian or emergency response:

1. Complexity: Is it just a simple major drought or is the drought in a war zone?
2. Phasing: Is it a sudden and unexpected and clearly defined earthquake event, or a drought that evolved over many months and was finally declared a disaster?
3. Timeframe: Are efforts needed for immediate lifesaving or is basic sustained survival the priority?
4. Resources: How much will be needed financially and materially to respond? Where will the resources come from? And will they be sufficient to satisfy needs?
5. Technical: What specialized skills will be needed (e.g., water infrastructure, emergency medicine, acute malnutrition response, mass food delivery)?
6. Logistical: How can people and materials best reach affected populations? Are constraints to access caused by physical barriers, or by insecurity and conflict?
7. Competition and timing: Resources are limited and always insufficient, and timely response is needed to assist people. The faster an organization can prepare a high-quality funding proposal the more likely they are to obtain resources. Usually, this process is a matter of days, or at most weeks, between disaster declaration and assessing, designing, preparing, and submitting a funding request.
8. Specialized coordination: Is there central coordination from the UN or the government (or among responding agencies themselves) helping to define common strategies and tailored standards?

Response Program Planning

We need to use the standards and universals and appropriately align and merge them with the realities, assessment, and local current information. Accessible, sufficient, appropriate information, baseline data, sociocultural knowledge, and metrics about the disaster and its effects and magnitude of needs is usually poor quality, but are instantly needed for key decisions that will affect people's well-being and lives and have long-term consequences for them, as well as for agency success in obtaining funding.

It is difficult to conduct good assessments with limited time, capacity, focus, and tools. It is like scuba diving at night: you see only what you point your light at and can only use what light you have. Initial information often comes from proxy sources, local nongovernmental organizations, displaced persons, and volunteers with varying degrees of observation and interpretation or analysis skills. The information is often filtered through perspectives influenced by specific agendas or biases and conditions and can therefore change very quickly.

But as quick-and-dirty and shallow as it often is, assessment information needs to lead to a real project design built on best assumptions about the feasibility of activities and veracity of targets. Those targets become promises to beneficiaries and commitments to donors, and they also guide the way projects are implemented. It is at this point that the standards and principles initially meet reality: in the process of program design, while using the input from the assessments.

Gaps

Certainly, there will always be gaps. This is true regardless of whether the standards are those focusing on ethical and behavioral guidance or those recommending technical targets. In some settings our ideals of neutrality, or impartiality, or equity cannot be fully fulfilled. For instance, from the author's personal experience, trying to apply commonly agreed (within the humanitarian community) universal ethical, moral, humanitarian principles while also, for example, working through local village leadership in rural Afghanistan, and demanding that elders treat all villagers impartially, and to provide goods to women equitably, is nearly impossible, and presents obstacles requiring culturally informed and innovative solutions. As anthropologists we know the constraints in this challenge.

As anthropologists we also recognize, and often struggle with, the understanding that we are working within primarily externally defined theories of what constitutes universal ethics, morals, rights, needs, and well-being; theories that are defined by the cultural orientation of those who promote those constructs as universals. While nonetheless built on significant and broad empirical evidence, we may still question how truly universal these theories are that define and guide humanitarian response principles and actions. From this perspective, any variation between the universal theories and humanitarian principles being applied and local realities could constitute a gap.

The aid worker should also know the potential risks of applying such universals. A key risk is in trying to enforce an external principle and thus alienating the local structures that depend on the provision of aid. Another important risk is making allowances (bending the principles) in order to provide some level of aid instead of being entirely denied access and opportunity by the local traditional gatekeepers, decision makers, and power brokers. These are very difficult situations that require significant negotiation skills, diplomacy, and flexibility.

Similarly, and also from personal experience, with specific universal metric targets to meet, it is very different at times to ensure twenty liters of water per person per day to an earthquake victim in urban Haiti and a drought victim in rural Ethiopia. On our side we see the logistical circumstances and budget realities. From the recipient's side they see their needs through the lens of status quo norms, cultural practices, and household priorities. Although providing twenty liters of water a day— which we had to tanker truck in—according to targets derived from minimum standards and on-the-ground capacity of aid agencies under severe drought conditions in the Somali region, Ethiopia, in 2011, it turned out that the people were using only ten liters for human needs, and giving the other ten to their animals to prevent the loss of their only livelihood assets. Yet while this countermanded the aid agency standard, it fit the people's cultural need.

As one aid worker colleague commented, "The standards are necessary, but in regard to humanitarian principles, reality cannot ever be totally impartial or neutral."

Many Gaps

There are also different kinds of gaps. The inability of the standards to address all the circumstances as they occur locally is a common gap. Insufficient knowledge of local context and baseline data is another. Aid practitioners' unclear understanding of, and inconsistent use of, the standards frequently occurs. There are also often clearly identified gaps in the simple management of resources, response capacities, and skills.

I noted earlier in this discussion that the standards are derived from real experience and needs on the ground. But they are often articulated, debated, finalized, and agreed at the higher policy levels away from front-line aid workers. The issue then arises whether the standards filter back down to the field operations where they are needed the most. In addition, humanitarian veterans are concerned that a great many of the newer aid workers, either those entering aid from other disciplines or those starting their careers, are not well trained or deeply familiar with the standards and approaches and principles. One infor- mant noted a recent sizable survey conducted by an NGO consortium that found that about 38 percent of aid workers were not familiar with the Sphere minimum standards. In another survey, this one by Building a Better Response (www.buildingabetterresponse.org) of a hundred organizations, most of which were NGOs, the majority of respondents said they were only "*somewhat* knowledgeable" about the international humanitarian system.

The gap in awareness can also be particularly true of volunteers and within disaster-affected countries for personnel of existing or new national or local NGOs, civil society organizations, and government offi- cials. This has been clear in such responses as the Haiti earthquake and the Syrian conflict. We could presume that somewhere in the disaster response program cycle—(1) assessment, (2) design, (3) implementation, (4) evaluation—senior or veteran aid professionals know the standards and build them into those four steps. Thus, in one sense, the real gap is not with the standards themselves but with the focus and sufficiency of ensuring that aid workers know, and use, them.

Humanitarian workers agree, including those specifically interviewed for this discussion, that, most importantly, the standards are not ideals. They are, in fact, the *minimum* thresholds we are aiming for. Also, stan- dards provide invaluable targets that are fundamental to our work. Every principle, every code, every metric-based standard provides clear aspira- tional guidance. The targets embodied in the guidance, whether behav- ioral or material, create a structure around which to work, plan, design, and implement. They motivate action by providing a starting point, and they enable us to break down a very complex, often chaotic, urgent, and stressful situation into a set of categories of response—that is, they pro- vide a framework that sets the foundations of our response programs. Put most simply, that framework sets up blanks that we can fill with the reality we find in each particular disaster or crisis. The framework also points to what we need to look at, look for, and pay attention to during our assessments, design, implementation, monitoring, mid-course adjust- ments, evaluations, and lessons learned.

Looked at another way, the standards exist to provide a template and allow us to see where we can and should fill in the gaps in a particular situation to design a more locally adapted and appropriate and effective response.

Gap Consequences vs. Expediency

As stated early in this discussion, the aid provision standards are intended to be guidelines, outlining minimum criteria, targets, and aspirations. In such dynamic, fluid, and complex circumstances as disasters and crises, with the exception of the ethical and moral codes, they were never meant to be strict and immutable rules. But they have been interpreted as such in multiple troublesome ways, causing gaps in both technical and operational areas or for social, cultural, or political reasons.

In one sense, the gap matters significantly. When funding is granted, it is based on detailed program objectives with quantified targets, based on standards and benchmarks. Funding agencies demand precise measurement indicators, assessed through careful monitoring and evaluation processes. They understandably seek means and metrics to measure and assess NGO program design, performance, impact, accountability, quality assurance, and effectiveness of programs, both with respect to victim well-being and dignity, and in terms of financial or material investment and cost effectiveness. Implementing organizations depend on demonstrating their achievement of these targets. In the reality of limited resources and a very competitive funding environment, the ability of an NGO to achieve promised outcomes in a particular response program affects its reputation (i.e., its track record), and influences subsequent funding decisions in that country, or even globally.

In another sense the gap is sometimes an unavoidable result of reality and circumstances. While personnel on the ground understand that reality intervenes, that reality can seem to contradict the program targets derived from global standards and mandates that are often perceived to be inflexible operational rules. In one example from my personal experience, in Ethiopia during the drought of 2011 we were thrilled that we were able to meet the global target of twenty liters of water per day per person, for many thousands of people through delivery to dozens of far-flung villages from water tanker trucks covering large distances. But we learned that people were sharing their water with their goats in order to preserve one of their primary livelihood assets, thus halving the amount per person. But we could not deliver more than we were already providing. Another personal experience involves the design of transitional shelters (which is housing meant to be inexpensive but sufficiently strong to suffice for two to three years) for Haitian households following the earthquake of 2010. Sphere minimum standards recommend at least 3.5 square meters (37.7 square feet) of covered floor space per person for such shelter. However, household sizes vary culture to culture as does the use of public space and private space, and there are also differences between rural and urban settlement patterns. Therefore, in this case, as often occurs in collective on-the-ground coordination between agencies, an agreed estimate was established through the Shelter Cluster of UN Office for the Coordination of Humanitarian Affairs. It stipulated that there be five people averaged per household for Haiti for the purposes of this response. Our transitional shelter design was engineered to use the least amount of materials and construction time, and therefore the least cost, while accommodating to the best of our ability the Sphere minimum standards. It was based primarily on using full-size 4x8 foot sheets of plywood and full-size sheets of corrugated roofing material, and thus required the least amount of time and wastage from cutting. Our final funding proposal totaled several million dollars for 3,500 shelters, but the design was a total of one-half meter less than full Sphere minimum standards for five people. Instead of strictly fulfilling the standards, with 17.5 square meters (188.4 square feet), our shelter would be 17 square meters (182.9 square feet). Nonetheless, our donor required a design revision to precisely meet the standards. The new design required considerably more wood and significantly more cutting and construction time. Because funding was limited, we were forced to reduce the number of shelters we could build. The flexibility that was originally intended along with the adaptation to realities was lost, and at considerable expense. Ironically, since we were using a designated average as the family size that was a planning model in this disaster, even had we reached 100 percent of the standard, those with fewer household members would have had excess space while larger families would have had too little. We also undertook the design without knowing what the local shelter norm was pre-disaster per person for living space.

Very often, too, with any of the precise targets, and with always limited resources, we must make difficult allocation choices. If we have only half the food/water/tarps/toilets/and so on available, do we provide 100 percent of the target to half the people affected, or half the inputs to all the people? In the Haitian case, for temporary shelter, the standard was providing two plastic tarps per household, so with limited immediate supply, and for logistical and operational reasons, we chose to provide half the recipients with both tarps, because we expected a resupply soon. Later we learned that many recipient households had given one of their tarps to households that had not received any in the first distribution. Far beyond the scope of this discussion is the extremely complex and highly charged challenge of targeting and determining levels of need and priority.

The same is true also for the less material standards and more behavioral ethical codes, such as equity, impartiality, and neutrality. For example, in Darfur, Sudan, trying to ensure equity in aid for all displacement victims, and, particularly, protection for women, creates friction with government authorities that can, and has, led to agencies being evicted from the country. In such very difficult and delicate situations the choices are to devise creative ways to mediate and soften some of the codes so they appear less objectionable to local powers, or to risk losing the capacity (through departure) of helping anyone at all.

As a result of these multiple realities, aid agencies consistently fall short of targets. Sometimes they "fudge around the edges," as one colleague commented, in order to address the dilemma and satisfy the strictest observers while needing to be pragmatic and practical on the ground.

So, does the gap matter? Yes, because it helps define how close we are coming to the ideals. At the same time, the gap also often matters operationally for agency funding and reputation. More importantly, what matters in disasters and crises is preservation of life, ensuring the least degree of suffering and danger, and striving for the highest level possible of well-being (as broadly defined). Targets will never be entirely achieved and ideals will never be fully met, and gaps will never be entirely filled. If we do not reach twenty liters of water per person per day, but we have saved lives by supplying ten liters, we have, at minimum, fulfilled our mission, while at the same time we realize it is at minimum.

Filling the Gaps: A Possible Chinese Menu

The gap can, and is, filled in several distinct ways. First, the overall universal mandates, standards, codes, theories, approaches, and guidelines are constantly being examined, researched, analyzed, adjusted, and improved to provide a clearer and more effective and useful framework with which to enter a particular humanitarian response. It is primarily humanitarian practitioners and the aid community itself that provides this input and revision. Second, the systems, structures, methods, and understandings of how to adapt and adjust at the local, episode-specific level are constantly advancing. Third, as depicted in the examples above, is having sufficient resources, operational capacity, skills, and political support to fulfill those standards and targets that do appropriately apply to a particular situation, to the greatest degree that circumstances will allow.

Of course, ultimately the real goal in all the work on standards, mandates, and principles is conducting better disaster response and relief,
with the highest level of humanity. There then follows moving on to
better recovery along with making preparedness, resilience, and mitigation stronger. Since aid professionals are fully aware that the gaps exist,
the standards, guidance, and tools are constantly evolving, being debated
and tested, and being revised, with new ones being created according to
emerging understanding or needs.

An example of this evolution at the global level is how, after the successful introduction and adoption of the Sphere minimum standards in
1998 (updated in 2011 and 2018) that were focused only on human victims,
in 2009 the Livestock Emergency Guidelines and Standards was introduced
and is now almost universally used to preserve livelihoods. Likewise, as
the transfer of cash and use of vouchers has become a key strategy in
disaster response in recent years, the Cash Learning Partnership initiated
in 2005 came about with guidance and training. This increasing approach
supports local markets and commerce, enables recipient independence,
embodies greater dignity for disaster victims, often facilitates more-rapid
overall societal disaster recovery, and is financially, logistically, and operationally more efficient for NGOs.

A crucial and urgent need that has emerged in recent years, particularly in Haiti, has been the development of understanding, guidance, and
standards for disaster and humanitarian response in urban settings where
more and more crises now occur, and where the humanitarian community
has found that existing rural-oriented approaches do not fit well. Shelter
and settlement, sanitation, water supply, and a host of other factors, as
well as social patterns, are significantly different and demand different
models and approaches.

One important trend has been the advent of consortia of organizations
that combine forces to fine-tune the effectiveness of disaster response.
For example, Building a Better Response is a partnership between
International Medical Corps, Concern Worldwide, USAID, and Harvard
Humanitarian Initiative. Likewise, the Assessment Capacities Project
(www.acaps.org), initiated in 2009, is operated by the NGOs Action
Against Hunger, Norwegian Refugee Council, and Save the Children.

By far the largest and most comprehensive and conscientious effort ever
mounted was the World Humanitarian Summit, which convened in May of
2016 and was under preparation since at least 2012. The summit was an
acknowledgement that there remains still significant improvement needed
in disaster and humanitarian response, particularly with respect to reaching shared and unified agreements on principles, practices, definitions, and
approaches between and among nations and aid agencies. Tremendous

amounts of money, material, and human resources are spent in multiple disasters and crises responses and our results should be much better, as demonstrated in the ongoing continued suffering of crisis victims in Haiti, Sudan, Somalia, and Syria, among many others. The ultimate results, utility, and effectiveness of the World Humanitarian Summit in facilitating better aid will need to be judged later, but the preparation has sought to include the broadest input possible, including multiple consultation events with diaspora, donors, international and local aid agencies, the global North and South, researchers, analysts, and decision makers. Distilling all this into practical, useful, guidance and new standards will be the challenge.

There is also a dilemma here. While all the many standards, codes, principles, tools, strategies, and approach are welcome, help fill the gaps, and increase our effectiveness, they also present somewhat of a Chinese menu of choices for those working on a particular response. It may as well be a menu with some pages hidden from view. The guidance becomes a mosaic and we have to know where to go for what, and which guidelines may supersede others. We are also charged with the question of who is responsible for knowing all the guidance and for choosing which of it fits best with a particular case at hand.

At the level of a particular disaster or humanitarian response as it unfolds that entails the means and capacity to apply the standards, the need to identify the real gaps and how to accommodate or tailor them, and the duty to conduct the best possible and most effective specific disaster response, it is vitally important to understand the local social, cultural, political, and resource baseline context, including the present situation in the face of the disaster or crisis. For example, in order to achieve equitable aid, it is crucial to discern what is the local resource control and distribution system. To provide twenty liters of water per person, what is the current access to water using accustomed local coping strategies? If we know the capacity of these coping strategies to satisfy part of the needs, we can calculate what degree of gap our external assistance must provide to reach minimum standards. All the while, as noted and depending on the existing knowledge base of the local and foreign aid workers, additional information may not be readily or quickly available, and some key decisions will likely need to be made with insufficient data or analysis.

This is one point at which the keen insights, analytical abilities, contextual assessment, and observational skills of practitioner anthropologists working for aid agencies are highly relevant. In the middle of a disaster response, their nuanced insight in shaping the design—and, equally importantly, in the later revisions and fine-tuning—of response activities can be key. Anthropologists can particularly focus on providing input and guidance to ensure that aid does no harm, and to ensure that aid is not provided needlessly or in excess of real needs (wherein coping strategies supply some of the needs), through faulty or incomplete information or interpretations of a situation.

Such aid assessment, analysis, and planning efforts are often addressed collectively by aid agencies under particular structures or mechanisms. One key aspect of this process is supposed to be led by the United Nations Office for the Coordination of Humanitarian Affairs (UNOCHA) that, in large-scale international emergencies, coordinates a predefined set of clusters organized by sector topics such as shelter; water, sanitation, and hygiene; health care; food security; and logistics. The cluster system is activated in-country at the beginning of each major response. Each sector is supposed to be co-led by a specific UN agency. For instance, the United Nations World Food Programme leads the food security cluster, along with affected country government specialists from an appropriate ministry or department; that cluster then comprises all aid agencies (including UN agencies) that are planning to work or already working in that particular sector. The clusters also usually include representation from external donor governments. Often an international NGO will be a member of multiple clusters during a response depending on their organization's technical expertise, plans, and resources at the time. Participating agencies in each cluster are assumed to be aware of, and to have entirely endorsed, the global principles and standards, and it is hoped that individual staff representatives of each cluster are also sufficiently knowledgeable about these fundamentals, though as noted elsewhere in this discussion many are not.

It is often in these clusters that the universals meet the local realities. It is here that rapid assessments are shared, joint appraisals are planned, and locally adapted collective standards are discussed, and are in part or wholly adopted. Although the process and its effectiveness in practice is often less than ideal and less efficient or meaningful than desired, it is one way that the gaps are collectively acknowledged and, to some degree, jointly accommodated. In the Nepal earthquake of 2015, the shelter cluster decided, along with the Nepal government, that the collective response of aid agencies for housing would be the provision of corrugated galvanized iron roofing sheets and a modest tool kit, as opposed to tents, tarps, or other building materials. Local housing style, local capacities, cost, and logistics led to the decision. In retrospect, and from my direct personal experience, this decision was entirely valid.

Multiple other mechanisms operate on the ground in any crisis. international NGOs themselves very often have their own inter-agency coordinating body that seeks to jointly problem solve. There are initiatives such as the Assessment Capacity Project, mentioned earlier, that conducts analysis to fill in the information gaps and leads to more-effective programs.

All these are greatly aided in every aspect of their efforts by increasingly better technology such as access to the internet, geographic information systems, vastly improved communications, and portable data and computing systems.

Back On the Ground, and Temporal Realities

Along with all else, there is a temporal element to the gap. Emergency and humanitarian responses unfold in phases and over time. In the early days or first phase of a response, the aim is coverage, not perfection, and aid givers are forced to work ad hoc and rapidly. We must "kick into operational mode and just do it," as one colleague noted during an interview. As the response proceeds, we become more systematic, analytical, and strategic. In the early stages, the endeavor is about life saving, and so we make compromises even on key issues such as gender, equity, and sufficiency. As long as early (the first days) decisions and actions, based on generic strategies, standards, and templates, are aimed at doing the best possible under the time, knowledge, and resource limitations, and in light of the inevitable initial chaos and urgency, activities can be revised and improved in the following weeks or months based on analysis as we learn more about the specific situation.

During the massive floods in Pakistan in 2010, during which 1.5 million people were displaced and 20 million were affected, many families found shelter in the compounds of host households who were sometimes relatives, but often just good Samaritans. Often two to four families comprising up to 20 people or more would stay in a single compound, tremendously overloading their facilities. Aid agencies, including the one I was working for, in responding to Sphere minimum standards, provided extra latrines to host households, all separated for male and female use. Upon return follow-up visits, we learned that the men had all moved to the local mosque to sleep. Cultural norms dictated against the type of cohabitation embodied in household sharing. As a response in keeping with cultural integrity and in satisfaction of Sphere minimum sanitation standards, we then provided additional latrines to the mosques. This adjustment enabled the people to keep their dignity, and at the same time fulfilled the required physical elements of a humanitarian response. In short, we started with the comprehensive standards as operational target and added in cultural knowledge and current situational status, thus deriving a solution that addressed all sides: the codes, culture, and circumstances.

In this way that humanitarian standards can be most effectively used, integrated, and adjusted to create the ultimate goal of overarching guidance, and to facilitate the best outcome in response to the worst situation. Herein lies the answer to the questions posed in this discussion. The ideal resolution depends on merging guidelines, information, and cultural knowledge, and on improving and adjusting a particular response as it evolves.

Still, the efficacy of a response yet depends on perhaps one of the most critical elephant-in-the-room realities of the standards and mandates and targets—that of resources. Reaching aid-giving targets is very expensive, and resources are always very limited. Always lurking behind every situation is the cost of fully achieving the standards. Aid agencies never have enough funding to meet needs. There are many arguments (and accusations) about how needs are financially calculated, which governments and entities (including affected countries themselves) pay for needs, and how the funds are ultimately provided and spent, and by whom. Those issues are sufficiently complex and politicized to merit their own discussion. However, from the perspective of on-the-ground aid agencies, the net result is that resources are almost always insufficient to meet needs based on both minimum standards and realities. In many emergencies, overall financial response targets, in terms of what is provided to the collected efforts, is between 50 percent and 80 percent of estimated needs. In some protracted humanitarian crises, such as those in Sudan, Iraq, and Syria, where funding struggles with other competing emergency needs elsewhere, and where there is increasing donor fatigue, it is even less. Aid agencies are therefore left to do the best they can do with what they have, and hope it is enough.

Conclusions

Although they label themselves and their work as pragmatic or applied research and analysis (and sometimes deliberately disavow any academic inclinations), the aid professionals developing the guidelines discussed here are developing theoretical constructs of universal human needs in the same way that academic anthropology builds theory from empirical and experiential evidence and data. They are positing what minimum standards are for material well-being, as well as intangibles such as dignity and rights. Whether they frame it this way or not, with every new disaster, by applying the universals first, they are testing their theories of humanity. Because these theories have been formulated externally to a particular disaster, they may well not be readily applicable, as discussed throughout this chapter. The local population had no input into the development of the theory being applied to them, and may well disagree significantly with it. Part of the theory itself is that local specificities will usually diverge to varying degrees. Nonetheless the gap between theory (the universals) and present circumstance can be examined and analyzed as input to refining the theories, as is done in academic theorization, and humanitarian research.

The gap is OK; it is even good. All told, the existence of a gap in the planning and implementation of humanitarian aid stems from having targets, aspirations, and goals that cannot be met. But without the carefully developed principles and targets to which organizations are held accountable, the organizations would be free to do or establish their own "good enough" targets with no overall consistency or agreed interorganizational structure and framework. In assessing the achievement of the benchmarks against the goals set by the standards, the gaps in any situation can be identified and measured. Consequently, the gaps themselves become a kind of target to be filled, and they push aid forward to improve and close those very gaps. Existence of the gaps forces discussion, analysis, and new strategies. The holes also point to needed local adaptations and local relevance and to how to obtain the most impact in the specific situation within the convergence of the many factors that comprise any given disaster or crisis. Therefore, indeed, the gaps are good. They ensure that responses are adapted and appropriate.

The standards and guidance established and shared by aid giver agencies are thus indispensable and serve a fundamental purpose in structuring and guiding emergency responses. The principles and guidelines and standards keep us focused and heading collectively in the same best direction. They increase the speed of planning, efficiency, equity, and professionalism of aid activities. They most probably save lives.

The standards are not, and cannot be, rules. They are targets against which aid workers focus their goals in a specific setting. If and when they become considered rules, they can become obstacles that serve only to detract from efforts. Financial donors and other decision makers must recognize this so they can accommodate justifiable programmatic flexibility.

Gaps demand and facilitate ongoing programmatic assessment and monitoring. They lead to ongoing revisions, refinement, and improvements in aid effectiveness, case by case. Such assessment and analysis has become a recognized part of response, and the methodologies and tools continue to improve.

Adam Koons has thirty-five years experience, including twenty years overseas, specializing in international and domestic humanitarian and disaster response and recovery. His career spans senior and executive management roles, program design, and field analysis, with large international NGOs, with private contractors, and with several disaster-focused agencies of the U.S. government. He has worked in thirty-five countries, on disaster response management and recovery field teams, ranging from Haiti, Indonesia, Nepal, Pakistan, the Philippines, Ethiopia, South Sudan, and Guinea to Mississippi and Puerto Rico.

References

Core Humanitarian Standard (CHS Alliance), and Sphere Project. 2014. *Core Humanitarian Standard on Quality and Accountability.* CHS Alliance, Group URD and the Sphere Project. Retrieved December 2016 from https://corehumanitarianstandard.org/files/files/Core%20Humanitarian%20Standard%20-%20English.pdf.

International Federation of Red Cross and Red Crescent Societies (IFRC) and the International Red Cross and Red Crescent Movement (ICRC). n.d. "The Code of Conduct for the International Red Cross and Red Crescent Movement and Non-Governmental Organizations (NGOs) in Disaster Relief." Geneva, Switzerland: International Federation of Red Cross and Red Crescent Societies (IFRC) and the International Red Cross and Red Crescent Movement (ICRC). Retrieved December 2016 from https://www.icrc.org/en/doc/assets/files/publications/icrc-002-1067.pdf.

International Federation of Red Cross and Red Crescent Societies (IFRC) and UN Office for the Coordination of Humanitarian Affairs (UNOCHA). 2011. *Model Act for the Domestic Facilitation and Regulation of International Disaster Relief and Initial Recovery Assistance.* Geneva, Switzerland: International Federation of Red Cross and Red Crescent Societies. Retrieved December 2016 from http://www.ifrc.org/PageFiles/88609/Pilot%20Model%20Act%20on%20IDRL%20(English).pdf.

Livestock Emergency Guidelines and Standards Project (LEGS). 2014. *LEGS,* 2nd ed. Rugby, UK: Practical Action Publishing.

McIllreavy, P., and Nichols, C. 2013. *Collective Responsibility: The Perceptions, Expectations, and Realities of NGO Coordination in Humanitarian Leadership.* Washington, DC: InterAction. Retrieved May 2019 from https://www.alnap.org/system/files/content/resource/files/main/collective-responsiblity-ngo-coordination-in-humanitarian-leadership-june-2013.pdf.

Sphere Project. 2018. *Humanitarian Charter and Minimum Standards in Humanitarian Response,* 4th ed. Rugby, UK: Practical Action.

U.S. Agency for International Development (USAID). 2005. *Field Operations Guide for Disaster Assessment and Response.* Version 4.0. Washington, DC: U.S. Agency for International Development.

CHAPTER 5

Disaster Theory versus Practice?

It Is a Long Rocky Road: A Practitioner's View from the Ground

JANE MURPHY THOMAS

Introduction

This chapter makes two main points: that it can be a very long, rocky road between disaster knowledge and practice, and that practice depends on a lot more than the knowledge. Such a myriad of factors can stand in the way of putting theory into practice that it is impossible to list them all, but this chapter tries to identify some of them, at least from one practitioner's point of view. If only it were as simple as just going out and applying ideas everyone agreed were correct!

The question at hand is this: Considering the decades of research, experience, and lessons learned from disasters around the world, why is there such a huge gap between this knowledge, policy, and practice? Why is there still such an uneven application of knowledge about disaster prevention and mitigation? What are the influences and barriers to applying knowledge? Moreover, what fosters and perpetuates the gaps, do they have functions, and what are the effects of the gaps? Can anything be done about them? The significance of these questions cannot be overstated: "The impact of aid on the recipient is sometimes helpful, sometimes as damaging as the original disaster" (Anderson and Woodrow 1989, 1).

After this introduction, there are three sections in this chapter. After looking at what being a practitioner means in the first section, the chapter identifies the kinds of gaps and their causes in the second section, then, in the third section, presents five cases studies from my own practitioner experience that help illustrate the above challenges or obstacles. These cases represent both negative and positive examples where gaps were created or even enlarged, or in other cases where gaps were bridged and the knowledge applied.

This chapter refers only to work in the reconstruction and recovery phases of a disaster, after the emergency phase has ended. Also, it does not differentiate sudden catastrophic events (earthquakes, hurricanes, floods, avalanches, wildfires, etc.), from the kinds of repeating catastrophic events of conflict or war. As explained elsewhere, the content of this chapter refers only to my experience. Some of this content may apply elsewhere, and some may not.

Section 1: What Being a Practitioner Means

While being a practitioner can mean many different things and take many different forms, I think of it from my thirty years living and working full time at the grass roots in international development assistance programs in Afghanistan, Pakistan, Bangladesh, and Azad Jammu and Kashmir (Pakistan-administered Kashmir). My work is in areas in particularly complex situations, in traditional cultures where varying degrees of conflict already exist. This chapter will focus on my experiences in these regions after a so-called natural disaster occurs.

As a social anthropologist specializing in community participation, I have worked as a self-employed independent consultant on contract with agencies, applying my specialization to a range of sectors including community and local nongovernmental organizations (NGOs) development, health, education, water management, flood control, earthquake reconstruction, forestry, transboundary water issues, micro finance, refugee camp management, agriculture, land acquisition, and so on. The agencies are international NGOs, United Nations (UN) agencies, donor and local governments, and international consulting firms carrying out projects for which I have been engaged over periods from a few weeks to a number of years.

Most of this work has meant living in the rural project areas with frequent trips for work at project head offices located in main cities. For each project, my work is to hire and train staff, carry out participatory analysis, mobilize the communities to participate then act as intermediary or broker, facilitating for information, planning, evaluation, requests, and plans between the community and project management.

From this frontline position, it means daily or even hourly contact with the people who are the intended beneficiaries of the project, through existing community-based groups or new purposely formed groups. To start this work with the community that continues throughout the project, I put together a community mobilization or social team. It is my practice to employ only those people who are from and currently living in the same districts as the disaster recovery project. In this way, we are able to build local capacities and to save much time and trouble. Besides being disaster victims or survivors themselves, these project staff members know at least some of the many local languages, already deeply understand the local culture, and already know who is who. In many places where I have worked, NGOs have never operated, so there have been no previous opportunities for locals to learn about such organizing and community mobilization. This being the case, those who we hire and train as social mobilizers are those who can at least give examples of what they already do on their own to help their own communities, and already know the challenges there can be to do so at times.

Once a project's social mobilization team is assembled, I lead members through an ongoing, participatory iterative process to analyze and decide approaches, including approaches to understand the local power structure and stakeholder groups, to understand the needs of project counterparts, and to understand what the team needs to do to support the counterparts' work. Depending on the sector of the project, this means social mobilizers learn how to work with other specialists such as engineers to facilitate in earthquake reconstruction, agriculturalists to improve vegetable growing, or teachers to get parents involved at schools. The social mobilizer's role is to know and help steer the project through the sociocultural context of the immediate area, ensuring the project can work with that cultural flow, and knowing who holds and uses or misuses power, who are the vulnerable people, how to protect local people from any loss, what are the people who lead in the connections or divisions, who are the leading connectors and dividers, and where is there conflict or risk of conflict. All these facts guide the project, then, to be most culturally and conflict sensitive: how to do no harm and still meet project objectives, schedule, and so on.

As typical of many developing countries, most people who are the intended beneficiaries from these projects where I work live in poverty on income of less than $2.00 per day as subsistence farmers, petty traders, or laborers. If they have any land at all, it is less than one acre. Millions simply are trying to make sure their families survive every day, with most living close to or on the edge of different kinds of disaster all the time.

Section 2: The Kinds of Gaps and Their Causes

What makes putting theory into practice so difficult at times? As this chapter's case studies in section 3 indicate, in disaster work, and referring only to the early reconstruction and recovery phase, there can be huge gaps—or chasms—between theory and practice.

What Gaps and Why?

So what are the gaps and why do they exist? While there are probably innumerable obstacles, using my own observations and experience, the main ones can be listed as behavior in organizations; power, politics and money; project design, management, and funding; phases of disaster response; whose knowledge and what knowledge; contexts, culture, and history; lack of local participation and local knowledge; siloing of issues and disciplines; accessibility to the research, knowledge, and practice; and anyway, just who are the practitioners and others who we expect will apply the knowledge?

Behavior in Organizations

Suddenly, a disaster happens—an earthquake, tornado, flood, wildfire, volcanic eruption—causing widespread loss of life and destruction. Local people are the very first responders, doing what they can. They are usually followed by waves of assistance from across the country and sometimes from around the world. Who are those disaster responders in the second wave? The largest-scale disaster responses are carried out by organizations with their work facilitated or hindered by factors within the organizations and their own environment, customs, beliefs, practices, and values. Those responders might be universities, governments, multilateral agencies, NGOs, businesses, and so on. While each may have the mandate to assist in such emergencies, they also have their own internal vulnerabilities and capacities to act in the most effective ways.

Like organizations or companies doing any kind of work anywhere, much can stand in the way of putting knowledge into practice to get successful results. Even in humanitarian agencies, there can be differing visions and priorities, internal power plays, or office politics: dictatorial management, poor communication, oppressive workloads, and personality conflicts. Sometimes people get along with each other and sometimes they are in tension. Hiring practices may be weak with recruiters not understanding needs of the job so people are hired who may or may not have the skills needed. There can be lack of will or entrenchment in the status quo. People with the knowledge may simply be ignored, their knowledge never put into practice, and so on. And as in any company or organization, humanitarian agencies unfortunately occasionally employ people who are simply incompetent, self-centered, ethnocentric, or unsuitable in other ways.

For professional recognition value in many organizations there is also the tendency to look for new ideas, and not just repeat what has been done in the past. Unfortunately, this can mean rejection of what has already been done and learned, even from successful models, leading to reinvention of the wheel, waste, and unsuccessful results.

In other words, main barriers to effective practice may be weaknesses inside the knowledgeable and expert organizations themselves. All these internal matters can have serious effects on their external matters. Weaknesses within, competition, or a sense of superiority or aggressiveness with other others renders these organizations unwilling or unable to go along with "the knowledge."

Power, Politics, and Money

"Humanitarian emergencies are always political events and play a role in national and international politics" (Middleton and O'Keefe 1988, 15). Whether acknowledged or not, both disaster and development assistance, especially for longer-term reconstruction and recovery, is heavily based on power, political relations, and resources among governments, institutions, and individuals. Humanitarianism, of course, is a strong factor, but close behind that, or in some cases at the front, are the disaster politics between provinces or states and between them and their federal governments, among them and other countries, and among multilateral agencies. The political nature of such assistance can be based on a wide range of factors such as bilateral political reinforcement or favors, as a reward for compliance to other agreements, to build future relations, or to get voter approval at home. When disaster strikes, and is broadcast into every home on television, everyone is familiar with politicians pledging large amounts of assistance, but in the end the public often does not know what, if anything, politicians actually provided or if their assistance actually helped.

These realities alone can create huge chasms in the needs and types of assistance and what is made available. As earlier stated, practice depends on far more than knowledge. Possibly the biggest gap, and by far the most complex factor between theory, policy, and practice may be power at both the macro and micro levels. The aid agencies, such as NGOs, donors, governments, and multilateral organizations, simply may not call on those with knowledge, theory, policy, and practice of previous disasters and/ or the people and cultures for whom help is intended. As a result, these people have no power, authority, influence, or means to use what power, authority, or influence they might have. Even if authorities call on the specialists, their specialized knowledge may still be rejected. Two of the case studies included are good examples of this: the Pakistan earthquake reconstruction of health facilities, and the Afghanistan customs and excise case.

Even with outside aid agencies, there are two particular factors that can facilitate or interfere with carrying out work known to be based on knowledge or best practices, but that are often overlooked. Of most relevance is to answer these questions: What is the local power structure? Who are the influential and powerful people, how do they use or misuse their power? When a disaster has happened, what roles will they play? Will they ensure assistance is provided to those who need it most or will they divert it to their own interests? Will they use their influence in their own communities to organize and manage their recovery affairs? Will they demonstrate their own knowledge to get their own best practice? Even agencies that know better (i.e., that know the theory and the practice) may not have a choice when overruled by local power sources.

The second factor determining longer-term sustainability of assistance is also a matter of power—or, more precisely, the rearrangement of power. When assistance comes from sources outside the disaster zone, whether from within the same country or from foreign sources, that help often brings with it a great amount of control or power, but as the outside assistance is temporary so is the external power. Once the outside source withdraws, the old local power structures and their ways of operating revert to what it was before. As a result, knowledge gained may or may not be put into practice. A common example of this is how frequently community organizations formed in projects cease to exist once a project is completed. No matter that skills have been developed to run a new community group, once the outside influence is gone, the old power structures take over again and the new group ceases to operate.

Power, politics, and money might be some of the significant factors creating the chasms between theory and practice. No matter when best practices are well known in some agencies, the reality is that all need funds to operate. This sometimes means taking short cuts, building territories, or overclaiming expertise just to get the grants.

Project Design, Management, and Funding

Again, why is there such a gap between what is known and what is done? Part of this problem is the actual ways assistance is provided. At least in long-term reconstruction and recovery, work is planned in the form of projects: undertakings with set schedule, budgets, activities, objectives, and funding availability based on the donor's time lines. Projects are designed with too little information, or with a lack of knowledge. By their nature, projects are usually of short duration—twelve months, two years, five years. They are not based on the time needed to apply the knowledge or to get practice established, or on the time needed for the all-important follow-through.

Too often, projects are decided and designed with too little information, especially information about the background, conditions, and ideas and needs of the intended beneficiaries. Such gaps of information result in weak or failed projects.

Phases of Disaster Response

Different phases of disaster response require different approaches to get the most-effective recovery. Longer-term reconstruction can be strongly affected by what happens in the early first response phase. While assistance in the emergency is needed to save lives, it has a reputation of being uncoordinated, rushed, and provided top-down. The early free handout style often creates expectations on what comes next, negatively influencing people's sense of self-reliance and creating dependency. The early rapid data collection but lack of follow-through in that early period can create mistrust. Dependency and mistrust are strong barriers to putting best practices to use.

In the two case studies on an earthquake reconstruction and recovery project in Pakistan, at first some people in the project area were reluctant to accept the project. Some wanted to know what the project would give them as free handouts, or do for them individually, having received aid this way in the emergency phase. In the project team's first visits to communities, there was great skepticism among the people that the project would even happen, as many explained: So many other organizations had come there already and asked so many questions, then went away and never returned. "Why should we trust you?," they wanted to know. In this case, the social mobilization team's work to apply their knowledge had to start with promoting self-reliance and establishing trust, beginning with forming a partnership. Knowing what practice is appropriate for what phase of recovery is essential.

Whose Knowledge? What Knowledge?

When we talk about there being a gap between knowledge and practice, we need to ask, Whose knowledge are we talking about? And what knowledge is it? Who are the knowledge holders? Whose knowledge is valued more? Some assume it to be a pool of knowledge at the top, in academia or among experts or researchers and that it should trickle down to be used in practice at the bottom.

It might be too obvious to state, but this alone is one of the main answers to the problem, the gaps between the people who know and the people who do. The academics, researchers, and experts tend to be people who live and work elsewhere, while the practitioners live and work on the spot. These are two very different perspectives, but even more important is the third perspective, that of the people themselves. How much of their ideas, opinions, customs, culture, and knowledge is included in the practice? Is their knowledge relevant? Compatible? Like other top-down approaches, more-useful knowledge can be gained by turning the situation to be more bottom-up.

When we talk about knowledge, what knowledge do we mean? Using the term "knowledge" implies there is a neat package of widely agreed information to make available to policymakers or practitioners, but that is not the case. The knowledge involved is so diverse, covers so many subjects, and is researched and practiced by so many people in different situations and institutions in so many places, from so many perspectives, and through so many different cultural lenses and biases, as to be over-whelming and contradictory. This factor alone could be the main reason for the huge gaps.

Case studies in this chapter include several examples of contradictory or even clashing sets of knowledge and how these create the gaps between knowledge and putting it into practice. Different sources have different knowledge, as well as ideas and options that contradict each other. It comes down to who has more power and the means to use it.

Part of the problem comes from gaps in education, where some disciplines are not exposed to other disciplines and have little sensitivity to them. For example, in their professional practice engineers, economists, or health professionals might not be taught or expected to listen to local people or even know anything about them. Such content might not have been included in any course work, and that missing work then is reflected in their approaches. Their practice, then, may be far harder or less success-ful because they do not have the local knowledge needed.

Context, Culture, and History

Project A from Country B arrives in Country C. How much do the people in Project A know about the place they have just arrived? How much do they know about the contexts: the people, their ideas, opinions, social structure, beliefs, practices, languages, preferences, and all the history (including other disaster response) that has come before? Without such knowledge, implementing the work is like working in total darkness, just hoping for the best, but in so doing can be creating its own obstacles.

In the aid business, the term "culture" is sometimes misunderstood as something superfluous, a luxury not really needed or important to know to be able help. A common attitude is, "All that these poor survivors need is food, water, and their houses and schools rebuilt. We can help with that, so that's all we need to do. They should be happy we have come to help!" Then outsiders often wonder why their projects run into trouble.

Lack of Local Participation and Local Knowledge

From the early days of the first international aid following World War II and establishment of the UN, many lessons were learned from the far too common project failures. Even that far back, development agencies began realizing, at least in theory, they had to understand the local cultures and see development as people being involved in the decisions that affect them, hence the community participation movement that grew out of that time. However, since local knowledge and "participation ascended to the pantheon of development buzzwords, catchphrases and euphemisms" (Leal 2011, 70) over the recent decades, donor agencies required community participation as a condition of funding, and agencies jumped on the bandwagon, so much so that it has become difficult to find a project that does not claim to have community participation. There is so much confusion and watered-down versions of community participation that the development literature now widely criticizes its poor practice.

From my many decades of experience working in this specialized field, the reasons for gaps between extensive knowledge and effective practice of community participation can be summarized in a few points: lack of knowledge about how to do it, lack of conviction of its importance, and a reluctance or refusal to shift or share power with the people. Aid agencies often ignore, reject, or disavow the existence of local strengths and knowledge. Such aid agencies, whether from the same country or abroad, often bring their own knowledge to the places they work, sometimes assuming that what they know and how they do things is superior. "Why would we ask the people? We already know everything we need to know!" Such attitudes reflect differences such as the urban–rural divide, levels of formal education, class, culture, nationality, wealth, and so on.

Siloing of Issues and Disciplines

From the practitioner's point of view working on the ground in long-term development where conflict and disasters are also factors, it appears that many related specializations are divided into they own boxes or silos. Knowledge is kept more or less separate from practice, disaster from long-term development, relief from development, participation from development, and other sectoral separations such as health from education, crime from homelessness, or culture from building design. Add to this picture how sociocultural factors of the people being assisted are often treated as separate or ignored by other disciplines such as engineering, economics, finance, education, logistics, and even health. As a consequence, when aid agencies ignore other disciplines, their own work may be less effective.

Even the term "disaster" appears to be treated completely separately from the term "development." Development efforts may already have been under way for decades in certain locations when disaster strikes, but some aid agencies ignore some of this development. Unfortunately, without this continuity or without using development approaches in times of disaster, responses often completely contradict and damage earlier development efforts, especially when they create dependence.

Accessibility or Inaccessibility of the Research, Knowledge, and Practice

Both academics and practitioners do research but, unlike the scholarly work that academics write up and publish, if practitioners write up research at all, they and their employers treat it as internal to their organizations. For the scholars, research and writing may be a full-time occupation, while practitioners, by the very nature of their work, have far less time (and possibly far less inclination) to write for publication. Unfortunately, this also leaves a large gap in practical, reality-based knowledge sharing, and leaves each set of knowledge—academic and practical—relatively inaccessible to the other.

The published knowledge, then, comes more out of academia, and academic language can be incomprehensible to others. The value of the research may also be unclear when such documents tend to be detailed descriptions and analysis but often avoid making conclusions or recommendations, leaving the significance of the research unclear. The practitioner reader is left with the question, "So what? What should be done about this? Is there something here that I could put into practice?" Now, with the internet, access to such information is vastly increased but it tends to be located in unknown websites requiring extensive searches and subscription, which are themselves restrictive.

Ironically, one of the main barriers to applying learning is lack of access to or use of lessons learned. From the many decades of disaster response, emergency relief, recovery, and development, vast amounts have been learned and documented. Most project final reports end with a section on "Lessons Learned." Never mind for the moment how relevant they are to different places and times, the question stands, Where are all those valuable documents?

Who Are the Practitioners?

Leaving it up to academics to do an analysis of themselves and their own spheres of work, I look instead at others involved in disaster response work. Most people working in disaster situations have expertise outside of the field of academic disaster studies. In my experience, some important questions are, Who are the practitioners or front-line workers? What do we do and what influences us? Just who is working in all the fields dealing with disaster mitigation, emergency relief, reconstruction, and recovery?

Speaking only about the aid programs in countries where I have worked, that still means hundreds of international, national, and local agencies or institutions involving thousands of staff and volunteers, mandated to provide a myriad of services to millions of people. Many organizations exist specifically to respond to disasters and arrive at the earliest phase and often leave again in the first few months, as soon as the life-or-death emergency is over. Those organizations that remain are often those that have been present all along and are not disaster specialists, and instead are concerned with poverty reduction, community-based development, social justice, and so on., These organizations work in different sectors such as health, education, infrastructure, agriculture, water, and others. Each organization has its own administrative and management levels with people who are the influential decision makers, but extremely few staff members may have any exposure to disaster research and most may not even know such a field of study exists. Most of these organizations probably have no specific mandate to work in disasters, even if they are based in disaster-prone areas, hence they have no specialist disaster practitioners. The number of agencies that specialize in disaster, or employ those who know the theory, probably are only a tiny fraction of those in the overall field of humanitarian assistance.

The non-disaster specialist organizations (e.g., NGOs, UN, local government programs) are those carrying out the long-term reconstruction and recovery. They are run and staffed by professionals whose work demands that they draw on multiple diverse skills: business, management, technical, financial, horticulture, engineering, health, education, livelihoods, agriculture, water, and so on. Maybe they are health-care workers who know how to show people how to reduce waterborne diseases, or agriculturalists who help people restore their damaged apple trees, computer operators who can also fix photocopy machines, drivers who know how to maintain their assigned vehicle, or managers who can write succinct reports. They may be highly skilled, dedicated professionals, with high levels of altruism or empathy, who enjoy working out solutions to very complex situations. In reality, though, the vast majority may have no formal knowledge at all about development, disaster studies, or community participation. They might have no idea of what is a good practice, or what has already been well proven to be a bad practice.

As mentioned already, front-line people may have no time or inclination to look up research that has been done, especially if that research is too theoretical or impractical. Faced with a disaster, such workers put their own other strengths and much needed, very important, highly welcomed expertise to use.

Section 3: Five Case Studies

From afar, we tend to see disasters and their results in summarized ways, as the media cover them. There may never be any detailed explanation to the public about what is involved, and if people have never worked on the ground it might be impossible for them to imagine just how many minute details are involved. The following case studies, taken from my own experience, illustrate a few such points where reality-based challenges come in, some of which show where there are chasms between theory and practice, and others showing where those gaps were overcome.

Bangladesh: Dampara Water Management Project

This case study is an example of where strong gaps existed between knowledge and practice at first, but then were brought together in the project. In the northeast of Bangladesh severe annual river flooding repeatedly caused loss of life and wide destruction of houses, crops, livelihoods, and infrastructure. In 1999 the Bangladesh Water Development Board (BWDB) decided to have a thirty-kilometer-long flood control embankment constructed; to plan for it, the BWDB sent out its own cadastral survey crews to measure and determine the alignment for the embankment. Although the project would affect about one hundred thousand local people, the BWDB did not consult them on any aspects of this plan.

In the meantime, as part of international agreements between Canada and Bangladesh, the Canadian International Development Agency (CIDA) agreed to fund the construction of this embankment, but on the condition it was not going to be "just another construction project." Because such embankments crisscross almost every area of Bangladesh, and because many are not maintained and need repeated reconstruction by government, CIDA and BWDB agreed that this embankment would be based on a whole new approach using strong community participation. Would that produce a sense of ownership among the people? Could participation motivate them to protect it? Toward those ends, CIDA took the unprecedented step of requiring their contractors, who were responsible for supervising construction, to hire a sociocultural and community participation expert to manage the project, and not an engineer as would be normal. The funder especially wanted to know what difference could it make if such a specialist was the overall project manager of such a large construction project, which would of course require supervising all the engineers and construction contractors. I was the one contracted to do the job.

Immediately upon start-up of the Dampara Water Management Project, it was in trouble. In the two years since the above cadastral survey had been done and alignment decided, opposition to the project had grown sharply. On our first trip to the project area, as the project's chief engineer and I arrived by boat on the river (because there were no roads), we were met by dozens of fully armed men brandishing various kinds of weapons watching our arrival from the riverbank. Their first words were threats shouted at us, over the din of our boat motor necessarily kept running to deal with the river's swift current. If we took one step on to the shore, they shouted at us, they would use the weapons because they would not allow us to build this embankment.

Clearly this was a serious situation. It was apparent the BWDB had arrived earlier with an eminent domain attitude, practically invading and deciding which land to take for construction, with no input from the people, and that had resulted in the armed resistance. We had two choices, go back to the city and tell CIDA and the BWDB that this was too dangerous, or try to establish dialogue. I chose the latter route, introducing myself and the chief engineer with me (a senior Bangladeshi), and explaining we had come today only to talk with the people and hear what they had to say. Discussion continued at a distance until they invited us ashore. Sitting down together with them and several under community members under the trees for a few hours of talks resulted in a long list of their specific complaints and what they wanted the project and government to do about each. I offered to take this information away and have it considered, but emphasized that no promises were being made. If they agreed, we would come back again in a week for more discussion. They accepted that plan.

To me, all the complaints and requested solutions seemed reasonable, but back in the office our chief engineer flatly rejected all the requests.

As he reasoned, the BWDB had already decided on the plan, and because they were the authority, we had to do what we had been told to do and build on the alignment decided. This then became a friendly but real matter of power. This very capable but by-the-book chief engineer had no experience with listening to the people at project sites. I reasoned with him that we absolutely needed to find alternatives and convince the BWDB of the need for alternatives. If we failed to do so, either the opposition could make it impossible for us to build anything or the BWDB would abandon the project for the same reason. The chief engineer and I then set about to consider the reasons for the people's complaints and determine if there were any possible options for proceeding.

By far the biggest issue from the people's point of view was the planned alignment, because it would destroy prime crop land. Although there was little dispute over the need for an embankment, it was con-sensus among the landowners and other community members that it should be built instead on a full-length strip of unproductive, public land closer to the river. The second biggest issue was that if people agreed to forfeit their land and assets on it, then the normal government land acquisition process would mean they would be expropriated and have to wait forever for compensation or never get it (a far too common reality), and in addition be forced to take endless trips to town to settle land ownership, pay bribes to get documents, and face other expenses of time and money.

The area was also rife with land issues. Because land holdings are so small and are co-owned by several family members who have inherited the land jointly, and land survey and records had not been kept up-to-date, the exact boundaries were often not known or were already in dispute. Constructing any embankment more than thirty kilometers long would mean dealing with about 320 tiny plots of land, co-owned by more than a thousand people.

There were also major engineering issues. The project engineers at first were sure allowing local people to choose a new alignment would put the earthen embankment dangerously close to the river and it would simply wash away—a highly valid concern—so I asked the engineers to select their own criteria that would satisfy technical concerns for ease of construction and long-term sustainability of the embankment. If the project would allow local people to choose the alignment, what would the minimum criteria need to be so that a new embankment would be protected? The engineers decided that the only way they would go along with a new alignment was if it, like the originally planned alignment, also respected river behavior. They decided that a new alignment would have to be a certain distance from the river, a distance that they would determine in each place (since bends or straight stretches had different erosion considerations, putting embankments at risk); and that all land owners connected to the proposed change had to agree on the choices.

At the next public meeting, accompanied by the project's newly formed social mobilization team, I had the engineers announce their willingness to consider a new alignment, but that it had to meet the above criteria. That day, over several hours of discussion and walking on the riverbank with villagers to start new measurements, the atmosphere completely changed to one of full cooperation. We divided the thirty-kilometer stretch into eight sections, each about five kilometers long, and asked for committees of land owners to form in each section.

Over the next few months, the project set out in each of the sections, going plot by plot with negotiations involving the co-owners and other community members to decide the new alignment. In each location, the project handed responsibility to each committee, to have landowners negotiate in public with each other (and not with the project) to choose the exact alignment, based on the new criteria. Since most people urgently wanted the embankment, these public negotiations resulted in much trading, bargaining, and helping among members that would not have happened if any outside source had dictated the choices.

Once the alignment was fully agreed, the project engaged other local expertise to reduce or eliminate the other problems villagers had dreaded. Instead of land owners having to make endless trips into town, the DWMP established land camps along the length, with the project bringing government officials to the project site to do new land surveys and install visible boundary markers, assess assets, update and legalize all land records, and make payment on the spot before construction started. According to local leaders and authorities, the combination of full payment, ownership transfer, legalization, and new documents issued on the spot before construction started was unprecedented in the country.

In the end, the new alignment suited everyone and the construction was completed in the planned three years. Engineers deemed it technically sound. Almost no crop land was lost, no resettlement was needed, and owners received cash payment and up-dated land ownership records right away. Communities received not only flood protection, but also a road: the top surface of the embankment had been designed to serve as their first and only road. In the end, it was the community participation, government cooperation (reluctant or otherwise), and donor approval that made it work. As the project ended and the chief engineer and I visited the communities for our farewells, the man who had led the armed resistance on the day we had first arrived tearfully thanked us, saying, "Back then, we could not imagine someone listening to us so we were surprised you did listen to our ideas and concerns and we are grateful for that." He knew all too well about the frequency of abandoned construction and other losses, caused simply by authorities who think they know better.

Conclusions

This case study is an example of two sets of knowledge that could have clashed in their practice. The engineers' knowledge and top-down practice is to build the most structurally sound infrastructure, while the sociocultural specialist's primary concerns are bottom-up: respect, social justice, and protection of the people and their wishes. But in this case, the concerns of both were met by fulfilling each other's objectives.

As discussed later, in the case study on Pakistan earthquake health facility reconstruction, power can be a strong factor in the application or rejection of knowledge. Realistically, because I (a sociocultural specialist) had the project manager role, I was in the most powerful position. That position gave me the authority to ask engineers to reconsider. Had an engineer been the project manager, as is far more common, it is highly questionable how much influence a sociocultural specialist would have had, if in fact one would have been on the project at all. It was the power of the donor, then cooperation within the project, that made this management arrangement possible.

Afghanistan: Customs and Excise

This case study is an example of where big gaps of knowledge and practice existed, might have been filled, but were not.

An international consulting firm was engaged to draft Afghanistan's long-term plan for customs and excise. I was one of the consultants and had several years of experience in Afghanistan, so the firm gave me the role to facilitate among government agencies and related business people from around the country to help develop the plan. To start, I worked with the consulting firm's overall country director who oversaw many other projects while also filling the role of acting manager on this project; the actual project manager was temporarily absent for other purposes. The country director enthusiastically helped to envision the participatory process to design and lead a process to set up workshops in four areas of the country. These workshops were to involve the main stakeholders who normally treat each other as enemies: cross-border traders and the government customs and excise officials. This was going to be a delicate and risky subject to raise, let alone address.

Afghanistan has often been described as a country whose history is based on smuggling, where borders are unmarked in most places due to the mountainous and desert terrain. In the decades of war, movement of people and goods across these largely invisible lines had grown exponentially and now government was ready to clamp down on it. Outside pressure on the then-new Kabul government to generate its own revenues and customs and excise was one way to do it. Early in the consultancy, government and the trader entrepreneurs at first refused to have any talks with each other, blaming each other for mismanagement, corruption, coercion, and other criminal behavior, based on their long-standing mutual impressions.

To start the process, I held separate discussions with the main stakeholder groups: those in government departments, others among the cross-border traders, transporters, and related service industries. While there was general trader agreement that the new government needed to be able to get up and stand on its own feet and Afghan industry should help, the government people remained extremely skeptical that traders would own up to this need. Finally, after I had spent much time talking with individual groups, the ice thawed a little and, with all parties agreeing not to get into insults, shouting matches, or worse, they agreed to start meeting to try to solve the problems. I scheduled the first big workshop.

Government people and traders arrived from around the country and all border crossings. After introductions, myself and an Afghan interpreter and facilitator conducted an exercise to have government people sit together and make a list of their complaints involving traders, while the traders did the same about government practices. For the plenary session, each side chose their most diplomatic representative to read out their list of complaints to the whole crowd and to do it in a friendly way. Later, discussing the matters point-by-point, there was finally some willingness to listen to each other; participants agreed to a series of workshops to follow up. Many enthusiastically agreed that this process was needed if they were going to contribute to any hope for the future of the country.

At this point, the actual project manager who had been absent from the country for some months returned and was furious with what I had done although facilitation was part of my terms of reference, and had been directed and approved from his superior. He dismissed all the consultation and workshop results and, along with the plan for the series of workshops, tossed out the idea of any further participation in the national problem solving. He was an economist, and argued that the subject of customs and excise had to be worked on first on more widely recognized economic principles before they were applied to Afghanistan. He wanted to lay out the theory first, with almost no knowledge about the immediate problems on the ground. He could not see the significance of solutions to these now practically being handed to him on a silver platter.

Unfortunately, the stakeholder participation stopped at that point. For months after, I received emails from government and business people, pleading for me to return and continue with the process. My contract was not renewed as optioned. All the momentum and potential were lost.

Conclusions

The gaps in this case study can be attributed to common problems and behavior in organizations: differing visions, internal power plays, workloads, and so on. Here also was the entrenched idea that formal policy needs to be in place before appropriate practice can happen. The returning project manager was furious that his power might have been usurped by key people getting together to solve immediate problems without him. Unfortunately, he did not see this situation as the unprecedented opportunity it was. Not only had the historic enemies jointly produced a detailed problem analysis and were eager to continue to work out solutions, but also this process and its results could have been the foundation on which to build the most relevant policy and see it put into practice.

Pakistan: School Reconstruction after Earthquake

This case study is an example of where theory and practice, even involving disciplines that are often contradictory, can come together successfully. In this case it meant that almost all the project's reconstruction of seventy-seven facilities were completed on or ahead of schedule.

In 2005 an earthquake of 7.6 magnitude struck northern Pakistan, killing seventy-three thousand people, leaving 3.5 million homeless and almost all public buildings destroyed. An international donor organization responded with emergency aid and a major project to rebuild sixty-one large schools and sixteen health facilities destroyed in the quake.

To carry out this project, the donor contracted with a major international engineering and construction firm to manage the project, and to hire and supervise local companies to design and construct the new facilities.

Part of the donor's requirement for this project was that a sociocultural expert was to be hired as a member of senior management, part of the project team to design and lead a wide community participation component to support construction. Later, donor personnel explained that such sociocultural content was included because they had learned some lessons from unsuccessful school construction in other countries. They had realized too late that in those other cases that many problems had arisen because there was practically no local participation.

By the time the project team arrived to start one year after the quake, many other projects by various agencies were under way for reconstruction and most were having problems or already had stalled construction, common in construction in Pakistan even outside of disasters. Again, as the sociocultural specialist I observed that many of these problems were caused by lack of local consultation or participation and lack of knowledge of the culture and aspects of it that had implications for construction, for instance issues about land and water ownership and access to either, heterogeneity of local communities, the history of government relations with locals, and more.

My work began with the formation of a social mobilization team to work as facilitators and to form a partnership between the project and the communities. School committees were formed to facilitate construction, to prevent or solve the construction-related problems involving community members, and for the first time ever, to draw parents and community members into involvement at the schools to help improve education. At the same time, the social mobilization team worked with project engineers to coordinate in a step-by-step process before, during, and after construction. Planning for and working with all these local sociocultural realities, along with strong construction management, resulted in almost all of the project's construction to be completed on schedule, which is practically unheard of in Pakistan and some other countries.

Such involvement to achieve the large construction tasks also served as strong institutional building in each location. School committees were so motivated and empowered by their experience they went on to make contributions far in excess of expectations. Besides helping to speed along construction by preventing local problems over land, water, social divisions, and other issue, these poor communities found ways to lend or donate a wide range of resources such as cash, land, water, electricity, hospitality, and thousands of hours of meetings. They also activated parent involvement in the schools and in their children's education for the first time, and got wide community support. One of their most visible, permanent achievements was to start their schools' first-ever libraries; to do so they held the first-ever book fairs in Azad Jammu and Kashmir.

Throughout the project, many local people said that no one had ever asked them to participate in such an infrastructure project like this before, and they very much liked participating. Perhaps most importantly, as many committee leaders often stated, their participation had helped restore the hope they had lost when the quake happened. While destruction was still visible throughout the area, wherever a big construction project like this was under way, they saw it as a symbol of recovery for the whole community. At inaugurations of many of the new buildings, many leaders said "Now, with how this project has challenged us and we have been successful, we feel like we can achieve anything."

Conclusions

Stakeholders including the local government, participating communities, construction contractors, the media, and other donors attributed the successes of the project to its strong construction management and community participation. Credit is due also to the donor organization for requiring a strong community participation component, for it to be led by a sociocultural specialist, and for this position to be integrated in the four-person senior management team. Since completion in 2013, the project has been used somewhat as a model, including by this donor for their response to the devastating earthquake in Nepal in 2015. A similar project to rebuild schools and health facilities was introduced there by this donor in 2017. I am an adviser on the Nepal project, and work with the Nepali social mobilization team and engineers to adapt our previous project's approaches for the Nepal situation. While internationally there is so much to criticize for the lack of applying lessons learned and theories developed, this is at least one rewarding example of where it is, in fact, happening.

Pakistan: Health Facilities Reconstruction

In this case study, power and levels of power determined the outcome. As mentioned above, part of the earthquake reconstruction project in Pakistan was to rebuild sixteen health facilities that had been destroyed in the 2005 earthquake. While the region was already under-serviced by health facilities, the quake had greatly complicated medical and health conditions and added many more demands. When this project team arrived a year after the quake, virtually all hospitals and clinics were operating in what remained of heavily damaged buildings or in tents or open air. All were struggling to provide the services needed and in far higher demand following the quake.

Although the donor organization had specified that the project was to be heavily based on community participation, as noted above, the department of health (DOH) refused to allow such participation when it came to health services. The project team had proposed that in each construction location, community members would form a committee to work with the project, helping to facilitate construction. At the same time, with department of health training and leadership, this committee could be community promoters in preventive health measures, a new idea in this area.

While another project carried out some health training, and the department of health did allow temporary construction committees, in this project, the DOH forbade community initiatives in preventative health matters. As the district health official stated, "What do those village people know about health? They know nothing. We (doctors) are educated and know all about health, so there is no use for villagers to be involved." Besides, the district official explained, some years earlier another donor project (Project X) had required community participation but it had caused "nothing but trouble." Suddenly villagers had started showing up at the clinic demanding to know where the health worker was, or why certain medicines were not in stock. "They started thinking that they were the bosses. So we just stopped all this participation idea." Clearly, the official put far more value on his own knowledge than on that of the uneducated villagers.

Unfortunately, while the school committees (discussed in the case study above) contributed greatly to school construction and thrived in getting community and parent involvement in the schools for the first time ever to improve education, no such activity or benefits occurred for health.

This health facility reconstruction project could not trace the earlier, pre-earthquake Project X, but it appeared to be one that led people to raise their voices to get entirely valid change needed. The strategy that project's organizers had used may have been too insistent on imposing participation, so that the attempt backfired. It simply had not dealt with the realities of local culture and its power structures and could not find possible ways for the power structures to support and not compete with each other.

Likely not taken into account were the common differences over ethnicity, caste, sect, political affiliation, and how those differences alone affect and determine who has what power to command over others and where revenge might have been a factor. Now, newly built facilities would miss all the opportunities of having community leadership and buy-in to preventive health measures.

Conclusions

This case study provides a good illustration of the many levels of power that can be the cause of gaps between theory and practice: the department of health official who forbid people's participation because it would interfere with his own power; the power of an earlier donor's project that may have encouraged locals to take the power without being sensitive to how that could backfire; the power of the next foreign project to suggest such participation should happen (again); and, finally, the power of the department of health to still say no.

Afghanistan: Rural Development Training Center

As indicated above, according to my experience, recovery needs, challenges, and processes are similar whether they stem from a sudden catastrophic event or from the kinds of repeating catastrophic events of conflict or war. In both situations the many challenges are not only to rebuild the physical infrastructure, but also to re-establish what may be shattered institutions, vision, policies, plans, communications, records, staffing, capacity building, and relationships among people. An underlying priority need is to encourage and support cooperation and trust-building, and to do so requires developing a process, as process is as least as important as the product.

Such a case occurred when, after 9/11, the fall of the Taliban government, and the start-up of the new Afghan government, I was engaged to help conceptualize and plan the country's first rural development training center, intended mainly to train government personnel. The idea was to turn the government's historic extreme top-down approaches with rural people into more participatory, bottom-up ways of working. For this assignment, I had been recruited by the then-Minister of Rural Reconstruction and Development, who knew my work from several years earlier in the Afghan refugee camps in Pakistan. At that time, I was facilitating the start-up of the first Afghan NGOs, and he then was an employee of an NGO. Having first led a slow, deliberate trust- and capacity-building phase with those Afghans to start their NGOs, the Minister saw a similar need in the new government. Accordingly, I set about to establish a similarly paced participatory process with the respective stakeholder government ministries and departments, having them envision their roles and needs, and consider how this training center could serve their needs. After a few months of this process with government, a sense of ownership was developing among the departments and ministry people: this center would belong to them and they would lead it.

However, just as this joint progress was noticeable, the Minister received notice from two universities in Europe that they were about to arrive in Kabul to set up their own rural development studies programs, with their own processes and curricula. Stunned, the Minister set about to find out how this had happened. Two years earlier, foreign consultants conducting needs assessments had him and others list what they thought was needed in Afghanistan, and he had listed, among many other items, the idea for such a rural development training center. No discussion on the idea had been held then or since, but suddenly now these gifts were being parachuted in. The university programs would teach; Afghans would learn. University representatives arrived, dismissing the efforts already started. They essentially were setting up shop in Kabul, transferring top-down, made-elsewhere programs, and the Minister had no real choice but to cancel having his own Afghan-owned, Afghan-led concept, instead accepting these gifts as part of the pledges from the respective countries.

Conclusions

In organizations' quest to stake new territory and compete for new sources of funding, especially in such large-scale disaster or war reconstruction efforts, many of the basics including best practices in policy, process, consultation, and local participation can be thrown out the window. Even when institutions and their officials know better, their own self-interest and survival may also be at least part of their agenda.

Jane Murphy Thomas, MA (Social Anthropology of Development, School of Oriental and African Studies, University of London), is an independent, self-employed practitioner, consultant, and social anthropologist specializing in community participation in international development, disaster, and conflict areas. She has more than thirty years of experience in this field, living and working full time in Afghanistan, Pakistan, Bangladesh, and Azad Jammu and Kashmir. She works with local communities to develop their own initiatives, and carries out assignments for various government organizations and NGOs, multilateral agencies and consulting firms, for whom she facilitates intended beneficiaries' participation in projects. The projects cover a range of sectors from earthquake reconstruction, to community organization, health, education, water management, forestry and land acquisition.

References

Anderson, M. B., and P. J. Woodrow. 1989. *Rising from the Ashes: Development Strategies in Times of Disaster.* Boulder, CO: Westview Press.

Leal, P. A. 2011. "Participation: The Ascendancy of a Buzzword in the Neoliberal Era." In *The Participation Reader,* edited by A. Cornwall, 70–81. London: Zed Books.

Middleton, N., and P. O'Keefe. 1988. *Disaster and Development: The Politics of Humanitarian Aid.* London: Pluto Press.

Part II

SITUATIONS AND EXPOSITIONS

Plights, Problems, and Quandaries

Since the mid-1970s, geographers, anthropologists, and sociologists who research disasters have made the case that disasters are by no means natural events, but instead are processes engendered over long periods by human practices that enhance the materially destructive and socially disruptive capacities of geophysical and hydro-meteorological phenomena, technological malfunctions, and epidemics. Furthermore, these social scientists have made the case that the human practices that give disasters their form and magnitude also inequitably distribute the effects of catastrophes along socially created differentiations of gender, class, race, ethnicity, and age. Consequently, issues of social difference and inequity should be central to the analysis of disaster and mitigation policy. All the chapters in part II address the topic of difference in one way or another. Furthermore, all the case studies presented here also highlight critical gaps or disjunctions between scientific knowledge about the social dimensions of disaster and established mitigation policies and practices on the part of governmental agencies and nongovernmental organizations (NGOs) in a variety of settings, from the United States to Haiti.

Shirley J. Fiske and Elizabeth Marino begin by reviewing the challenges confronted by people and communities being directly affected by anthropogenic climate change-related sea level rise in their chapter, "Slow-Onset Disaster: Climate Change and the Gaps between Knowledge, Policy, and Practice." A key issue here is the way climate change effects manifest in a different temporal scale (e.g., the slow manifestation of a drought or sea level rise) in comparison to other kinds of disasters whose emergency phase has a more dramatic (e.g., the immediate effects of an earthquake or a tornado) onset. The slow-onset qualities of climate change in governmental agencies and the public at large being less likely to recognize a state of emergency and provide affected people and communities with the resources and assistance they need to mitigate the disaster at hand. Most importantly, communities that are hardest hit by climate change are also communities with long histories of experiencing environmental injustice, racism, and class-based discrimination. The social positionality of climate change–affected people and communities, then, compounds the sociopolitical effects of sea level rise, requiring policymakers and researchers to be specifically mindful of the ways marginalized populations voice their self-defined needs and visions of disaster vulnerability reduction.

A key dimension in the social distribution of risk is that of gender. In chapter 7, "Disrupting Gendered Outcomes: Addressing Disaster Vulnerability through Stakeholder Participation," Brenda D. Phillips provides a thorough discussion of the diverse ways gender differences manifest in the varied cultural contexts people live in and the ways gender difference comes to matter in disaster vulnerability and risk reduction. Phillips's main points are that policymakers and disaster response experts must recognize (a) how location-specific circumstances shape gender roles, (b) the way gender becomes interrelated with local systems of production (e.g., livelihoods and subsistence patterns) and social organization (e.g., kinship, class structure, and political organization), and (c) the manner in which disaster risk is differentially distributed and experienced as an effect of the relationships involved in items (a) and (b). Phillips's observations also resonate with the core messages of Adam Koons and Jane Murphy Thomas in chapters 4 and 5: Although disaster response experts and policymakers may at times have ideal values concerning gender (or universal humanitarian standards or project objectives), institutional ideals must always be negotiated with location-specific particulars. Flexibility and resourcefulness in the face of contingency are qualities to cultivate and encourage when bridging gaps between knowledge, policy, and practice.

Another major theme that concerns disaster researchers and policymakers is that of community relocation. As Anthony Oliver-Smith details in chapter 8, "Resettlement for Disaster Risk Reduction: Global Knowledge, Local Application," whether it is done as a measure of preventive mitigation or as part of recovery following a catastrophic event, the movement of communities from sites that are considered at risk of geophysical, hydro-meteorological, or climate change–related hazards is a practice that affects millions of people around the world every year. Despite the frequency and lengthy history of community relocation in disaster risk reduction—which has resulted in a significant body of academic knowledge on the topic—we continue to see problematic outcomes in state- or NGO-sponsored projects that address disaster risk through resettlement. A key concern here is that the movement of communities in space can disrupt relationships between people and their broader surrounding environment. These relationships are incredibly important for household and community livelihoods and also transform social institutions and structures in ways that prolong the social effects of disasters. A key dimension of the gap between knowledge, policy, and practice in these instances is the collection of relationships involving power and knowledge between relocation experts (e.g., architects, urban planners, and government officials) and the members of affected communities. Oliver-Smith echoes the observations by Fiske and Marino that those who are displaced by disasters are often members of socioeconomically disadvantaged communities whose voices, knowledge, and practices are not considered as valuable or valid as those of the experts assigned to assist them. This differentiation in value becomes a wedge that breaks apart what should be an important partnership between project managers, planners, and beneficiaries, resulting in relocation programs that disrupt the life-ways, kinship patterns, and social organization of those who are displaced by disasters. Like many of the chapter authors in this book, Oliver-Smith emphasizes the importance of flexibility, empathy, and a willingness to work in partnership with disaster survivors on the part of disaster management and response professionals.

The word "disaster" often evokes images of catastrophes set off by geophysical or hydro-meteorological hazards, but disaster researchers have also recognized the importance of including technological accidents and epidemics within their analyses for quite some time now. Ryo Morimoto's contribution to this volume, the chapter "From Nuclear Things to Things Nuclear: Minding the Gap at the Knowledge-Policy-Practice Nexus in Post-Fallout Fukushima," brings our attention to the combined Fukushima tsunami and nuclear power plant meltdown and beckons us to consider the gap between knowledge, policy, and practice from a new vantage point. Whereas chapters 1–3 in part 1 focused on gaps and disjunctures, Morimoto's chapter 9 makes a compelling argument that the social challenge of radiation cleanup and containment is not as much a matter of missed connections as it is a matter of unexpected emergent meanings that lie beyond the purview of governmental and scientific knowledge, policy, and practice. Very much like nuclear radioactive fallout that does not preexist a power plant meltdown, the meanings people attribute to the material objects that are produced through radioactive cleanup do not preexist a meltdown disaster nor are they anticipated by existing knowledge and policy. What is more, these meanings come to play a vital role in how residents of nuclear disaster–affected areas make decisions about returning to their homes and participating in recovery programs. Furthermore, existing knowledge, policy, and practice lacks the analytical and methodological resources to document and grasp the development of such meanings. Consequently, Morimoto's chapter highlights the relevance of anthropological and semiotic analyses in disaster research and recovery.

The final chapter of this part brings us back to concerns about voice and inequity. In chapter 10, "'Haitians Need to Be Patient': Notes on Policy Advocacy in Washington Following Haiti's Earthquake," Mark Schuller shares his experiences as a public anthropologist working in collaboration with an umbrella organization of Haitian NGOs in the aftermath of the 2010 earthquake. This case study demonstrates how issues of power and representation play out at the highest levels of governmental and NGO administration, especially when non-Haitian U.S. government officials appropriate the right to speak for the needs and concerns of disaster-survivors from this Caribbean nation. Schuller's chapter resonates with Oliver-Smith's observation that effective, culturally sensitive, and locally relevant disaster recovery requires a partnership between disaster survivors and NGO and governmental agency staffs where discretion over the spending of budgets and use of resources is equally shared. Together, part II's chapters provide readers with a review of the sensitive issues of difference and inequity that complicate disaster mitigation, response, and recovery, and offer invaluable insights for practitioners interested in enhancing their ability to address these themes in their work.

CHAPTER 6

Slow-Onset Disaster

Climate Change and the Gaps between Knowledge, Policy, and Practice

SHIRLEY J. FISKE and ELIZABETH MARINO

Introduction

Most disasters, such as those brought on by hurricanes, floods, and torna-does, have a relatively recognizable window of existence and a so-called "shelf life," whether anticipated or not. Although the underlying conditions of vulnerability that often predicate disasters and determine who is placed directly in harm's way have long time horizons, the meteorological, atmospheric and hydrological expressions of most publicly recognized disasters develop quickly and are treated as temporary aberrations from normal conditions of everyday life. Thus, rapid-onset disasters stemming from floods, hurricanes, and tornados become the prototypical model of disasters on which most of the emergency management and disaster response procedures and policies in the United States are based.

A slow-onset phenomenon such as climate change, however, is global, gradual, and cumulative over time, and alters the underlying environmental baselines on which disasters occur. While the social preconditions that underlie most disasters often accrue over long periods, climate change is unique in that it continuously shifts ecological baselines through subtle and insidious sea level rise, warming oceans and land areas, increasing erosion, and declining sea ice and snowpack, among other things. The changes are almost imperceptible at any one moment, but lead to permanent changes in the ecology and landscape that will render some homes, communities, and cities uninhabitable. Not only does exposure to climate change hazards put individuals, families, and communities at greater risk, but also in the United States, we argue, there is little statutory or regulatory basis for responding to slow-onset disasters, leaving people and communities to fend for themselves as these changes occur. States, cities, and counties are struggling to identify ways to ensure public safety and constrain property loss, but to date disaster policies focus on the immediate rapid-onset disaster event and its aftermath, and climate change policies have focused almost entirely on mitigation (reducing carbon dioxide (CO_2) emissions) rather than on adaptation (Urwin and Jordan 2008). Therefore, we suggest there is a gap between our knowledge of the impending effects of climate change-related phenomena and policy and institutional infrastructure (Bronen and Chapin 2013) to protect even those already at the most extreme risk.

The slow-onset nature of climate change establishes a new baseline for disasters in two ways. First is the global incidence of higher sea levels, a warmer atmosphere, decreasing sea ice in the Arctic, and changes in fringing reefs and mangroves in tropical areas, all of which add up to greater vulnerability to people and communities onshore and new meteorological and atmospheric thresholds for storms. These facts mean that slow-onset climate change exacerbates the formation and impacts of rapid-onset disasters, such as the storms that pound Shishmaref and other coastal communities in Alaska.

In 2012 East Coast communities in the United States experienced the catastrophe of Hurricane Sandy, with higher sea levels (from sea level rise) along with high tides that brought greater devastation to downtown Manhattan (New York City) and the New Jersey shoreline communities than expected. Similarly, along the Arctic shorelines of Alaska the communities of Shishmaref, Newtok, and Kivalina (among others) have repeatedly suffered disastrous winter storms and erosion that have eaten away at their communities for the past twenty years. These calamities are enabled by rising sea levels and declining sea ice, which in previous centuries added a protective barrier to erosion.

Second, the invisibility of climate change with its changing baselines may lead to an acceptance among the general public (if not coastal communities) of the risks of increased and repeated flooding, erosion or hurricanes as the new normal. This gradual shift in belief systems, in effect, regularizes expectations of more frequent catastrophes and allows decision makers to continue to define slow-onset disasters as non-disasters. Therefore, while rapid-onset disasters are recognizable to emergency management and hazards officials, and while policies and programs exist for planning and mitigation in response to rapid-onset disasters, the disasters that stem from slow-onset climate change phenomena are more or less invisible in terms of policies and programs, and create critical challenges to pre-existing disaster management planning (Schipper 2009).

Slow-Onset Disasters and Their Effects on People

Despite their seeming invisibility, slow-onset disasters really are disasters: the gradual shifts are likely to result in the same panoply of human trauma and loss of property that that are brought by rapid-onset hurricanes, fires, floods, and tornadoes. Effects of slow-onset disasters on people include death and diaspora, loss of property and community, loss of cultural icons and way of life, and loss of livelihoods. Wisner and colleagues (2004) describe "slow-maturing" disasters as "unfolding" rather than "happening" (147) and include such widespread disasters as drought and famine, and pandemics such as HIV-AIDS, in addition to climate change. They note that, similar to rapid-onset disasters, the root causes of slow-onset disasters form over a long period (see Oliver-Smith 1999), but there are fundamental differences between them. First, there is a temporal difference: despite long-term genesis, most rapid-onset disasters can be measured in minutes, hours, or days, as in the case of hurricanes or flooding.[1] Slow-onset disasters, on the other hand, unfurl over longer periods and the length of trauma and destruction are cumulative, but nonetheless consequential. Wisner et al. (2004) argue that prime examples of slow-onset disasters such as drought and famine account for the vast majority of all human deaths due to "natural hazards" (87 percent), followed by floods, which account for less than 10 percent (3).

A second fundamental difference between slow- and rapid-onset disasters is the chronic and increasing ecological impact of the former on people's livelihoods and community habitability. Climate hazards will increase in intensity and frequency, and make some livelihoods and living situations less sustainable (Wisner et al. 2004, 83), especially for indigenous populations who depend on subsistence hunting and fishing, such as communities like Shishmaref, Alaska (this chapter), communities in the bayous of Louisiana (Maldonado 2019; and this chapter), or Maryland's Eastern Shore fishing communities (this chapter). Moreover, Wisner et al. argue, "Slow-onset disasters require careful analysis of social adaptation, the emergence of new rules of inclusion and exclusion regarding networks of support and changing access qualifications for new and existing income opportunities" (109). Slow-onset disasters, over time, encourage out-migration and gradually erode social networks and income opportunities as employment steadily withdraws from the region. Communities become more isolated, and access to resources becomes even more tenuous, as has happened in rural communities facing sea level rise on the Eastern Shore of the Chesapeake Bay among other island and shore-based coastal communities in low-lying areas across the world. Communities that once had sustainable economies are vitiated with repeated hazards assaults, out-migration, and loss of jobs.

Slow-Onset Disasters and Environmental Justice

Anthropological and social science studies of disaster have taught us a great deal about the intersection of social and environmental systems during disaster events, including slow-onset disasters. For the past two decades social scientists have been examining so-called natural disasters and their aftermath, along with the precursors to differential loss of lives and property in these disasters (Faas 2016; Hoffman and Oliver-Smith 2002; Oliver-Smith 2013). Anthropologists have argued that natural disasters are in fact socially constituted and have shown the need to pay attention to social context, as well as historic and economic preconditions in assessing risk, in dealing with disasters, and in emergency management planning (Hoffman and Oliver-Smith 2002; Wisner et al. 2004, 4–5).

In addition, research on environmental justice makes explicit that environmental risks are distributed unequally across classes, race, and ethnicity (Mohai, Pellow, and Roberts 2009). To understand the differential effects of disasters, whether rapid- or slow-onset, we must pay attention to underlying vectors of social stratification, poverty, and access to resources and social networks. We must understand power and power relationships at multiple scales, as they map across social classes and race, and, in turn, as they map across environmental risks (Pellow 2007). Historic contexts such as colonization, housing segregation, forced relocation, isolation, and enclosure, are critical for understanding climate justice today. As Marino and Ribot (2012) observe, "Climate change is redistribution. It alters the timing and intensity of our rains and winds, the humidity in our soils, and the sea level around us. As redistribution, climate change is also a matter of justice—it is about who gains and who loses as change occurs and as interventions to moderate change unfold" (323).

Following this line of reasoning (questions of power and social stratification) we would be remiss if we did not call attention to the well-known fact that slow-onset disasters are even more likely to parse the heaviest damages on the most marginalized areas and people, often those who live in low-lying delta areas, on small islands, at high altitudes, and in high latitudes (see Fiske et al. 2014). As the Intergovernmental Panel on Climate Change (IPCC) puts it, "Climate change will amplify existing risks and create new risks for natural and human systems. Risks are unevenly distributed and are generally greater for disadvantaged people and communities in countries at all levels of development" (IPCC 2014, 13).

We will provide six cases of threatened towns from three geographic areas in the U.S. that document how communities are falling into the gap between knowledge and policy to address slow-onset disasters. The chapter will highlight the ways in which climate change outcomes are being experienced by human communities as both slow- and rapid-onset disaster events, given the new baseline of climate change, and what communities are doing about the accumulating disasters. We will highlight communities already experiencing hazards and ecological shifts linked to climate change and will examine their risk mitigation options in contemporary policy environments including migration, accommodation, and adaptation. We will take into consideration the role of the U.S. Army Corps of Engineers (USACE), the Federal Emergency Management Agency (FEMA), local and state governments, and national policies as applicable. The chapter observes that climate change exacerbates existing environmental justice gaps and that U.S. policy and programs leave much to be desired in addressing social justice issues (see Fiske et al. 2014). We note that it is the *lack of policies and programs* for slow-onset disasters such as climate change, that widens the gap into which rural and low-income, and marginalized communities and people fall. We will discuss how the gap was created and maintained politically, the implications for social equity, the need for climate adaptation policy in the United States, and whether the gap can be closed. Finally, we will discuss the ramification of facing climate change outcomes from a policy standpoint and make suggestions for future policy and research areas.

The Eastern Shore, Maryland

Eleven counties, nine in Maryland and two in Virginia, comprise the Eastern Shore of Chesapeake Bay, the southern portion of which is one of the most vulnerable regions to sea level rise in the mid-Atlantic (Boesch et al. 2013, 7). Lined with low-lying marshlands, shoals, and islands, interdigitated with necks of land, this idyllic shoreline and its waterways are home to heritage groups such as Maryland and Virginia watermen who harvest crabs, oysters, and finfish, and farmers descended from those who have lived and worked the water and lands close to the water for hundreds of years. On some islands, such as Tangier Island, families trace their ancestry back to the descendants of John Smith's voyages up the Chesapeake in the 1600s. Their way of life and livelihoods have been immortalized in books such as William W. Warner's *Beautiful Swimmers* (1976), John R. Wennerston's *The Oyster Wars of Chesapeake Bay* (2007), and James A. Michener's *Chesapeake: A Novel* (1978).

Heritage and Environmental Justice Communities at Risk

In addition, the Eastern Shore is haven to hundreds of small communities established by descendants of freedmen and slaves from plantations after emancipation (in Maryland in 1864), communities in Talbot, Dorchester, Somerset Counties, and other tidal counties where plantations once lined the creeks and waterways. These communities, or really clusters of homes, were generally settled in the flat, forested marshlands along the creeks and rivers that open into the Chesapeake Bay, such as the Choptank River, Fishing Creek, and Blackwater River. Oral histories and recollections, combined with recorded deeds, reported in Miller Hesed (2016), indicate that Smithville, in Dorchester County, for example, was established in 1886, and that early on people sought work in local sawmills, as farm laborers, in seafood processing (blue crabs and oysters), or developing livelihoods as watermen as the regional economy changed and evolved over the next century (77).

Many such clusters of residents dotted the necks of land between the creeks and provided spaces for small groupings of families, generally centered around a church as the hub of social and spiritual life. Most of the clusters were too small to be designated as villages and therefore lacked representation in local government (Miller Hesed 2016, 77), leading to political and economic isolation from the 1870s (Jim Crow era) through the 1940s (see Wennersten 1985). Today, what remains of Smithville are a few trailers and clapboard homes, some abandoned, a graveyard, and a wooden frame church, still in use, sitting on the edge of the tidal creeks, in sight of the rising water and vulnerable to frequent flooding.

From the mid-twentieth century onward, many of the children of families in villages such as Smithville moved to larger regional towns or urban centers such as Baltimore in search of employment and wage labor opportunities, with the net effect of helping to depopulate coastal rural counties. Small and large businesses have been exiting the region from the 1900s until today. Agriculture and seafood processing have been mechanized, as the region has become integrated into regional and national economies over the course of the twentieth century. The same economic constriction typifies all of south Dorchester County and other low-lying tidal communities in Maryland.

The focal institution of many of the remaining African American towns remains the United Methodist Church, still actively attended although by a dwindling flock. Churches have been the nucleus of strength and resilience for the social life of many of the Eastern Shore small communities for generations, safeguarding their history and family members for more than one hundred years (Miller Hesed 2016, 59–60). Ironically, these same churches and their associated graveyards replete with families' relatives are particularly at risk from sea level rise and flooding; assistance in saving them is difficult to find from public and private sources unless the funds are raised from the parishioners themselves.

Because the social preconditions that underlie most disasters often accrue over long periods, it is worth noting the historical context surrounding these communities. The lands that were settled by African Americans in the post-emancipation era were conspicuously distant from the main towns of Cambridge or Easton in areas that plantation dwellers, merchants, and farmers did not claim or had no use for. They settled on the less desirable lands, typically at more risk for flooding or environmental hazards. This settlement pattern is a recurrent theme not only in the United States, but also across the globe where the stigmatized and often legally segregated poor are left to their own resources to live in places with risky exposure to natural hazards. "The fact that many of the rural African American communities today are in close proximity to water bodies and are at high risk of flooding from sea level rise is directly related to historic settlement patterns during times of intense racial discrimination" (Miller Hesed 2016, 59–60). The pattern of settling on marginal lands is rooted in the historical context of segregation and discrimination toward African Americans after emancipation (Andersen 1998; Wennersten 1985).[2] The history of discrimination and residential segregation has the long-term effect of putting African American communities at risk under almost all scenarios of climate change (Miller Hesed 2016, 77).

Starting some sixty years ago, the opening of the Bay Bridge in 1952, linking the Eastern Shore to the metropolitan areas of Baltimore and Washington, DC, brought another demographic community to the Eastern Shore. The Eastern Shore's inviting shoreline and tranquil waterways, with ample room for private docks and boating, began to attract wealthy "amenity migrants" to its coastline and municipalities. As they generally have higher income and education levels, the "come-heres," as they are known locally, flocked to buy waterfront properties along the creeks and rivers of the Bay, seeking a more rural, smaller-scale, and more recreationally oriented lifestyle. While these residents are also at risk to sea level rise, they do not have the vulnerabilities that characterize African American and fishing-based communities along the shorelines, who have many fewer resources to adapt to or resist slow-onset disasters. The newcomers likely have greater adaptive capacity to withstand changes brought by changing climate and sea level rise.

Changing Baseline of Sea Level

The Eastern Shore is a poster child of slow-onset climate change, with gradually eroding islands and necks of land, widening creeks and "rising tides," as farmers refer to it, and imperceptible change from geological subsidence.[3] The region is largely low-lying and flat, with the highest point in Dorchester County only fifty-seven feet above sea level (Maryland Geological Survey 2014). The shorelines are expected to recede, gradually turning working farmlands, forests, and waterfronts into marshlands and broader reaches of water. This is especially true in the lower Eastern Shore counties (Dorchester and Somerset), where sea level rise is expected to encroach inland creating wetlands and open water covering thousands of acres.[4] One can already see that roadside ditches, built originally for drainage and mosquito control, contain standing water that rises and falls with the tides. Residents are now used to driving on their roads covered by a few inches of standing water on sunny days with no storms in sight.

Changing shoreline conditions were important enough that the state of Maryland and counties bordering the Chesapeake Bay enacted legal statutes to mitigate slow-onset environmental problems such as erosion and declining water quality that mandate set-back zones of a hundred feet to act as a buffer for streams, wetlands, and shorelines. But the Chesapeake Bay Critical Area Law of 1984 has not stemmed erosion nor of course rising tides and sea levels that cumulatively add to the deteriorating habitability of the region (constant flooding that threatens homes, drinking water and septic systems), and slowly chip away at economic viability (loss of property value, out-migration of businesses) on parts of the Eastern Shore. Adding insult to injury from the declining economic activity, now sea level rise has created a new urgency for considering strategies of adaptation or migration.

The impacts described above will become even more commonplace if sea level rise accelerates, which, in fact, appears to be happening. In 2012 Sallenger, Doran, and Howard published a report demonstrating that the Eastern Shore lies in the middle of a thousand-kilometer-long sea level rise hot spot, where sea level rise increases were approximately three to four times higher than the global average. Additionally, a report by Boesch and colleagues (2013) provides evidence that sea level has been rising faster in the region than elsewhere along the Atlantic Coast (7) and highly intense storms are projected to become more common during the twenty-first century (14). The 2014 National Climate Assessment states that the region "will experience an increase in coastal flooding and drowning of estuarine wetlands" (Melillo, Richmond, and Yohe 2014, 381).

The Knowledge-Policy Gap

In Maryland, scientific, governmental, and public knowledge levels about climate change projections, especially sea level rise, are quite high, and coastal residents are climate-aware. In a recent survey, George Mason University's Center for Climate Change Communication reported that 55 percent of Marylanders support protecting coastal areas and low-lying shorelines from sea level rise (Akerlof and Maibach 2014). Furthermore, Marylanders support changes in regulations, such as zoning laws and set-backs, tax incentives for property owners, and the use of government funds to buy natural areas as buffers. Most Marylanders also expect hotter weather and more frequent severe storms (Akerlof and Maibach 2014, 1–2).

Because of the cycle of yearly spring high tides, regular tidal flooding, and episodic Category 4 hurricanes, erosion and flooding are familiar issues throughout the Chesapeake Bay region. Hurricanes and their storm surges arrive at regular intervals, inundating the counties and com-munities. Local environmental knowledge holds that hurricanes come every twenty to thirty years, and people can recite the names of the big ones: an August hurricane in 1933, Hazel in 1954, Isabel in 2003, and Sandy in 2012. The general view of people born and raised on the Eastern Shore is that flooding is an expected, normal event that they have seen before, planned for, and seen earlier generations survive them. They do not see flooding as a source of vulnerability the way that state managers and environmental groups perceive it. Instead, south Dorchester County people see flooding, occasionally up to three feet high, as a periodic but normal occurrence and more of an annoyance than a disaster. People feel they have been adapting to coastal flooding for centuries, literally, and that they are prepared for it (see Fiske, Paolisso, and Clendaniel n.d.; Johnson 2016).

At the state level, Maryland has been a leader in climate change planning and adaptation, which accounts in part for the high level of public aware-ness of adaptation strategies cited earlier. In 2007, then-Governor Martin O'Malley established the Maryland Commission on Climate Change, which resulted in climate change adaptation programs and an action plan to reduce Maryland's contribution to greenhouse gases. One of a minority of states enacting a CO_2 emissions program, the state passed the Greenhouse Gas Emissions Reduction Act of 2009, designed to reduce emissions by 25 percent by 2020, primarily in the transportation and energy–utility sectors (71 percent) and the agriculture–forest conservation sector (12 percent).[5]

Yet despite the high level of knowledge of climate among policy and decision makers and the fact that Maryland is proactive in climate change planning and adaptation, there remain systematic gaps in the institutional safety net to address the needs of low- and moderate-income families who are linked to their coastal communities on the land, shorelines, and islands in the Chesapeake Bay.

The following paragraphs detail some of the challenges for state and local government, and for communities, because slow-onset disasters will play out under current disaster policies, that lack climate adaptation policies for slow-onset changes like sea level rise.

To start, a recurrent theme in this chapter, is the concern that disaster and hazard mitigation policy is focused on individual property owners, whether commercial or residential, rather than on community survivability, resilience, or sustainability. States and counties, particularly when it involves land use policies, provide home owners and property owners with guidance on how to keep their property habitable. For instance, in Maryland property owners can bulwark their shorelines, raise their houses and churches, or try the softer approach recommended by the state to reduce the land erodibility—planting sea grasses and stabilizing the shoreline with naturally growing vegetation as the sea moves inland. Right now, the property owner typically bears these expenses. While owners of million-dollar properties can take out second mortgages, pay cash, or get loans to pay for raising their homes on stilts, multi-generational working-class families working the water on the Eastern Shore have to mortgage their boats, borrow or depend on social capital from their networks of friends.

Some of the residents on Eastern Shore's islands and peninsulas are "amenity migrants" who vacation there and use their homes for only several months a year, but the majority of residents retain close ties to the land or water. They are families of watermen harvesting oysters and crabs, individuals who work in seafood processing, marina owners, or those otherwise involved in services to the fishing industry and shoreline trades who have lived there for generations. In the African American clusters of homes, the long history of their settlements is chronicled by the graveyards and headstones of ancestors, relatives, and friends adjacent to the churchyards. What kind of support will be available to help restore, protect, or move churches, graveyards, and other private structures?[6]

Meanwhile, basic questions of rights and property ownership are evolving and remain unclear. As the mean high tide line moves landward, who owns that property, which is now a marshland or underwater? Do people simply lose their property and its value, and the property becomes part of the state's tidelines below the mean high tide line? Will the state compensate property owners? Who absorbs the loss in value?

For the past several years, the state, in conjunction with FEMA, has been systematically updating Flood Insurance Rate Maps. The Digital Flood Insurance Rate Maps are used to determine whether a home-owner resides in the hundred-year floodplain, which then determines the availability and rate for federal flood insurance as set by FEMA. More homeowners will find themselves facing high insurance rates or becoming ineligible altogether as FEMA adjusts the boundaries of the floodplain inland. The implications for low-income homeowners are that they may not be able to afford flood insurance and will have to face the risk of flood damage or worse on their property (Goldman 2011). Making the National Flood Insurance Program economically sustainable and simultaneously affordable for homeowners has been almost impossible and has resulted in changing legislation between 2012 and 2014. In order to make the National Flood Insurance Program sustainable, the Biggert-Waters Flood Insurance Reform Act was passed in 2012, but when rates among some homeowners threatened to spiral upward in 2014, Congress amended the language so that rate increases to property owners were less substantial. The problem of legislating an economically viable insurance program that homeowners can afford is still an unsolved problem. The issue threatens both the long-term viability of the National Flood Insurance Program and the ability of low-income coastal dwellers to find affordable flood insur-ance, especially under conditions of rising sea levels (see Wriggins 2015).

Dorchester County has been forewarned that it may have to abandon the upkeep of roads that are flooded on an increasingly regular basis, leaving communities on small islands or necks of land lacking regular ser-vices via roadways. The reason for this is simply the cost of maintenance for infrastructure such as roads and bridges, including continually raising the road, which is the extant policy for dealing with recurrent flood-ing. The county's dilemma is compounded by gradual de-population, a declining tax base, and limited resources at the county level (Cole 2008). The observation of this researcher among others is that the systematic, planned abandonment of infrastructure in low-income communities and properties in the Eastern Shore will become a wide-spread environmental justice–climate justice issue.

An example of the contentiousness of costs imposed by slow-onset disaster and the county's need to encourage residential and business construction to maintain a tax and population base is the question of "freeboard" regulations: Should the County require people to build their houses higher off the ground? The question of regulating new construc-tion to be built at a higher level to accommodate flooding has been under discussion at the county level since the 1990s. Dorchester County has been aware of the need to enact freeboard regulations into their building codes, but every time the issue comes up it is defeated. Freeboard regulations would require new construction to have three feet of freeboard (above base flood elevation) on new construction and on renovated construction as well. It has been proposed before county officials four times, and four times it has been defeated, due to concerns citing increased costs of construction (Goldman 2011).

Quintessential and consequential questions remain: "Who is responsible for absorbing the loss when homes become surrounded by the daily tide? The property owner, who may be elderly and not have the income to start over? The property owner whose solvency is tied to their property equity? The employee whose employer must relocate the business many miles away, leaving a workforce behind who cannot sell their flood-damaged properties?" (Cole 2008, 54). Many of these unanswered questions about property ownership and sea level rise will come down to zoning and land use that are typically the purview of states and counties. Possible responses to sea level rise include conservation easements, property acquisition programs, setbacks, flood hazard regulations, density restrictions, and building size regulations. Two of the most extensively used are "rolling easements" and Transferred Development Rights (TDRs).

Rolling easements are land conservation regulations that generally prohibit property owners from erecting hard structures to protect their property from flooding (e.g., with a seawall) but allow owners to undertake any other kind of use or activity on their land, including building or re-building their house on their property. The easement "rolls" landward as sea level rises. Rolling easements put the risks and costs of continuing to live on the land squarely in the hands of the property owner, an income-regressive policy, since the private property owner is essentially managing the risks of sea level rise (see Titus 2011). As of 2012, California, Hawaii, Maine, Maryland, North Carolina, Rhode Island, South Carolina, and Texas are using some variation of rolling easement policies in their coastal areas to manage sea level rise adaptation (Cox 2012).

TDRs are zoning regulations that allow for the "severance" of development rights from one property, say, in a coastal flood zone, to another tract of land in a less risk-prone area. The property owner is selling their right to put houses or other developments on their coastal property, but they can still use it for other purposes. These zoning and land use mechanisms (TDRs) are usually undertaken by a county, and work best for undeveloped land at larger, regional scale rather than individual properties. The development rights are generally bought by developers via the real estate market. TDRs appear to be a disadvantageous mechanism for individual property owners in small, rural communities where there is no demand for nearby development and the value of development rights will be minimal.

In return for the easement or TDR, what does the property owner receive? The answers vary by locality: an owner can donate the rolling easement to the state or municipality; the owner can sell the easement to a municipality, county or nonprofit if the latter have funds; or a property owner can sell their right to coastal development on their property if a TDR zoning program exists (see Cox 2012).

These persistent questions of rights, ownership and responsibilities, and the relative paucity of programs to re-coup lost property value with sea level rise, signal "climate injustice," as some term it, where the gaps between our knowledge of the future and what it will bring, and the policies we have in place to assist the public have vast lacunae that disproportionately affect lower-income, working class, and rural families and communities. Since erosion and sea level rise are not considered to be disasters, there is little recourse through FEMA and even the USACE. And since most adaptation models in the U.S. are built on individual property owners' responsibilities, not community-based engagement, these communities will suffer the indignities of their homes and villages becoming un-livable, un-viable and un-sellable.

The Rural Coast of Alaska

As in the Chesapeake Bay, rural coastal Alaska has also been identified as a poster child of climate change, where persistent changes, especially permafrost thaw and erosion, have led to deteriorating conditions and habitual flooding in some communities. When infrastructure and livelihoods are compromised irreparably, some communities will consider relocation/migration as the only possible adaptation strategy. It is important to note that most migration linked to climate change will take the form of temporary migrations allied with increasingly challenging conditions in which to earn a livelihood. In such cases, migration patterns are likely to follow traditional economic and labor migration routes (Raleigh and Jordan 2010). However, in other cases climate change will make areas that are inhabited no longer viable places to live for anyone, at any time. In the most extreme cases, land will be inundated by water permanently. Rural Alaska, and especially the coast of rural Alaska, presents all of these situations.

In 2002 and again in August 2016, the community of Shishmaref, Alaska, voted by a narrow margin to relocate their rural community from their island home to the mainland due to extensive erosion and flooding. Since 1986 there have been eight state-declared flooding events in Shishmaref, and the community continues to lose land every year during fall storms.

The island is approximately three miles long and a quarter of a mile wide, meaning that storms that cause erosion significantly impact the vulnerability of the island (Department of Commerce, Community, and Economic Development 2016). The increase in flooding and erosion over time has been linked to climate change (Azelton 2009; Smith and Levasseur 2002; USACE 2009; U.S. General Accountability Office 2003, 2009) and other complex interactions among infrastructural and geomorphological processes (Jordan and Mason 1999; Mason et al. 2012). Sea level rise, continued permafrost thaw, erosion, and larger storms in the Chuckchi Sea are likely to continue to threaten the island with flooding for the foreseeable future. In 2009 USACE estimated that the community had only between fifteen and twenty years before it became uninhabitable (USACE 2009).

In Shishmaref, residents have many concerns about climate change, but they also have concerns about the policies available to them to help craft successful relocation as an adaptation strategy for a future under a new ecological regime. Residents have expressed worry that relocation will not occur before a major disaster unfolds (Marino 2015, 69–71) and that voting to relocate will inhibit minimum infrastructure development, including basic sewage and water projects, on the island for the foreseeable future (Department of Commerce, Community, and Economic Development 2016; Lazrus and Marino 2015). The interaction between these two concerns—the lack of realistic planning, policies, and funding for relocating and the lack of infrastructure investment in place if residents discuss relocation as an adaptation strategy—is incredibly problematic. Without a realistic plan for moving forward, disinvestment from current infrastructure, coupled with continued erosion, could lead to the deterioration of living conditions on the island. In other words, the inability to plan under conditions of persistent change is creating a situation where both policy protocols *and* climate change outcomes are simultaneously threatening the viability of the community for the foreseeable future. Again, similar to the rural East Coast, we see that abandoning maintenance of community infrastructure in the face of extreme risk creates its own critical outcomes. It is essential to remember that the community of Shishmaref has been discussing relocation with the state since at least 1974 (Marino 2015, 61–62).

Another community in Alaska, Newtok, has moved forward with relocation with some success. At a new site, called Metarvik, the community of Newtok has successfully begun infrastructure development in anticipation of relocation (Newtok Planning Group 2016), although no one in the community has moved to the site year-round (Semuels 2015). A brief analysis of the processes that Newtok has gone through demonstrates the lack of streamlined policy available for climate change–displaced populations.

In 2006 the Newtok Planning Group formed; it was a collaboration among two local (Newtok) political organizations, nine state organizations, twelve federal organizations, and six regional organizations. Also in 2006, Newtok residents built three homes at the new site at Metarvik, establishing and demonstrating the site selection for relocation. The weight of this decision to construct three new houses at a site where no one lives should not be overlooked. In rural Alaska new homes are scarce and heavily coveted. Here, up to fourteen people can live in a three-bedroom home. Putting new houses at Metarvik, knowing that no one would live in them in the near future, was a bold statement.

Additional infrastructure in Metarvik has been added through an ad hoc suite of policy opportunities (Bronen and Chapin 2013, 9323). One of the most inventive was the extended use of the U.S. Department of Defense's Innovative Readiness Training Program, which brought military personnel to Newtok to assist in the construction of a boat landing, a rock quarry, and a road at the new site. This program exists in order to give military personnel "real world training opportunities." The use of the Innovative Readiness Training Program by the Newtok Planning Group was seemingly successful for all parties involved, and yet we can ask, Why is a military training program the best (or only) option for disaster mitigation under conditions of climate change in the community of Newtok?

Watching the Newtok Planning Group and the community of Shishmaref work toward relocation over the past decade has been a study in improvisational policy manoeuvring. As leaders in the climate change–mitigation process, these communities, and the state and federal agency employees who partner with them, are perpetually working with limited funding and limited personnel capacity. The formal organizations and coalitions that have worked on climate-induced relocation in Alaska have morphed from the Immediate Action Workgroup in the Governor's Subcabinet on Climate Change (2007–9), to the Alaska Climate Change Mitigation Program (2009 to the present), to President Obama's decision in 2016 to put the Denali Commission in charge of the relocation of Alaska rural communities threatened by climate change (2016). Likewise, positions within the community have had to stretch to build local capacity. For years, a funded position of transportation coordinator was the point person for relocation issues in Shishmaref because there was no position for a climate change relocation and erosion protection coordinator.

What climate-induced relocation planning in Alaska points to is that procedural templates, local and state institutional capacity, formal policy, or what Robin Bronen calls an "institutional relocation framework" (Bronen and Chapin 2013, 9323–24) are lacking. When communities such as Shishmaref vote to relocate to avoid catastrophic flooding, the pragmatic result to date has been a vote to decrease the likelihood of important community infrastructure (Marino 2015; Marino and Lazrus 2015). This adding of insult to injury demonstrates the lack of policy equipped to handle climate change–induced relocations and is infuriating to community members and community allies alike.

Policy changes that could benefit rural Alaskan communities in this situation are likely many. We will address three possibilities here. First, erosion is a slow-onset disaster not recognized under the Robert T. Stafford Disaster Relief and Emergency Assistance Act of 1988 (known as the Stafford Act), which governs FEMA (Bronen and Chapin 2013). If erosion were recognized, then storms that cause land loss or threaten critical infrastructure could trigger disaster policy procedures. This could free up funding for responses that cause risk but do not immediately inundate homes and other infrastructure with water. Second, while there is some leverage within FEMA to buy back high-risk and repetitive flooding homes, that leverage has not been effective in Alaska to date, for a number of reasons (Marino 2018). Funding that is dependent on cost-benefit analyses are challenging for rural communities to qualify for, particularly in the Arctic where the cost of building is high and populations are small. Also problematic is the pervasiveness of the nuclear family or household as the foci of disaster mitigation and response programs in the United States, as mentioned earlier. In Shishmaref and Newtok researchers, advocates, and community leaders discuss the community as a collective, including community relocations. Buy-back programs, however, and much disaster mitigation work in general, is focused on household-level decision making and the process has applicability only to buy back individual family homes. These limitations on who is visible (i.e., home owners and "households" vs collectives) compromise community-oriented solutions and do not provide a clear way forward to fund relocation or reconstruction of community-based infrastructure. Finally, there need to be infrastructural frameworks (Bronen 2008; Bronen and Chapin 2013) for communities to follow once the decision to relocate has occurred. Any institutional framework for relocation needs to include capacity building within communities and within institutions tasked with carrying relocations forward. We see again that rural communities in particular are at a disadvantage simply by not having the personnel and institutional capacity that larger communities would have to deal with both risk itself and the bureaucracy of risk mitigation.

Southeast Coastal Louisiana

Facing some of the most rapid rate of land loss and relative sea level rise in the world, southeast Louisiana is home to both tribal and nontribal communities who have dwelled in place for generations. Their lifeways are "shaped by their subsistence and traditional practices and spiritual connection to the land and water" (Maldonado and Peterson 2018, 289). Here, in Louisiana bayou country, among the residents of Isle de Jean Charles, for example, the palpable sense of "community" and importance of place stand out clearly. The Isle de Jean Charles band of the Biloxi-Chitimacha-Choctaw Tribe (the Tribe) is a state-recognized tribe that in the late 1700s and early 1800s evaded forced removal to the West during the period of Indian Removal, by escaping to the southern bayous of Louisiana.[7] For nearly a century they lived in relative political and economic isolation, which led to a strong sense of independence, community, cultural identity, and livelihood sustainability (Maldonado 2019, 57–79; Peterson and Maldonado 2018).

The community has faced multiple, cumulative assaults to their cultural sovereignty and sustainability through erosion and loss of wetlands from oil and gas drilling, building of levees for flood control, and relative subsidence of the region from extraction. The community was once located geographically well inland from the Gulf of Mexico, but today is face to face with the Gulf waters themselves. The tenuous state of their land is compounded by climate change–induced sea level rise, greater exposure to waves and storm surges, and more-rapid erosion from continued storms. By 2017 after perpetual flooding, lost homes, and "accumulating disasters," only seventy-eight houses and 325 people remained on the Isle de Jean Charles. The island had shrunk by 98 percent from 33,000 acres to an estimate of less than 500 acres (Maldonado 2019, 29).

The Tribe has been determined to find every pathway possible to restoration and remaining in place on the Isle de Jean Charles, and their decision to relocate did not happen instantaneously. It developed over time, as their land disappeared from under them, and residents were forced to move after continued flooding and loss of their homes, connecting roads to the mainland, and schools. One critical event, however, changed their deliberations and feelings about the course of resettlement. An ambitious federal-state-parish plan (the "Morganza-to-the-Gulf" plan[8]) to better protect southern Louisiana communities from hurricanes and flooding was revised and redrawn in 1998, and subsequently left the entire island and community of Isle de Jean Charles outside of the boundaries of the Gulf of Mexico Hurricane Protection System (Maldonado 2014; Maldonado 2019, 115–116). That decision to revise the boundaries essentially sacrificed the island and its people, ostensibly because the cost-benefit ratio for USACE and other decisionmakers was not high enough to justify the cost. Close on the heels of the USACE decision were a continual series of disastrous storms, each one eating away at the Tribe's narrow spit of land.[9]

At this point tribal leaders proactively decided that in order to keep residents safe, their community together and culture intact, the only option left was to relocate *together* on a new site of their choosing, away from the perpetual flooding and land loss but close enough to still access culturally significant sites and subsistence waterways. The path to tribal-led, community-based resettlement has taken almost twenty years of effort by leaders and community members, a direction and quest that Maldonado and Peterson call a "community-based model for resettlement" (Maldonado and Peterson 2018, 292–295).

The saga to find support for proactive relocation takes on a similar tone with other communities facing slow-onset disasters: it is ad hoc and opportunistic, and there is no body of dedicated laws, agencies, or programs to which the Tribe can turn. Furthermore, the search is lengthy and expensive and requires extensive outreach and many alliances with other communities, organizations, and allies. In Louisiana the Tribe has appealed to levee districts, their parish (equivalent to a county), and the state government; casting wider nets, they have appealed to the United Nations (UN) and the U.S. Congress.

In January 2016 the Tribal Council's persistence resulted in a landmark decision by a federal agency, the U.S. Department of Housing and Urban Development (HUD) in partnership with the Rockefeller Foundation, awarding a $52 million grant to the Tribe to support resettlement. The award for the Tribe's resettlement plan is designed to allow resettlement as a community, as opposed to individual residents, and is one of the Tribe's critical priorities for resettlement. The plan also includes restoration and protection of the Isle de Jean Charles under stewardship of the Tribe, and is described as being a courageous "path away from cultural genocide" (Maldonado and Peterson 2018, 295).

In yet another ironic setback, tracking the progress of the grant, the public-private grant (part of a larger grant to the state of Louisiana), has been delayed due to conflicting interpretations regarding resettlement for a tribal group in light of the Fair Housing Act, among other delays. As of this writing 2019, no funds have been officially received by the Tribal Council, according to Maldonado (2019, 122–123).

Resist, Accommodate, Relocate:
Responses to Slow-Onset Disasters

These cases illustrate ways that communities are experiencing and coping with slow-onset disasters, including sea level rise and erosion. In the most extreme expression, communities are forced to consider relocation and to find assistance wherever they can in a very bleak landscape. Dealing with agencies, state and local government, and program officials takes time and resources to travel and present cases, always based on volunteer efforts and no money. Some communities have a longer time horizon to deal with the advancing sea levels, but all of them face a dim prospect seeking assistance in this ad hoc process.

Coastal communities in the United States generally face a series of adaptation responses and choices, and often resort to a combination of strategies. Sea level rise and climate change adaptation strategies can be thought of conceptually as three major types of approaches, including (a) some form of shoreline protection, or in-situ *resistance* to encroaching waters; (b) ways to *accommodate* to rising sea levels, where people move out of harm's way but still continue to inhabit the shoreline; and (c) the *relocation/resettlement* of people and structures, basically migration and resettlement in another location (Nuckols et al. 2010; Titus 2011).

The first approach entails resisting sea level rise generally by using rock and concrete seawalls, jetties, or levees and dikes (often called hardscapes) to keep flooding and erosion from inundating public and private property. Metropolitan areas such as New York City and Miami have looked at seawall and dike systems offshore to stem storm surges and flooding (Rosenzweig et al. 2011). Smaller municipalities that are also at risk such as Norfolk and Hampton Roads, both in Virginia, have undertaken studies to look at options to protect their historic centers and vulnerable areas by building floodgates and seawalls, and raising the roadbeds. In beach communities where tourists want to have sandy beaches, resistance to slow-onset disasters is often done by continual beach nourishing and widening with dredged sand. In these scenarios, environmental protection decisions are often based on cost-benefit analyses by the state or municipality, a process which raises social justice questions regarding protection of more wealthy communities and tourist destinations vis a vis poorer ones (Cooper and McKenna 2008).

Some of the hardscape solutions in resistance adaptation to slow-onset disasters may work well in the short term, but in the long term they are often overcome by severe storms and rising sea levels, or unintended consequences (e.g., intensifying erosion in adjacent areas). The limitations of hardscape strategies, for example, have been demonstrated again and again in coastal Alaska. In 2006 the USACE built a seawall in the town of Kivalina, which was compromised by a storm *the day following* the dedication ceremony (Bronen 2008). In Shishmaref a cement block revetment (barricade) was installed along the seaward shore of the island in 1982, but failed in the first year of its construction (Mason et al. 2012). Many Shishmaref community members acknowledge, however, that seawalls and other revetment options may be a necessary interim step on the way to relocation.

The use of riprap, bulkheads, and seawalls to armor shorelines from eroding and keep flooding at bay in the state of Maryland is legal on private property in designated zones and widespread around the Chesapeake, but it is seen as environmentally destructive and expensive, and is not encouraged by state policy. Maryland's climate change policy on protection and resistance is to encourage municipalities, counties, and private property owners to use *softscapes* (i.e., restoration of wetlands, planting of sea grasses, creating vegetative buffers) to mitigate erosion and to moderate the increase in wave action, tides, and storm surges that accompany sea level rise as well as to provide additional habitat for fish and wildlife. One of the main tenants in Maryland's "Climate Action Plan" is, "Protect and restore Maryland's natural shoreline and its resources, including its tidal wetlands and marshes, vegetated buffers, and Bay islands, that inherently shield Maryland's shoreline and interior" (2008).[10]

The USACE has a role to play to control erosion and flood damage, but its role in mitigating slow-onset disasters ahead of time is more limited than what it can do *after* there has been flood damage. The USACE has limited dollars that Congress appropriates, and the decision on projects often takes years. It starts with with a feasibility analysis, which itself can take years, and an assessment of the cost-benefit ratio. If the cost-to-benefit ratio is small, there must be a strong rationale that there is a "national interest" in a particular project, particularly if it is sited in a small, rural community. The USACE also depends on the ability and willingness of state and local politicians to earmark funding and support individual projects.

Cost-benefit benefit analysis is well-documented to be insensitive to questions of environmental justice, since it is difficult to translate people's sentiments, emotional attachments, and dependence on community and local subsistence resources into strong dollar-and-cents economic terms. Ultimately the cost benefit analysis works against small rural communities with projects in remote areas, despite the fact that these projects are often critical to the community's needs for sustainability. The cost of protecting or saving a small community results in a very large cost-to-benefit ratio when figured on a per capita basis. As a result, the USACE is directed to do certain projects that have a high political visibility and support in the appropriations process. The implications of this authorization and appropriation process are that almost all small, rural communities in United States coastal areas will predictably fail to meet the standards the USACE has for restoration projects to protect communities and community infrastructure. Under conditions of wide-scale sea level rise, this could mean the exposure and neglect of the needs of many or most low-lying, rural coastal communities.

An example of a counter-intuitive, yet ultimately successful restoration project in the Chesapeake Bay (Maryland) sheds light on how the process can prioritize highly important urban areas and disadvantage small coastal communities. The case is Poplar Island, located near the deep shipping channel in the upper Bay, and reduced in acreage by continual erosion to virtually nothing (five acres in the late 1990s); also, no one had lived on it for seventy years. It was not as if the project would restore a land mass with heritage communities where people desperately wanted to remain on their island and maintain their fishing livelihoods. The high cost-to-benefit ratio of the $1.5 billion project was justified by the argument that restoration of Poplar Island was crucial to protecting Baltimore Harbor (Gertner 2016) from severe storm surges, and thus in the national interest. The argument of national interest, in addition to two powerful Senators in the U.S. Congress responding to political pressure from a major constituency (Baltimore City), meant the Poplar Island project received the funding.

In contrast, saving Tangier Island in the southern Chesapeake Bay (Virginia), where a community of fishing families lives and has sustained livelihoods for six generations, would take only $20 million to $30 million, according to a USACE feasibility study (Milman 2015). The total is much less costly than restoring the uninhabited Poplar Island to a height of up to 25 feet, higher than most of the inhabited islands in the Bay. Where is the environmental or climate justice in that decision? The irony is not lost on the residents of Tangier Island. We will come back to Tangier Island when we discuss their responses to relocation as an option for slow-onset disaster.

Accommodation is a second general type of adaptation; in this case public entities and private property owners allow increased flooding to occur without resistance as described in the earlier paragraphs (seawalls, jetties, riprap), but enact zoning and easement requirements on existing properties to allow people to continue to live in their homes and use the structures, albeit at their own risk. Accommodation is an apparently preferred adaptation for cities, states, and metropolitan areas since it limits their expenditures to maintain public infrastructure, such as beach recreation areas, roadways, and bridges. Private property owners like it because it does not restrict them from using their property or force them to relocate and sell their property (at a loss), and allows to repair their houses and docks after a hurricane or severe storm.

In North Carolina along the Outer Banks (i.e., the barrier islands between the open ocean and the mainland), houses sit on stilts already, and current property owners are paying to have their homes raised a full story up from the ground so that storm surges and waves can wash underneath. On the Eastern Shore in Maryland, some property owners have constructed earthen berms to keep tidewaters and storm surges away from particularly valuable (e.g., historic) properties. Other property owners have used FEMA funding to raise their home four to five feet in the aftermath of a hurricane disaster such as after Hurricane Isabel (2003), while still others pay out of pocket to have their houses raised and freeboard installed. All these options are expensive and outside the realm of possibility for most long-time residents with lower incomes in rural counties. As expected, most of the financing for accommodation strategies is the responsibility of the individual property owner, and not vested in with a community-level entity or other communal-oriented organization.

The third general strategy in the face of sea level rise is relocation and migration. As we see after every major hurricane, and pointedly heard from many residents of the Jersey Shore after Hurricane Sandy, most people do not want to move from long-standing communities and believe they can "ride it out." Heritage communities are no different, unless pushed to the brink of losing life and homes as in Shishmaref and Isle de Jean Charles. This tenaciousness to place and community is a widespread sentiment and behavior that befuddles meteorologists and hazards planners.

Belief systems among multigenerational farming families and watermen families about their homes, and the bonds of livelihood and family on the Eastern Shore are very similar. Farmers and watermen are very resolute about staying where they are and staying in their homes, believing they already have been and can continually adapt to floods for many generations. Farmers, in particular, believe that they have been adapting to environmental changes for generations and are prepared to handle climate changes. They have survived centuries of hurricanes and flooding and believe they will still be able to survive the next big flood. Some farmers are accommodating the changes in the environment, and those who have the resources have relocated their homes inland from where their grandparents lived and raised their newer homes above flood level. Farmers often have enough acreage to allow them to move inland and still stay on their own land (Fiske, Paolisso, and Clendaniel n.d.).

The Alaska villages have mixed views regarding relocation. Community members in Shishmaref are divided on whether to stay on the island, but are resolute in insisting that any relocation occur within their subsistence territory. In other words, if the relocation occurs within traditional geographies, allows for the continuance of culturally embedded human-ecological relationships, and maintains tribal sovereignty, then a majority of community members favor relocation. If relocation means moving outside of traditional subsistence territory such that embedded human-ecological relationships are severed, then research indicates that 100 percent of community members interviewed resist relocation (Marino 2015). One observation in comparing these cases may be that relocation, particularly for heritage populations, is not seriously threatening in and by itself, but is resisted when it becomes seen as a catalyst for the eradication of a way of life. If, conversely, relocation is predicated on rebuilding sociocultural capital and human-ecological relationships—in other words, community—then relocation may present a much more palatable adaptation solution to a multitude of communities.

Relocation efforts offered primarily by public agencies can be rebuffed and often are. An early offer by the USACE to individuals on Isle de Jean Charles to relocate individual families was rejected in a contentious process (Maldonado 2018). However, over time and with greater erosion and repeated storms, the tribal council changed their views on relocation as long as it met their priorities, including, resettlement as a community-based process (Maldonado 2018).

In Maryland and Virginia, several highly vulnerable island communities have rejected offers of relocation. Reflective and symbolic of the emotional meaning that the term "community" holds for people, as well as ties to their livelihoods and homes, the residents of Smith Island, Maryland, rejected an offer of relocation in the face of sea level rise, erosion, and ever more-severe storms. Pursuant to Hurricane Sandy (2012), the state of Maryland offered Smith Islanders $2 million to relocate from their island to another location in the state. The residents refused the offer, preferring to stay where they have been living for more than three hundred years despite the fact that they remain vulnerable to flooding during severe storms and have already lost much of their island to erosion. They have since drawn on their strong sense of heritage and developed a vision plan for their future, which is intended to keep themselves and their historic island culture and livelihoods sustainable. Again, we can see that a sense of community is vital to well-being, whether staying-put or relocating. The U.S. Fish and Wildlife Service subsequently agreed to invest $8.5 million on the south side of the island to restore the marshlands, hopefully stabilizing it and stemming erosion in a wildlife refuge and in the nearby town of Ewell (Wheeler 2015).

Returning to the example of Tangier Island, Virginia, we find another example of people wanting to stay in their community of birth and livelihood despite dire experience and predictions in the face of rising sea levels. Tangier is a set of three, small separate islands in the middle of the Chesapeake Bay. The main residential area, called the town of Tangier, is on the southernmost island, where the main harbor is located. The town of Tangier is a heritage community settled in the late 1600s whose descendants still live and work there in relative isolation and with their own dialect of English. The elevation of the three islands is mostly at sea level with their highest point barely a meter above the water line. More than 33 percent of the island has disappeared since 1850. This slow-onset disappearance of the islands is not an uncommon occurrence for shoals and sandy islands in the Chesapeake Bay that face storms, hurricanes, and geological subsidence causing erosion, and now aggravated sea level rise. Climate change is an intensifier of existing vulnerabilities and low-lying coastal areas will face these challenges with more and more frequency.

Residents are stalwart, religious fishing families. The Methodist Church has consistently played an important role in their community. They are proud of their heritage, do not want to relocate, and are adamant about it. The residents of the island describe their home as the "soft-shell capital of the world," their most profitable product being blue crabs in molt stage and the famous Chesapeake Bay oysters. If they resettled off the island, they would face losing their identity, their repository of spiritual, social, personal legacy and, most likely, their livelihoods. They do not want to be separated from their way of life that is fundamental to their well-being and culture and where they have lived for over 300 years.

Tangier island is listed on the National Register of Historic Places and the plight of its residents has received much attention in the media and by the public, locally and regionally. They have received some support because of the heritage angle: the Commonwealth of Virginia has provided archaeologists to recover and transport bones from the earliest graveyards that were exposed after Hurricane Sandy to be analyzed and re-buried at residents' discretion. The Virginia Department of Historic Resources (https://www.dhr.virginia.gov/) is also surveying the land for potential listing on the state of national landmarks registry, in hopes of attracting funding and more tourism.

However, mitigation of erosion and the destruction of surrounding marshlands on the islands is a daunting and expensive challenge. USACE has undertaken a feasibility study of what it would take to stabilize the island, reduce erosion, and keep it habitable (Wheeler 2015). Estimates range as high as $20 million to $30 million for the USACE to construct a proposed system of 450-foot long seawalls and a jetty to protect the harbor and the island; to provide new wide beaches and sand dunes to buffer storms; and to build up the wetlands (Milman 2015).

Under the current disaster policy and lack of climate change policy, the primary authorization to save the islanders is the USACE, and, if it is declared a disaster area, FEMA. Pre-disaster work by FEMA is not financially covered when it involves primarily sea level rise and erosion. As reported in August 2018 in the ShoreDailyNews.com, the USACE has approved $3 million for a seawall and jetty to protect the island's harbor from erosion, starting in 2018 and finishing in 2019.[11] Although the islanders did not get the full $20–$30 million from USACE to protect for the entire set of islands, it appears they will have their harbor protected, which may help ensure their livelihoods in the short run. As of May 2019, there is no word on the completion of the sea wall and jetty.

Conclusions

In review, it is important to grasp the high vulnerability of specific geographies to climate change impacts, but perhaps it is more important to grasp the communities and lifeways of those people who call these vulnerable places home. In the Chesapeake Bay and along the Eastern Shore these vulnerabilities might be in the reduced viability of rural, coastal places and the entrenched desire to stay put in their homes and communities. A 2009 report, technical guidance to Dorchester County, Maryland, identified the single most detrimental impact of sea level rise to the Eastern Shore as the "loss of land mass to inundation and *loss of property value*," because it undermines the economic vitality of working families, businesses, and communities (Cole 2008, 13, emphasis added). Residents will not be able to sell their homes and property due to the risks of flooding and projected sea level rise, even if they want to relocate, and the county will lose revenue from property taxes at lower assessed values, even as the cost of road repairs and elevations increase. Flood insurance rates will increase dramatically; local businesses will move inland and with them, the few remaining jobs. People's lives and current livelihoods may be more difficult or even impossible to maintain. Most watermen live relatively close to where they keep their boats, which of course is at sea level, and they are not likely to move away from docking and off-loading and processing facilities—unless of course they have to move. This scenario most certainly describes a slow-onset disaster and shows the unique vulnerabilities on the Eastern Shore that unfairly put at risk low- and even middle-income families who cannot afford to move their home, or elevate it, or leave their livelihood.

In Alaska, relocation of entire communities is challenged by a lack of institutional and policy frameworks that can channel funding to aid in relocations. Adding to the obstacles has been the cessation of government infrastructural investment, such as HUD-funded housing or water and sewage systems in these communities, since their potential relocation is a possibility for the future.[12] The interaction of these policy gaps means that community members suffer and that Shishmaref can become a challenging place to live in the present as residents wait for relocation in the future. At the moment, there has not been significant out-migration in Shishmaref (Hamilton et al. 2016), but there is fear that compromised living conditions could create a slow population decrease, especially for younger generations. In this case, gaps in climate change policy themselves create a slow-onset disaster.

What we hope to have demonstrated is that, the ability of current public policy and nonprofit groups to address people's needs in resisting an insidious, slow-onset disaster such as climate change lags behind what is projected and already happening in climate change. People are going to need to relocate, to re-create communities, to save churches, and to relocate burial grounds, their lives, and social networks. As the dramatic recent flooding in Louisiana showed us in fall 2016, we may end up with coffins floating through residential areas. Climate change flooding dislodges even the dead.

Several insights emerge from the case studies discussed in this chapter. First is the ad hoc nature, and the time- and emotionally-intense nature of the work that is required by communities and their leadership. This constitutes a default approach, out of a lack of suitable paths, to seeking assistance for dealing with multiple and accumulating climate change disasters. In almost all cases, communities and families patch together funding and resources from a number of ingenious and inventive sources, such as defense training funds, the U.S. Fish and Wildlife Service, USACE, tourism councils, levee districts, historic preservation funds (for buildings, not people), HUD, and in some cases private foundations. The examples from Louisiana's eroding marshlands driving people, literally, to homelessness and relocation makes it very clear that each community is finding a different and sometimes desperate path. The lack of policy infrastructure makes vulnerable communities search for aid for whatever their desired outcome, whether to stay put or to relocate. This lack of policy infrastructure, the default approach, includes what Bronen and Chapin (2013, 9320) call the lack of an "institutional relocation framework," only on a broader scale. Transaction costs are one of the biggest deterrents to community viability, and, in the case of climate change impacts, the most vulnerable communities must bear the brunt of the monumental transaction costs that it takes to find appropriate funding to save their communities.

U.S. policy for disasters and for climate adaptation make the assumption that the appropriate unit is the nuclear family and/or the individual property owner. The role and centrality of community, and subsequently community infrastructure, whether social capital, labor for subsistence and sharing, or family ties, is not recognized. The individualism in this policy focus continues to reflect our Western European notions of private property and the historic tragedies of the commons. What emerges from these case studies is that the sense of community and its ties to livelihood, and resilience—and people's well-being—is crucial and needs to be acknowledged in policy. Those communities who do agree to relocate need a policy framework that encourages the community staying together, and which includes support for community infrastructure. Meanwhile, those communities who want to remain in place and accommodate slow-onset change also need a policy infrastructure that honors that desired outcome to the extent possible. U.S. policy does not, and never has, respected the value of community as a pathway for resilience.

What all the above points to is that climate change, through the slow-onset sea level rise, and increasing erosion, is creating a greater basis for storm-related disasters. Yet the most visible and tangible elements of climate change—erosion and sea level rise—are not considered to be disasters. Hence FEMA and the USACE cannot readily assist in the hazard mitigation phases. Climate change-related adaptation, including retreat and relocation, are being handled on a case-by-case, or municipality-by-municipality basis, as in rolling easements and TDRs. Furthermore, rarely is any targeted assistance directed at low-income or heritage populations and under the zoning and easement options mentioned, the risks of exposure to sea level rise and storm surges are borne by the individual or community property owners. Critically, there are no special mechanisms to address the historical constructions of vulnerability and, following that line of reasoning, to protect low-income or heritage populations. This patchwork approach to climate change policy clearly leaves gaps in dealing with rural and dispersed populations, isolated politically, and in fixed- or low-income communities. If climate justice is to be accomplished, communities such as the ones discussed in this chapter need policy protections that fit with their realities of slow-onset changes and the realities of disaster experiences, both social and ecological.

Acknowledgments

The authors are enormously grateful to Julie Koppel Maldonado, Ph.D., for her close reading of drafts of the manuscript and providing valuable and informative reviews that improved the chapter enormously. The inclusion of the case of the Isle de Jean Charles would not have been possible without her generous sharing of supplemental insights on the challenges and strategies of the people of Isle de Jean Charles, seeking support for relocation (based on phone interviews 25 October 2016 and subsequent conversations). The chapter also benefitted from the very helpful editing of the coeditors of the book, Professors Susanna Hoffman and Roberto Barrios, anonymous reviewers, and especially the close reading and very helpful suggestions of copyeditor Alison Hope.

Shirley J. Fiske is research professor with the Department of Anthropology, University of Maryland. She is an environmental anthropologist with twenty-five years' experience in executive and legislative branches of government in ocean, climate change, and natural resources management, policy, and governance. Most recently she coedited *The Carbon Fix: Forest Carbon, Social Justice, and Environmental Governance* (2017), addressing social equity concerns of global carbon offset policies. She chaired the American Anthropological Association (AAA) task force on global climate change (Changing the Atmosphere). Her work has focused on climate change and environmental and social justice. She was awarded the AAA Solon T. Kimball Award in 2016 for contributions to applied and policy anthropology.

Elizabeth Marino, PhD, is assistant professor of anthropology and sustainability at OSU-Cascades. She conducts research on issues of repetitive flooding, climate change displacement, risk perception, and the intersection of culture and the environment. Marino is an author on the fourth National Climate Assessment, and is a U.S. delegate of the Arctic Science Ministerial. Her book, *Fierce Climate, Sacred Ground: An Ethnography of Climate Change in Shishmaref, Alaska* was released in 2015.

Notes

1. Tornadoes are exceptionally hard to predict, and their duration ranges from momentary touch-down to ten minutes of roaring destruction. Even hurricanes last only a day (although the rainfall and flooding can last longer), in general, and earthquakes and mudslides are equally time limited. Forest fires and wildfires are time limited, but containment may take weeks.
2. Maryland enacted a number of Jim Crow laws after 1870 that were on the books well into the twentieth century. Although nothing prohibited living next to or

selling residential lands to African Americans, segregation was mandated on every other front (education, marriage, streetcars, steamboats, trains, and public accommodations) (see https://padlet.com/urielvictorinomartinez/2wwx9ov3d3xq for a listing of laws [last accessed 14 May 2019]) The municipality of Cambridge, in Dorchester County, was divided by a street aptly named Race Street, the name of which remains as a historic reminder of the residential divisions in the city during the era of segregation in Maryland.

3. Subsidence refers to the land mass sinking relative to sea level. Geological changes along the East Coast that started in the Pleistocene Era are causing land to sink along the seaboard. This fact exacerbates the effects of sea level rise, which has been occurring faster in the western Atlantic Ocean than elsewhere in recent years. See Boon, Brubaker, and Forrest 2010.

4. It is estimated that forty thousand acres of forest, wetlands, and farmlands will be lost, assuming a conservative three-foot rise in sea level over a century (Cole 2008, 18).

5. For more information on Maryland's climate programs, see http://www.mde. state.md.us/programs/Air/ClimateChange and http://www.mde.state.md.us/pro grams/Marylander/Pages/mccc.aspx.

6. Small, rural and poor communities face a daunting task in finding support for preservation of buildings and historic sites, such as churches and graveyards. The highly competitive national program, the Historic Preservation Fund (http:// ncshpo.org/issues/historic-preservation-fund/ [last accessed 14 May 2019]), whose funds are appropriated by Congress, will be hard-pressed to meet the needs of communities with threatened structures, sacred sites, and homes, even if they are listed as National Historic Landmarks. Private foundations and non-profits are other sources of assistance for historic preservation in the face of climate change, but the costs of finding and attracting such sources of support are high for such communities.

7. http://www.isledejeancharles.com/our-resettlement (last accessed 14 May 2019).

8. Please see the USACE site, https://www.mvn.usace.army.mil/Portals/56/docs/PD/ Projects/MTG/117.pdf and the Morganza site: http://www.morganza.org/morgan-za-to-the-gulf-description/ (both last accessed 14 May 2019).

9. Sequentially, the major flooding events were 1985 Hurricane Juan, 2002 Hurricane Lili, 2005 Hurricanes Katrina and Rita, 2008 Hurricanes Gustav and Ike, 2011 Tropical Storm Lee, and 2012 Hurricane Isaac.

10. Climate Action Plan, Executive Summary, August 2008, page 5. Maryland Commission on Climate Change. Maryland Department of the Environment. https:// mde.maryland.gov/programs/Air/ClimateChange/Pages/Reports.aspx (last accessed 14 May 2019).

11. https://shoredailynews.com/headlines/army-corps-of-engineers-approves-constr uction-of-tangier-seawall/ (last accessed 14 May 2019).

12. There are exceptions to this. In the past few years there have been some new housing, a paved road, and improvements to the health clinic. However, compared to other communities in the region Shishmaref lags behind.

References

Akerlof, K., and E. W. Maibach. 2014. *Adapting to Climate Change & Sea Level Rise: A Maryland Statewide Survey.* Fairfax, VA: Center for Climate Change Communication, George Mason University.

Andersen, M. L. 1998. "Discovering the Past/Considering the Future: Lessons from the Eastern Shore." In *A History of African Americans of Delaware and Maryland's Eastern Shore,* edited by Carole C. Marks, 101–21. Wilmington: Delaware Heritage Commission, University of Delaware.

Azelton, M. T. 2009. "Geophysical Investigations at Shishmaref, Alaska." In *Cold Regions Engineering 2009: Cold Regions Impacts on Research, Design, and Construction,* edited by H. D. Mooers and J. Hinzmann Jr., 8–15. Duluth, MN: American Society of Civil Engineers.

Boesch, D. F., L. P. Atkinson, W. C. Boicourt, J. Boon, D. Cahoon, R. Dalrymple, T. Ezer, B. Horton, Z. Johnson, R. Kopp, M. Li, R. Moss, A. Parris, and C. Sommerfield. 2013. *Updating Maryland's Sea-Level Rise Projections. Special Report of the Scientific and Technical Working Group to the Maryland Climate Change Commission.* Cambridge, MD: University of Maryland Center for Environmental Science.

Boon, J. D., Brubaker, J. M., and Forrest D. R. 2010. *Chesapeake Bay Land Subsidence and Sea Level Change: An Evaluation of Past and Present Trends and Future Outlook. Special Report No. 425 in Applied Marine Science and Ocean Engineering.* Gloucester Point, VA: Virginia Institute of Marine Science.

Bronen, R. 2008. "Alaskan Communities' Rights and Resilience." *Forced Migration Review* 31, 30–32.

Bronen, R., and F. S. Chapin. 2013. "Adaptive Governance and Institutional Strategies for Climate-Induced Community Relocations in Alaska." *Proceedings of the National Academy of Sciences* 110, no. 23, 9320–25.

Cole, Wanda Diane. 2008. *Sea Level Rise: Technical Guidance for Dorchester County.* Maryland Eastern Shore Resource Conservation & Development Council. Retrieved December 2016 from http://dnr.maryland.gov/ccs/Publication/SeaLevel_Dorchester.pdf.

Cooper, J. A. G., and J. McKenna. 2008. "Social Justice in Coastal Erosion Management: The Temporal and Spatial Dimensions." *Geoforum* 39, no. 1, 294–306.

Cox, K. E. 2012. "The Implications of Rolling Easements and Transferred Development Rights in New Hampshire and Rhode Island." *Sea Grant Law Fellow Publications.* Paper 59. Retrieved November 2016 from http://docs.rwu.edu/law_ma_seagrant/59.

Department of Commerce, Community, and Economic Development (DCCED). 2016. Shishmaref Strategic Management Plan. Shishmaref, AK: Department of Commerce, Community, and Economic Development. Retrieved December 2016 from https://www.commerce.alaska.gov/web/Portals/4/pub/Shishmaref_SMP_August_2016.pdf.

Faas, A. J. 2016. "Continuity and Change in the Applied Anthropology of Risk, Hazards, and Disasters." *Annals of Anthropological Practice* 40, no. 1, 6–13.

Fiske, S. J., S. A. Crate, C. L. Crumley, K. Galvin, H. Lazrus, L. Lucero, A. Oliver-Smith, B. Orlove, S. Strauss, and R. Wilk. 2014. *Changing the Atmosphere. Anthropology and Climate Change, Final report of the AAA Global Climate Change Task Force.* Arlington, VA: American Anthropological Association.

Fiske, Shirley J., Michael J. Paolisso, and Kathleen G. Clendaniel. n.d. "Climate Change and Environment among Farmers." Manuscript.

Gertner, John. 2016. "Should the United States Save Tangier Island from Oblivion?" *New York Times,* 6 July. Retrieved September 2016 from http://www.nytimes.com/2016/07/10/magazine/should-the-united-states-save-tangier-island-from-oblivion.html?_r=0.

Goldman, Erica. 2011. "Before the Next Flood." *Chesapeake Quarterly* 39, no. 4, 9–14. College Park, MD: Maryland Sea Grant College.

Hamilton, L. C., K. Saito, P. A. Loring, R. B. Lammers, and H. P. Huntington. 2016. "Climigration? Population and Climate Change in Arctic Alaska." *Population and Environment* 38, no. 2, 115–33.

Hoffman, Susann M., and Anthony Oliver-Smith, eds. 2002. *Catastrophe and Culture. The Anthropology of Disaster.* Santa Fe, NM: School of American Research Press. Fourth Printing.

Intergovernmental Panel on Climate Change (IPCC). 2014. *Synthesis Report Summary for Policymakers.* Retrieved June 2016 from https://www.ipcc.ch/pdf/assessment-report/ar5/syr/AR5_SYR_FINAL_SPM.pdf.

Johnson, Katherine Joanne. 2016. *Resilience to Climate Change: An Ethnographic Approach.* PhD dissertation, University of Maryland, College Park.

Jordan, J. W., and O. K. Mason. 1999. "A 5000 Year Record of Intertidal Peat Stratigraphy and Sea-level Change from Northwest Alaska." *Quaternary International* 60, 37–47.

Maldonado, Julie K. 2014. "A Multiple Knowledge Approach for Adaptation to Environmental Change: Lessons Learned from Coastal Louisiana's Tribal Communities." *Journal of Political Ecology* 21, 61–82.

———. 2019. *Seeing Justice in an Energy Sacrifice Zone: Standing on Vanishing Land in Coastal Louisiana.* New York and London: Routledge.

Maldonado, Julie K., and Kristina Peterson. 2018. "A Community-Based Model for Resettlement: Lessons from Coastal Louisiana." In *The Routledge Handbook of Environmental Displacement and Migration,* edited by Robert McLeman and François Gemenne, 289–299. New York: Routledge.

Marino, Elizabeth. 2015. *Fierce Climate, Sacred Ground: An Ethnography of Climate Change in Shishmaref, Alaska.* Fairbanks: University of Alaska Press.

———. 2018. "Adaptation Privilege and Voluntary Buyouts: Perspectives on Ethnocentrism in Sea Level Rise Relocation and Retreat Policies in the US." *Global Environmental Change* 49, 10–13.

Marino, Elizabeth, and Heather Lazrus. 2015. "Migration or Forced Displacement? The Complex Choices of Climate Change and Disaster Migrants in Shishmaref, Alaska and Nanumea, Tuvalu." *Human Organization* 74, no. 4, 341–50.

Marino, Elizabeth, and Jesse Ribot. 2012. "Special Issue Introduction: Adding Insult to Injury: Climate Change and the Inequities of Climate Intervention." *Global Environmental Change* 22, no. 2, 323–28.

Maryland Geological Survey. 2014. "Highest Point in Dorchester County." Retrieved July 2014 from www.mgs.md.gov/geology/highest_and_lowest_elevations.html.

Mason, O. K., Jordan, J. W., Lestak, L., and Manley, W. F. 2012. "Narratives of Shoreline Erosion and Protection at Shishmaref, Alaska: The Anecdotal and the Analytical." In *Pitfalls of Shoreline Stabilization,* edited by J. A. G. Cooper and O. H. Pilkey, 73–92. London: Springer.

Melillo, Jerry M., Terese T. C. Richmond, and Gary W. Yohe. 2014. "Climate Change Impacts in the United States." *Third National Climate Assessment.* Retrieved December 2016 from http://s3.amazonaws.com/nca2014/low/NCA3_Full_Report_0a_Front_Matter_LowRes.pdf?download=1.

Miller Hesed, Christine D. 2016. *Integrating Environmental Justice and Social-Ecological Resilience for Successful Adaptation to Climate Change: Lessons from African American Communities on the Eastern Shore of the Chesapeake Bay.* PhD dissertation, University of Maryland, College Park.

Milman, Oliver. 2015. "Climate Change Could Leave Chesapeake Bay Island Uninhabitable in 50 Years." *The Guardian,* 10 December. Retrieved Dec 2016 from https://www.theguardian.com/environment/2015/dec/10/climate-change-chesapeake-bay-tangier-island.

Michener, James. 1978. *Chesapeake: A Novel.* New York: Random House.

Mohai, Paul, David Pellow, and J. Timmons Roberts. 2009. "Environmental Justice." *Annual Review of Environment and Resources* 34, no. 1, 405–30.

Newtok Planning Group. 2016. *Timeline of Relocation.* Retrieved December 2016 from https://www.commerce.alaska.gov/web/dcra/PlanningLandManagement/NewtokPlanningGroup.aspx.

Nuckols, W. H., P. Johnston, D. Hudgens, and J. G. Titus. 2010. "Maryland." In *The Likelihood of Shore Protection along the Atlantic Coast of the United States. Volume 1: Mid-Atlantic,* edited by James G. Titus and Daniel Hudgens, 512–662. Washington, DC: Report to the U.S. Environmental Protection Agency.

Oliver-Smith, Anthony. 1999. "What Is a Disaster? Anthropological Perspectives on a Persistent Question." In *The Angry Earth: Disaster in Anthropological Perspective,* edited by A. Oliver-Smith and S. Hoffman, 18–34. New York: Routledge.

———. 2013. "Disaster Risk Reduction and Climate Change Adaptation: The View from Applied Anthropology." *Human Organization* 72, no. 4, 275–82.

Pellow, David Naguib. 2007. *Resisting Global Toxics: Transnational Movements for Environmental Justice.* Cambridge, MA: MIT Press.

Peterson, Kristina J., and Julie K. Maldonado. 2016. When Adaptation Is Not Enough: between Now and Then of Community-led Resettlement. In *Anthropology and Climate Change,* 2nd ed, edited by Susan Crate and Mark Nuttall, 336–53. New York: Routledge.

Raleigh, Clionadh, and Lisa Jordan. 2010. "Climate Change and Migration: Emerging Patterns in the Developing World." In *Social Dimensions of Climate Change: Equity and Vulnerability in a Warming World,* edited by Robert Mearns and Andrew Norton, 103–31. Washington, DC: The World Bank.

Rosenzweig, C. W., D. Solecki, R. Blake, M. Bowman, C. Faris, V. Gornitz, M. Linkin, D. Major, M. O'Grady, L. Patrick, E. Sussman, G. Yohe, and R. Zimmerman. 2011. "Developing Coastal Adaptation to Climate Change in the New York City Infrastructure-shed: Process, Approach, Tools, and Strategies." *Climatic Change* 106, no. 1, 93–127.

Sallenger, A. H. Jr., K. S. Doran, and P. A. Howard. 2012. "Hotspot of Accelerated Sea-Level Rise on the Atlantic Coast of North America." *Nature Climate Change* 2, no. 12, 84.

Schipper, E. Lisa F. 2009. "Meeting at the Crossroads? Exploring the Linkages between Climate Change Adaptation and Disaster Risk Reduction." *Climate and Development* 1, no. 1, 16–30.

Semuels, A. 2015. "The Village That Will Be Swept Away." *The Atlantic,* 30 August. Retrieved December 2016 from http://www.theatlantic.com/business/archive/2015/08/alaska-village-climate-change/402604/.

Smith, Orson P., and George Levasseur. 2002. "The Potential Impacts of Climate Change on Transportation." Summary and Discussion Papers. U.S. Department of Transportation (U.S. DOT), Federal Research Partnership Workshop, Oct. 1–2, 2002, Brookings Institution. Washington, DC: U.S. DOT, 151–161. Retrieved May 2019 from https://www.transportation.gov/sites/dot.gov/files/docs/workshop_0.pdf#page=158.

Titus, James G. 2011. "Rolling Easements, A Primer." *Climate Ready Estuaries.* Washington DC: Environmental Protection Agency. Retrieved September 2016 from http://papers.risingsea.net/rolling-easements-8-1.html.

U.S. General Accountability Office (GAO). 2003. *Alaska Native Villages: Most are Affected by Flooding and Erosion, but Few Qualify for Federal Assistance. Report to Congressional Committees.* Retrieved December 2016 from http://www.gao.gov/new.items/d04142.pdf.

———. 2009. *Alaska Native Villages: Limited Progress Has Been Made on Relocating Villages Threatened by Flooding and Erosion. Report to Congressional Committees.* Retrieved December 2016 from http://www.gao.gov/new.items/d09551.pdf.

U.S. Army Corps of Engineers (USACE). 2009. *Alaska Baseline Erosion Assessment.* Anchorage, AK: Elmendorf Air Force Base.

Urwin, K., and A. Jordan. 2008. "Does Public Policy Support or Undermine Climate Change Adaptation? Exploring Policy Interplay Across Different Scales of Governance." *Global Environmental Change* 18, no. 1, 180–91.

Warner, William W. 1976. *Beautiful Swimmers: Watermen, Crabs, and the Chesapeake Bay.* New York: Little, Brown, and Company.

Wennersten, John R. 1985. "A Cycle of Race Relations on Maryland's Eastern Shore: Somerset County, 1850–1917." *Maryland Historical Magazine* 80, 377–82.

Wheeler, Timothy B. 2015. "Smith Island, Threatened by Rising Water and Dwindling Population, Aims to Shore up Its Future." *Baltimore Sun*, 24 October 2015. Retrieved May 2019 from https://www.baltimoresun.com/news/maryland/politics/bs-md-smith-island-20151024-story.html.

Wriggins, Jennifer B. 2015. "In Deep: Dilemmas of Federal Flood Insurance Reform." *U.C. Irvine Law Review* 5, no. 6, 1443. Retrieved May 2019 from https://scholarship.law.uci.edu/ucilr/vol5/iss6/7.

Wisner, Ben, Piers Blaikie, Terry Cannon, and Ian Davis. 2004. *At Risk: Natural Hazards, People's Vulnerability and Disasters*, 2nd ed. London: Routledge.

CHAPTER 7

Disrupting Gendered Outcomes

Addressing Disaster Vulnerability through Stakeholder Participation

BRENDA D. PHILLIPS

Introduction

Historically, gendered disaster studies have focused on gender as a variable, typically distinguished through binary constructs of male and female. Such a bifurcation has ignored the complexities of gender, particularly the full range of ways in which gender can be and has been expressed. This two-sided gender coin has also masked the contexts from which gendered roles and relationships arise, including the impacts of broader systems influenced by ideology, politics, social arrangements, economic realities, or legal circumstances.

Indeed, gender cannot be understood as such a dualistic set of experiences with expected outcomes. A more contextualized approach becomes necessary to capture the fuller range of ways in which gender impacts and intersects with overlapping identity markers. Age, for example, complicates gendered experiences in disasters—from young girls at risk for human trafficking to older women thwarted amid frantic scrambles for food and water. Geographic location, such as being situated in a developing nation, further exacerbates such risks. Historic patterns of racial and ethnic relations may also marginalize women by demarcating occupations, housing locations, or educational opportunities that could otherwise enable and empower women to address deeply embedded patterns of prejudice and discrimination.

The purpose of this chapter is to explore gender and disaster through the lenses of several major theoretical perspectives: vulnerability theory, sociopolitical ecology theory, and intersectionality theory followed by a discussion of how gendered outcomes might be disrupted. These perspectives provide a basis for understanding why gendered vulnerability exists in disaster events and what strategies might be engaged by those at risk, their communities, responding governmental and nongovernmental organizations (NGOs) and other relevant actors. To start, a focused summary of disaster impacts will illuminate some of the challenges, identify gaps, discuss outcomes, and provide opportunities to understand ways in which engagement strategies might develop.

Gendered Disaster Outcomes

Numerous researchers and practitioners alike have observed differential vulnerabilities prompted by gendered roles, relationships, and social structures. Such outcomes have not altered over decades. To illustrate, consider:

- *Bhopal, India gas disaster, 1984.* A nighttime chemical release of methyl isocyanate claimed between twenty-two hundred and thirty-eight hundred lives, and injured thousands of others for life. Considered the worst industrial accident by many, the event sparked protests that continue to the present day. Led primarily by women, the effort focuses on justice, corporate apathy, and compensation (Thomas, 2018).
- *Hurricane Andrew, 1992.* In a classic, early study that helped to spark the rapid evolution of gendered disaster social science, low-income and single parents found themselves overlooked and disempowered following Hurricane Andrew. In contrast, they produced a strong organizational framework that demanded and brought about gender-specific post-disaster programs, including child-care and summer teen programs (Enarson and Morrow 1998).
- *The Indian Ocean Tsunami, 2004.* The tsunami that claimed nearly three hundred thousand lives included an 80 percent loss of life among women and children. Gender roles produced higher death rates: for instance, women waited on the shore in nations like India to process and sell fish (Oxford Committee for Famine Relief [OXFAM] 2005). Reconstruction neglected gendered roles that empowered women and redirected funding to men (Barenstein 2006).
- *Pakistan Earthquake, 2005.* Multiple disaster events, including the earthquake in Pakistan, revealed differentiated and gendered struggles for food, medical attention, and personal safety. In a disaster marked not only by gender but also by class, women widowed by the earthquake faced even more difficulties including shunning, being the victims of violence, and disappearing into human trafficking.
- *Haiti Earthquake, 2010.* In what seems like a never-ending lessons-unlearned scenario, women and girls competed for nutrition, health

care, and security after the massive earthquake. Responding organizations set up lighting for camps and established gender-specific security patrols. Such repetitive patterns revealing women and girls at risk have been observable for decades, yet efforts must be made anew in every disaster to protect those at risk (Enarson 1999; Jenkins and Phillips 2008; Phillips and Jenkins 2013, 2016).

Differential Outcomes: Gendered Vulnerability

Disasters reveal social problems, as indicated by these brief and non-changing examples given above (Barton 1969). The persistence of such problems over time reflect deeply embedded and socially institutionalized ways of influencing societies, communities, households, and personal lives that lead to disparate outcomes. In this section, I look at several themes around in which research has revealed such differential consequences of gender. The section also reveals the complex and nuanced ways in which gender is influenced by other socially, politically, and economically institutionalized ways that marginalize and imperil people in disaster. I begin by examining the most significant outcome of all: the potential to lose one's life as a reflection of gendered patterns of living, followed by a section reviewing the potential for increased personal risk of violence. This section is followed by several factors that influence such gendered outcomes, economic disparities, and legal circumstances. A final section helps us to understand how gendered vulnerability manifests.

Fatalities

It is not unusual to hear that women and children disproportionately bear the deadly consequences of disasters. The Indian Ocean tsunami of 2004, for example, is believed to have included an 80 percent mortality rate among women and children (Kottegoda 2007; OXFAM 2005). Their deaths occurred as women waited on the shore for men to return from the sea with their catches of fish. Along the coast of Naggapattinam, India, women's roles included cleaning and marketing the fish. A similar outcome occurred in Bangladesh when a 1991 cyclone and tidal surge claimed between 67,000 and 138,000 individuals. Mortality rates reflected that "gender-related vulnerability is deep-rooted in persistent inequalities" (Ikeda 1995, 188). Higher losses of life occurred among children, seniors, and women, with women aged twenty to forty-nine experiencing death rates four to five times that of men (Ikeda 1995). Although younger women had similar death rates to men, those older than fifty saw higher rates, yielding an overall death rate for women approximately twice that of men (Ikeda 1995). Causes of higher mortality for women included not being able to secure information in order to make life-saving decisions.

A larger study examining disaster effects from 1981 to 2002 spanned 141 countries (Neumayer and Plümper 2007). Given that women tend to live longer than men, the data revealed that disasters reduce that gender gap, which closes more rapidly in catastrophic events (Neumayer and Plümper 2007). One important variable influencing the gap arises out of socioeconomic status, revealing that disasters impact "the vulnerability of affected people, which can and often does systematically differ across economic class, ethnicity, gender, and other factors" (Neumayer and Plümper 2007, 561). Age complicates fatalities. The 1999 Taiwan earthquake, with 1,826 deaths, resulted in higher mortality rates for women and seniors (Chan et al. 2003), a finding similar to the 1995 earthquake in Japan (Tanida 1996). A somewhat different pattern occurred after Hurricane Katrina, albeit tied to gender and with the addition of race as a complicating variable: more African American men over sixty years of age died than white counterparts at a similar age (Sharkey 2007).

It is equally clear that gendered outcomes for males exist as well as for females in disasters, a consequence that appears to be context-specific. For example, gendered role expectations dictate that men stay behind to protect family farms during hurricanes. Striking Honduras, Hurricane Mitch storm caused significant loss of life among men who adopted such a traditional male role. Both women and men may feel compelled to make difficult choices whether to evacuate. In Australia a controversial stay-and-defend approach to wildfires (Haynes, Handmer, and McAneney 2010) may increase risks, particularly to males and traditionally male fire brigades. In Bangladesh women have died in cyclones when they stayed to protect their assets (Enarson 2000).

Fatality studies reveal that gender situates people in roles and locations that subject them to risk, as magnified at times by age, race, and other demographic characteristics. While teasing out such other factors remains complex, it is clear that socioeconomic standing emerges as another influential variable, particularly in comparing developmental standing by nation. The dual earthquakes of 2010 in Haiti and Chile produced significantly different numbers of deaths and damage. Though Chile endured a stronger earthquake, Haiti suffered grievous losses of people and resources. The primary difference stemmed from pre-disaster contexts (e.g., political corruption, lack of building codes) and insufficient resources that exposed Haitians to higher vulnerabilities. Disasters, as revelatory events, divulge how deeply embedded and discriminatory practices on the basis of gender, age, race, and livelihood opportunities influence the ability to survive and recover. Another such differentiated example can be found in studies of interpersonal violence in disaster contexts.

Violence

Disasters of all kinds, including terrorism, reveal significant vulnerability and potential exposure to a range of violence, particularly to women. From terrorist groups that kidnap and enslave girls and women to human traffickers that prey on survivors living in temporary shelters, there is a significant threat to their personal safety. Danger appears particularly gender-based with women and girls as frequent targets; females with disabilities are at even higher risk (Felten-Biermann 2006; Petersilia 2000). Disasters not only do not stop such targeting but also appear to increase the risks, such as was found in Sri Lanka and other locations when women and children disappeared from relief camps after the tsunami (Fisher 2005; Kottegoda 2007). Issues with personal safety arose again after the 2010 Haiti and 2016 Nepal earthquakes, suggesting that little had changed over the previous decade.

The range of potential violence covers physical, sexual, and spoken violence, as well as prostitution, enslavement, early marriage, and all forms of torture (Abirafeh n.d.; Phillips and Jenkins 2016). Factors that increase the potential for violence vary, but no social or economic class is exempt nor are members of any particular dis/ability, race, ethnicity, culture, or geographic location. One disaster-related factor that is relevant is displacement, which disasters certainly produce as do humanitarian crises (Horn 2010). Being disrupted from familiar settings and social networks that provide safety or oversight can render displaced people more vulnerable to violence. After the 2010 Haiti earthquake, for example, considerable challenges prevented women and girls from being safe. Three Haitian feminist leaders died in the earthquake, which also devastated the women's ministry. Shelter conditions left women and girls with limited institutional or interpersonal means for personal security (Abirafeh n.d.). Safety patrols had to be organized in massive tent camps simply to ensure safe passage for them to use toilets. NGOs brought in additional lighting. An array of experts, government, and NGO leadership altered camp layouts, created safe spaces, reconsidered aid distribution, and restored livelihood opportunities. They also worked to establish a means for women to voice their concerns and to influence providers (Abirafeh n.d.).

Nor are men and boys safe, although the context may vary. Being homeless, dealing with exposure to crimes, or assuming traditional roles with higher danger such as police work or military service may increase male exposure to disaster-related violence (Phillips and Jenkins 2013).

Specific disasters may also jeopardize men more than women. A study of 247 flood fatalities in Europe and the United States, of which two thirds involved drowning, found higher numbers of male fatalities (Jonkman and Kelman 2005). Attributing factors included men being more likely to drive into flooded areas, perhaps due to a more gendered socialization toward higher risk-taking. Age also appears to influence male mortality, with elevated rates among those under twenty-one or over sixty-five, as noted in several studies (Jonkman and Kelman 2005; Sharkey 2007).

Nor are support and advocacy agencies involved in gender-based violence ready to assist clients in disasters. Most fail to conduct any significant level of preparedness and planning for disasters and, historically, have been surprised by post-disaster increases in violence (Enarson 1999; Phillips and Jenkins 2016). Emergency managers and disaster officials are singled out for failures to integrate agencies that protect those at risk for violence in their own emergency response and recovery planning (Phillips and Jenkins 2013). An example from Hurricane Katrina revealed that agency workers in New Orleans improvised their own evacuation. When the storm destroyed all domestic violence shelters in a three-parish area of Louisiana, local providers struggled to safeguard clients and to provide services for years (Jenkins and Phillips 2008).

One means of escaping such violence has been personal resources, including income and social networks. Disasters also pull people away from their resources, leaving them without the means to escape violence, secure adequate protection, or find safety among those who have helped in the past. In disasters, already vulnerable people will compete with others who are more likely to secure resources (Peacock and Ragsdale 1999). Desperation can result when disasters further compromise already meager or nonexistent resources, leading survivors back into dangerous circumstances or onward into similarly threatening conditions. Because disasters threaten people, one effort that has been tried involves restoring livelihoods as a means of empowering women in their communities.

Economic Disparities

There are gendered wage disparities worldwide, undermining abilities to prepare for and recover from disasters. Such wage disparities neglect the broader array of women's work, including uncompensated and volunteer labor. A limited number of studies describe that continuum from traditional unpaid domestic work to compensated labor within a disaster context (e.g., see Enarson 2000, 2001; Krajeski and Peterson 1999). However, the bulk of women's work in disaster conditions appears to be largely unpaid, traditionally gendered, and with low to moderate social visibility

(Enarson 2000, 2001). Such a combination undermines the perceived value of women's disaster work, particularly in more-traditional locales.

To illustrate, disasters intensify gendered work burdens. Caregiver roles, for example, expand as women move to protect others, possessions, and homes, and to shoulder post-disaster recovery efforts (Enarson 2000, 2001). Volunteering and community organizing emerges as another way to work and, though unpaid, serves as a critical means to restore family and community and potentially increase women's influence, power, and localized leadership (Neal and Phillips 1990). Some evidence suggests that rewarding resources to women could undermine those who would misuse the aid distribution. Sri Lanka, for example, reported that men who received aid bartered it for alcohol or tobacco rather than using it for household and family needs like food (Kottegoda 2007). After the 2005 Pakistan earthquake, men secured food for themselves and their families while women widowed by the event went hungry.

Some unpaid volunteer efforts may merge into protest efforts, with women working to claim space and resources for themselves, their neighborhoods, and their communities; this was demonstrated after Hurricane Andrew in the United States, and in the Bhopal, India, post-disaster movement (Thomas 2018; Enarson and Morrow 1998; Enarson 2000, 2001). They may even shoulder volunteer work designed to rebuild entire communities, as women did when preparing for volcanic eruption in Montserrat (Soares and Mullings 2009). Though a male headed the primary committee, "women had more intimate knowledge of the village" and in reality led "planning and preparing" before the community-wide disaster (Soares and Mullings 2009, 253). Taking on these key leadership roles, officially or from the margins, not only increases women's roles but also can lead to a higher degree of political empowerment.

Post-disaster times can also offer paid work to all genders, from construction work to case management. Disasters, though, may compromise a woman's ability to return to work, such as because of the loss of child care. Post-Katrina, those returning to New Orleans and other affected areas found minimal child-care facilities and severely disrupted public transportation systems. Disasters may also rob women of their assets including work space, equipment, and other critical resources. Tornadoes, floods, hurricanes, and wildfires may take one's home-based resources, whether for paid or unpaid work. After Hurricane Katrina, the Mennonite Economic Development Associates opened child-care facilities in affected areas and provided micro-loans across the U.S. Gulf Coast. Disasters may also mean that women are widowed. In contexts where women lack inheritance rights, they may be forced to leave their homes where their livelihoods are located (Enarson 2000). In short, options for paid work may be bounded by a number of circumstances out of the control of those affected.

Business sizes and types also appear to matter. Very large firms tend to survive disasters more readily due to a broader and deeper set of resources (Webb, Tierney, Dahlhamer 2000). Conversely, small businesses—where women and minorities are more likely to establish enterprises—have higher failure rates. This includes home-based work from traditional crafting to child care or consulting, to small enterprises in kiosks, open markets, or storefronts. Corporate settings, where women have historically faced glass ceilings, may offer the safest post-disaster work environment due to their broader and deeper set of resources and abilities enabling business continuity.

Post-disaster, new work may emerge that is specific to the calamity. In a U.S.-declared disaster, work might be compensated through government programs or when voluntary organizations take on additional staff to meet new or increased needs. Internationally, particularly in developing and traditional contexts, paid work has been directed more readily to men. Organizations external to those contexts would be wise to observe local practices—as in the case of NGOs responding to India's tsunami experience and displacing women who managed local construction efforts (Barenstein 2006). Later in this chapter I detail some of the ways that women have become employed in post-disaster settings as a means of disrupting such gendered and inequitable post-disaster recovery assistance efforts.

Legal Issues

Several legal concerns also beleaguer women and their families in disaster contexts, some of which arise from pre-disaster customs and cultures. Areas that require a marital dowry, for example, may see issues arise when disasters affect assets intended to secure a daughter's marriage (MacDonald 2005). Post-disaster recovery practices may worsen the matter. After the 2004 Indian Ocean tsunami, for example, reconstruction funds went to males rather than to women who owned houses as part of their dowry, a significant asset of personal status and power (Gamburd and McGilvray 2010). Even in Sri Lanka, where women and men share constitutional equality, land ownership practices view men as the head of household. In areas where dowry practices pass land on from mother to daughter, state gendered land distribution after a catastrophe undercuts traditional ways that women historically care for their families. Some households, however, have used state distributed land to "offer a better basis for attracting a daughter's future husband" by subsequently transferring the land and house to the daughter on her marriage (Thurnheer 2009, 86). A related and significant concern occurs when, lacking land ownership as collateral, women cannot secure loans. State issuance of land then falls into male control and "by being deprived of equal access to land ownership, women remain vulnerable to eviction from the home and to domestic violence" (Kottegoda 2007, 21). Conversely, land rights, property ownership, human rights, and the recognition of such by post-disaster recovery programs have the potential to reduce post-disaster vulnerability (Kottegoda 2007). Representing a significant resource, such funds and programs could destabilize gendered inequities.

Loss of assets, family relationships, and safety networks can also lead to legal problems and significant risks. As a consequence of the 2004 Indian Ocean tsunami, for example, women and girls endured forced marriages and an increased exposure to violence (Kottegoda 2007). Women who shared custody of their children reported losing access after Hurricane Katrina when the U.S. evacuation prompted some separations, requiring extensive and expensive travel to share custody and fulfill custody requirements. If a child happened to be with a parent during the time frame that the disaster happened, the other parent might not be able to retrieve the child for some time. Some had to go into court, once such processes resumed, to restore custody, and child custody cases increased as courts tried to determine appropriate decisions (Fothergill and Peek 2007).

Legal concerns extend beyond women and children into concerns about lesbian, gay, bisexual, or transgender (LGBT) families. Lesbian families, for example, report difficulty in evacuating together, providing public comfort to each other, securing services, and qualifying for programs (D'ooge 2008; Eads 2002; Stukes 2014). Internationally, LGBT families face significant levels of discrimination, prejudice, and violence in the context of disasters. Coupled with gender, the intersection of sexuality with gender, age, income, race, or disability heighten the problems and uphill battles that LGBT populations face. While some countries offer legal protections and rights, those same rights have been under attack in other countries, which may continue to undermine people's abilities to be resilient in the face of disaster. Furthermore, not only do disasters disrupt and worsen personal circumstances, but they also may serve as the first step in an elongated process of recovery. Losing one's home also undermines one's social networks that can provide safety and resources (Gorman-Murray, McKinnon, and Dominey-Howes 2015; Stukes 2014).

Little is clear about people who live as transgender or in transition, suggesting that researchers and practitioners have largely failed to consider the continuum of ways in which gender can play out. Anecdotal accounts suggest that post-disaster experiences present dehumanizing challenges for transgender women and women in general, from shelter-ing experiences to personal hygiene and medical care. Post-Katrina, one transgender individual was incarcerated in Texas when attempting to use a bathroom. Rapid evacuations also mean that important resources can be left behind including medical resources, cosmetics, binding equipment, sanitary supplies, and related personal needs. Legal concerns include documentation by name and gender on official documents, which are required to secure aid. And, while nondiscrimination laws have increased, significant ideological discourse still remains to threaten access, use of resources, applications for aid, and case management (Mottet and Ohle 2003). Aid and relief agencies have been observed to deny resources on the basis of ideological beliefs, pushing transgender people into condi-tions of denial and disparity. Legal challenges to such have not yet been observed but may have new life in the latest round of cases in the United States regarding such rights. Advocacy organizations may lead the way in this. In the United States, the National Center for Transgender Equality offers preparedness information to contrast with such outcomes. The National LGBTQ Task Force Policy Institute and the National Coalition for the Homeless have also offered guidance on making non-disaster shelters welcoming (Mottet and Ohle 2003).

Psychological Impacts

One might assume, having read through the previous impacts, that recov-ering from disaster would prove debilitating in terms of physical, emo-tional, spiritual, and psychological impacts. Stereotypes do depict women, in particular, as victimized by disaster, seemingly helpless in the face of the event and unable to pull themselves from despair. However, such a disaster syndrome or mental health breakdown imagery remains largely unsubstantiated. Rather, disasters for the most part prove that women survivors are fairly resilient (Norris, Friedman, and Watson 2002; Norris et al. 2002).

A related stereotype depicts women as more negatively affected than men. Although some psychological studies report higher stress and post-traumatic stress disorder (PTSD) symptoms among women, several factors other than disaster may account for this (for a thorough meta-analysis, see Norris, Friedman, and Watson 2002; Norris et al. 2002). For example, women may be willing to report symptoms more readily than men. Furthermore, gendered roles traditionally permit women to be more expressive—which may be a healthy way to react to disasters and does not necessarily mean that negative psychological impacts may result. Another influential factor comes from the observable role overload borne by women who tend to shoulder more of the resource acquisition process, extra caregiving, and disaster volunteerism, which may increase stress-related symptoms (Enarson 1999).

Across psychological studies, exposure to traumatic or violent events tends to surface or exacerbate mental health needs (Norris, Friedman, and Watson 2002; Norris et al. 2002). This may be particularly true among women left to fend for themselves, under purdah for example, which increases their potential exposure to a disaster agent. Similarly, women and children not permitted to learn how to swim face significant peril when floods occur. Culturally specific or religiously designated clothing can also make it difficult to climb and increases the potential of the women becoming entangled in debris. It may also be true for men who are more likely to take on emergency response roles that expose them to danger (or in traditionally male-dominated military units.) As with other examples given in this section, multiple factors influence gendered outcomes. Hurricane Katrina, for example, resulted in PTSD appearing in 22 percent of a sample of 810 people in Mississippi (Galea et al. 2008). While female gender emerged as one factor influencing PTSD, findings were mediated by exposure, financial loss, stress, and social support.

One factor that does appear to influence psychological impacts has to do with prior traumas. Within populations more likely to experience traumas, such as interpersonal violence, the related pre-disaster exposure may increase a risk of post-disaster psychological impacts. While a clear solution is to address potential traumas for those at risk, such as domestic violence, the harsh reality is that gender-based violence will continue worldwide including after disasters. Intervention measures must then necessarily incorporate an understanding of the factors that increase gendered psychological impacts and address them in addition to limiting additional exposure to trauma. This calls to mind the need to secure economic resources and legal rights for those at risk in an effort to stave off potentially debilitating psychological consequences by empowering those at risk to leave.

Understanding Gendered Vulnerability

Understanding vulnerability requires attending to the originating sources or problems that cause differential outcomes. Historically, several theories have provided useful insights into why women experience higher rates of fatalities, violence, economic impacts, legal challenges, and psychological consequences than others. In this section, I briefly review the three most frequently used theories to understand why such differentiated outcomes occur: vulnerability, sociopolitical ecology, and intersectionality theories. Each provides not only explanations but also an eye toward the ways in which theory and research can be used to disrupt and alter gender-differentiated outcomes in disaster contexts.

Vulnerability Theory

Two perspectives have dominated discourse over why some populations bear the highest risks in disaster contexts. The first one, called the dominant perspective, assumes that nature causes disasters. The solution to vulnerability wrought by such acts of nature must be managed through feats of engineering and acts of humans, such as dams, safe rooms, or retrofitted structures.

Alternatively, the vulnerability perspective redirects attention away from the physical elements of nature and toward the social, political, and economic acts of people and the societies in which they live (Fordham et al. 2013; Wisner 2001; Wisner et al. 2005). Rather than argue that weather generates a hurricane with storm surge and high-velocity winds, vulnerability theorists propose that human arrangements place people in harm's way. Historic patterns of race-based segregation, for example, marginalize people into locales that bear repetitive losses and expose residents to significant risks (Cutter 2005; Phillips, Stukes, and Jenkins 2012). Similarly, gendered role expectations place both women and men into dangerous locations, such as when men feel compelled to stay behind and protect family farms or when women wait on shore to receive, process, and sell the fisherman's catch (Phillips et al. 2008). In short, this theory posits that the earthquake did not lead to fatalities or human trafficking—those outcomes occurred because human beings gendered work structures or became predators.

Vulnerability theorists would argue that other factors influence risk as well. Political practices that exclude people from jobs or participation alter disaster outcomes because policies and programs do not consider or incorporate gender-based realities. Institutionalized political realities may reflect the dominant or majority group (e.g., whites, males, or people without disabilities) and, through exclusion, fail to consider the very real needs of those who will ultimately pay a price. People with disabilities or transgender people may feel invisible or, conversely, believe that they are subjected to high levels of uncomfortable scrutiny in an otherwise straightforward search for safety. Similar contexts have impacted people of Latinx ancestry as well (Bolin and Stanford 1999). The 2005 earthquake in Pakistan revealed lack of access to recovery programs due to gendered interactions that afforded more opportunities for men to secure resources. To address the lack of health-care services, medical organizations created gender-specific programs and segregated health tents (Sayeed 2009).

Thus, vulnerability arises because of the ways in which society situates both physical and social locations, creates policies and programs that either include or exclude people from access to resources, or arranges organizations that fail to reflect the broad and deeply diverse realities of affected populations. A critique of vulnerability theory might suggest that it visualizes those affected as passive actors affected by their environment. In reality, though, vulnerability theorists posit that those at highest risk carry within themselves and across their commonalities a significant capacity to respond to and recover from disasters albeit within the restrictions and challenges of associated cultures. Indeed, cultural configurations arrange how we manage our environment as well as how people should, even may, behave within those environments (Hoffman 2017). Culturally specific norms will endanger people, sending them into conflict over scarce resources. Thus, an implicitly structured element of vulnerability theory concerns transforming the means to secure resources to protect and recover from disasters, which lies at the heart of an alternative perspective: sociopolitical ecology theory.

Sociopolitical Ecology Theory

From the perspective of sociopolitical ecology theory, dealing with disasters requires the acquisition of resources, such as food, water, shelter, medical support, and a means to ensure personal safety. As the above examples have demonstrated, though, securing such life-saving resources is troubled by gendered arrangements within a social system. When people have to compete for food, such as after the 2005 earthquake in Pakistan when women struggled to survive, not everyone will obtain what they need. The competition over scarce resources characterizes the heart of sociopolitical ecology theory, where winners and losers result from entangling over the means to survive or enhance resilience (Peacock and Ragsdale 1999).

Such fractious interactions over scarce resources occur because of pre-disaster inequalities embedded in the existing society. Social structures that prevent women from pursuing economic and educational opportunities ultimately set them up to fail in the battle to survive. Gender thus emerges as a considerable barrier, because it is embedded within institutions, organizations, and practices that either unintentionally or purposefully marginalize. In the example of hurricane Katrina, failure to send buses into areas populated by older African American men increased the probability of those men dying. Even intangible resources, such as personal security in Haiti, demonstrate the competition for scarce resources.

Also important to consider is that pre-disaster inequities, embedded throughout the human experience, can predispose some to languish during recovery times. Housing, for example, can be a scarce commodity and subsequently unavailable in the aftermath of a disaster—especially if affordability issues existed prior to the disaster. People who experienced marginal housing or homelessness will likely find themselves pushed farther back in the line for post-disaster housing as they compete with newly displaced people (Phillips 1995; Settembrino 2016). Those with more resources, such as funds that can be used for a rental, will be more likely to secure housing. State-run programs that dispense funds to traditional male households, as found in Sri Lanka, will increase housing insecurity and often in gender-specific ways.

Intersectionality Theory

Another perspective, intersectionality theory, compels researchers and practitioners alike to incorporate other deeply nuanced realities including those marked by gender. But intersectionality proponents urge us to wrestle more deeply with the overlapping, competing, and intersecting aspects of gender, age, race, ethnicity, disability, development, and other factors. Failure to consider the multivariate ways in which gender is influenced by and affects other ascribed roles is assumed to result in a misleading and disproportionate understanding of disaster effects. Gender alone, for example, cannot fully predict disaster outcomes. Indeed, a woman (or man or transgender person) living in a developed context, with a higher education degree and related income, who grew up in a world marked by white or gender-specific privilege—would presumably be even less able to prepare for and recover from a disaster.

But, a binary contrast between developed and developing nation contexts or male and female obscures gendered outcomes. Even within those realms of developed and developing, we find differential effects as found at the intersections of age, gender, and race in New Orleans or along the shores of Naggapattinam, India (Phillips et al. 2008; Sharkey 2007). Intersectionality theorists push us to think beyond single (gender) or binary ways (male/female or developed/developing) in which people become marginalized. They ask us to think in more complex ways, which demand us to think of each person as a unique human being marked by a variety of ascribed and achieved characteristics, all of which could produce differential outcomes in a disaster (Walby, Armstrong, Strid 2012). Intersectionality theorists argue that gender cannot be divorced from its context and, when considered in terms of real human complexities, helps to explain vulnerability and to target places for change. In short a "gender analysis alone is insufficient" (Hyndman 2008, 118; supported also by Enarson, Fothergill, and Peek 2007).

Disrupting Gendered Outcomes

Taken together, the currently dominant theoretical perspectives compel us to consider gender at a more nuanced level, involving multiple conditions, circumstances, and contexts. As evidenced earlier, a woman-specific agenda will divert attention from the realities faced by men and boys including the benefits and privileges of being male as well as negative consequences that do occur. Thinking in a binary manner also diverts attention from the range or continuum in which gender can be expressed including those who transgender. A gender-specific focus on women also obscures the multiple identities ascribed to or achieved by women through assignations of age, dis/ability, race/ethnicity, geographic location, or economic status. Thus, an essential departure point in terms of disrupting differentiated outcomes is to embrace gender as a matrix of identities that influence, sway, and endanger people through a complex, socially structured set of uncomfortable, even deadly, realities.

In this section I examine efforts that can disrupt such complex, gendered outcomes in an explicit attempt to fill gaps in knowledge, policy, and practice. While significant and widespread social, political, and economic change is clearly needed to permanently alter such gaps, a more realistic and practical approach is needed even as work continues on more macro levels. To start my consideration, I focus first on a key principle—to involve people at local levels to honor their lived experience and shared realities. Involving people in that way requires incorporating a practice of working with people rather than doing for them through participatory processes. Such a localized, stakeholder-driven process is essential. Indeed, such involvement reflects a clear and present need given that so many donor-driven, state-mandated, or organizationally arranged efforts have failed. Empowering stakeholders, particularly those at risk, inherently disrupts embedded power structures and can lead to not only heightened awareness of those in positions of authority but also to enhanced protections for those at risk.

Promoting Locally Centered Solutions

Local people are more likely to understand their own worlds, the ways in which gender both marginalize and reveal spaces for action, and how to navigate such challenging contexts. Though some may fear that locals endure socially imposed and gendered restrictions as a reflection of false consciousness, such is not necessarily the case. Empowering and involving local people, women, men, and transgender alike, can provide a broader set of insights into what happens, how and why it occurs, and what options might exist—particularly those that would be accepted and sustained. Post-tsunami construction and land distribution within India disrupted local practices and—as a consequence—failed to consider local preferences, many of which provided a basis for women to hold power. Emergency managers and related officials neither can or should attempt a universal solution without thoroughly understanding local customs, cultures, and preferences. Such a view has been consistently expressed for decades, for example, as when the United Nations (UN) stated, "The most effective relief and reconstruction policies result from the partic-ipation of survivors in determining and planning their own needs" (UN 1982, 4). Documenting the ways in which empowerment and participatory strategies work would enable a body of knowledge to accumulate, which those at risk and those in official positions can then further leverage to reduce risks.

State agents, relief organizations, and donors need all bear in mind what local people want, prefer, and need. In the aftermath of the Haitian earthquake, for example, a major donor conference collected and consid-ered opinions from tent cities, rural areas, urban enclaves, and a wide dis-tribution of people from myriad economic circumstances and livelihoods (Montas-Dominique 2011). Rather than being mired in false consciousness, those surveyed desired widespread, uplifting, and systemic change. They wished for and sought educational opportunities, environmental resto-ration, health care, civic empowerment, and job transformations. They wanted action to alter deeply embedded social inequalities, particularly through transforming their economic relationships to other nations. A widespread set of efforts ensued. One example, emanating out of a faith-based setting, connected those providing funds and work teams for solar-powered water distribution systems with locals. The participato-ry-driven approach started with identifying local community leaders who established an equitable partnership with the outsiders, found suitable sites for the plants, and agreed to a ten-year partnership. That decade of work would move the effort from jointly led to locally owned and sustain-able. From the donor side, training included both technical installations and cultural sensitivity training relying on participatory strategies. Nearly forty sites have been installed since the 2010 earthquake, an effort now expanded to Honduras and Kenya (see Solar under the Sun at www. solarunderthesun.org). Such locally driven and ultimately locally owned efforts exemplify best practices. In subsequent hurricanes, the solar units have fared well, suggesting that local, sustainable, low-cost solutions emerge as a viable practice, albeit one that came from the faith-based rather than governmental sector. Spreading such practices into political structures might prove equally valuable.

In comparison, efforts in Turkey failed when NGOs and contractors offered housing too far away from livelihoods, a situation that occurs too frequently after disasters. Shrimpers trying to survive Hurricane Katrina faced the same during an elongated evacuation. Their displacement across the southern United States kept them away from sustainable work situated within physical locations that provided cultural, environmental, and social support. This donor-driven versus owner-driven approach produced housing options, but they were options that failed to work for those who needed to earn a living (Yonder, Akçar, and Gopalan 2009). Survivors should not have to choose between livelihoods or roofs over their heads when they need both. Often attached to meaningful places, such ties generate significant resistance when outsiders make assumptions about what is needed (Cuevas Muñiz and Luján 2005; Davidson et al. 2007; Handmer 1985; Hummon 1990; Johnson, Lizarralde, and Davidson 2006). Such a clear gap between documented research and NGO or contractor practices is disturbing.

Indeed, asking people what they think matters. After a Gujarat, India, earthquake in 2001, an area NGO called Self-Employed Women's Association (SEWA) brought women together to relaunch local craft embroidery work (Lund and Vaux 2009). Members "began travelling around their own villages and nearby areas, and in many cases themselves became leaders of the whole community" (Lund and Vaux 2009, 217). SEWA noted, from villagers' reports, that the tents supplied by outside donors did not meet local needs. People required a place close to their destroyed homes that allowed them to keep an eye on their children and offered four walls, a roof, and a lockable door. "If donor agencies had understood this need, far more of these shelters could have been constructed" (Lund and Vaux 2009, 218). Bayou communities in Louisiana witnessed a similar transformation when women stepped up as leaders first after Hurricane Katrina and subsequently when the British Petroleum oil spill occurred adjacent to Native American lands. Such local knowledge, sometimes referred to as traditional ecological knowledge, "is derived from experience and shared from person to person" (Prober, O'Connor, and Walsh 2011, 12).

Participatory Approaches and Local Knowledge

Participatory strategies enable people to have more control over their lives, essentially bringing them together (including with outsiders) to define problems and solutions (Chrislip 2002). The social capital they produce serves as an intangible equivalent of money, a key resource generating ideas embedded in culture and history. Participatory processes work because they produce both sensible and innovative ideas as the community considers and discerns the most viable. Local stakeholders own the process, the problems, and the solutions. The method, typically designed as a collaborative and reflexive effort, can take place in a number of ways, from informal embroidery gatherings to formalized meetings or workshops (Natural Hazards Center 2005). Participatory approaches can range from a structured routine (Stringer 2013) to the less formal Highlander Folk School approach. The latter, which emanated out of grassroots coal miner organizing in the United States and transitioned into support of the civil rights movement, asks two questions: "What is the problem?" and "What would you like to do about it?" (Horton and Freire 1990). Or, as civils rights icon Ella Baker encouraged, facilitating group-centered leadership can lead to transformative moments and times (Grant 1999). Such an approach has been adopted nationwide by the New Zealand Ministry of Civil Defence and Emergency Management. Quoting local Maori culture, New Zealanders would state "he tengata, he tengata, he tengata" or "it's the people, it's the people, it's the people."

One strategy, in place for some time, is called the ladder of citizen participation (Arnstein 1969). The ladder situates people at the top of participatory strategies, firmly in control of decision making, to the lowest rung where officials simply inform survivors of decisions as oft-used donor-driven approaches. In between, various strategies of involvement can occur from the lower rungs of nonparticipation through tokenistic processes involving consultation and placation to more fully engaged strategies requiring partnerships, delegation to survivors, and survivor control. The ladder reflects a widely adopted recommendation from researchers, where "effective recovery can be achieved only where the affected community participates fully in the recovery process and where it has the capacity, skills and knowledge to make its participation meaningful" (Coles and Buckle 2004, 6). While such programs will not transform a society, when put into place early enough they have the potential to reduce mortality and improve the quality of post-disaster life. While broader change is certainly needed and required, it may pay to "think globally but start locally."

Despite calls for such local empowerment over the past several decades, one key and potentially transformative action remains under-used in addressing policy and practice gaps: the active participation of women in disaster contexts. Yet, from planning to recovery, women have demonstrated "they are potentially competent enough to overcome disaster situations" (Ikeda 1995, 189). For example, Algerian women took on reorganization of relief camps that were created in response to war and famine (Wallace 1994). Women in Bangladesh have worked collabo-ratively to address canals muddied with silt (Rodda 1994). Post-tsunami, some NGOs paid women for agricultural work in Sri Lanka, enabling them to earn a living and rebuild devastated industries. NGOs helped women by "promoting the leadership of female survivors in recovery processes ... and protecting and reviving the livelihoods of female tsunami vic-tims" (Joshi and Bhatt 2009, 311). Clearly, "vulnerability does not render them helpless" (305). Evidence certainly supports public participation as a means to promote sustainability in communities beset with repetitive losses and risks (Pearce 2003).

Empowering and incorporating participation is consistent also with a holistic recovery process that engages people in considering ways to not only move on from a disaster but also to do so in a way that embraces and outlines sustainable, transformative spaces in which to live (Natural Hazards Center 2005). Reflecting a comprehensively broad and deep view of what a social environment could offer, a participatory and holistic approach mirrors what Haitians wanted so fiercely after the 2010 earth-quake. A holistic approach requires participatory approaches, considers environmental concerns, promotes social and intergenerational equity, and addresses economic vitality. The Self-Employed Women's Association embraced a holistic approach in India, not only concerning themselves with a mission focus on employment but attending also to nutrition, health care, housing, child care, leadership, education, and self-reliance (Lund and Vaux 2009). Indian women initially resisted participating because they felt uneducated or thought men would disallow their involvement. Once exposed to the process of working collectively, they moved quickly into establishing multipurpose centers as they acquired land, organized labor, and managed community-focused construction (Yonder, Akçar, and Gopalan 2009). Similarly, an NGO in Turkey created first a public space in which women could talk and network, which led to peer learning and ultimately to economic self-sufficiency (Yonder, Akçar, and Gopalan 2007).

After the 2004 tsunami, Sri Lankan women used the informal economy to produce marketable goods including vegetables, handicrafts, and baby clothes (Ruwanpura 2008). They did so in contrast to pre-tsunami work that focused on coral mining disrupted by the tsunami. Women in Tamil

Nadu, India, also had to adapt to loss of income when the tsunami devastated offshore fisheries. NGO organizations provided crafting materials, including plastic drink bottles, which women cut and wove into baskets. In Turkey, post-earthquake efforts brought women into converted boxcars where they made toys. In contrast to the more home-based economic activities in Sri Lanka and India, though, Turkish women secured contracts with the government and eventually opened a productive toy factory (Yonder, Akçar, and Gopalan 2009). Women in Aceh, Indonesia, benefitted from pedal-driven sewing machines. Such a transition from displaced survivor to entrepreneur also transforms gendered relationships and livelihoods. Women in devastated disaster circumstances, marginalized by gender, isolated by age, and constrained by livelihood opportunities, proved adaptive as they stood together for themselves, their families, and their communities.

Conclusions

Solutions to gendered outcomes of disasters range from individual-level suggestions, such as teaching women how to swim, to broadly based calls for social change. Doing the latter requires significant and difficult efforts across cultures, ideologies, political systems, and time. Following 1991 Bangladeshi flooding deaths, Ikeda (1995, 189) recommended "that both men and women be given equal opportunities, resources and information in all phases of a disaster, and that they take part in major decision-making in and outside the household." In a culture marked by the isolating practice of purdah, such seemingly straightforward recommendations would prove complex and slow to take place.

OXFAM, a well-known NGO that operates in disaster contexts, summarizes well the approach promoted by vulnerability theorists: "The special needs of women—whether or not these needs are related to one of the biggest natural disasters the world has ever seen—must be urgently addressed by the international community" (MacDonald 2005). Identifying gender equality as the key, OXFAM joins others in working toward a different set of social, cultural, political, economic, and legal structures that jeopardize gendered people in disasters. Requiring widespread and systemic change, the achievement of gender equality will clearly remain a goal for generations to come—which leaves us with the question of how to address current needs.

Significant gendered vulnerabilities have become resistant to change over time. They have been put into place through centuries of institutionalized government policies, NGO procedures, and customs. Typically, disasters usually do not disrupt many pre-disaster patterns. To parallel their findings, if gendered relationships exist prior to an event, they will exist afterward. Isis International, a feminist organization, has called for structural change to undermine such disaster and non-disaster gendered contexts, an effort they have been pursuing for more than four decades (Kottegoda 2007). Unfortunately, as evidenced in this chapter, efforts that address gendered concerns in disasters have had a tendency to reproduce or worsen rather than to transform, undermine, and alter.

However, disasters do produce opportunities that could be leveraged more fully. The challenge is to know where to leverage opportunities effectively. While large-scale social, economic, and political change takes time, inroads can be made when disasters occur, as demonstrated by the examples in this chapter. Localized, focused interactions with people in the context of their communities—before and after—is a recommended solution. Funding from governmental and international NGOs represents one means when targeted specifically to women in all their diversity that could be placed within their hands. As shown in India, Turkey, Sri Lanka, and other places, empowering local people can lead not only to more-effective solutions but also to transformative moments when women move into their communities as entrepreneurs and empowered leaders.

Brenda D. Phillips is dean of liberal arts and sciences and professor of sociology at Indiana University South Bend. She authored or coauthored *Disaster Recovery, Introduction to Emergency Management, Qualitative Disaster Research, Mennonite Disaster Service, Social Vulnerability to Disasters,* and *Women and Disasters,* and has published extensively in emergency management journals. Phillips earned the Blanchard Award for excellence in emergency management education and the Myers Award for work on the effects of disasters on women. She is a member of the Hall of Fame for International Women in Emergency Management and Homeland Security.

References

Abirafeh, Lina. n.d. "GBV in Haiti: Prevention, Response, Coordination." PowerPoint presentation. Retrieved 8 June 2016 from http://www.usip.org/sites/default/files/GBV%20in%20Haiti.pdf.

Arnstein, Sherry. R. 1969. "A Ladder of Citizen Participation." *Journal of the American Institute of Planners* 35, no. 4, 216–24.

Barenstein, Jennifer D. 2006. "Challenges and Risks in Post-Tsunami Housing Reconstruction in Tamil Nadu." *Humanitarian Exchange* 33, 39–40.

Barton, Alan. 1969. *Communities in Disaster.* New York: Doubleday.

Bolin, Robert, and Lois Stanford. 1999. "Constructing Vulnerability in the First World: The Northridge earthquake in Southern California, 1994." In *The Angry Earth: Disasters in Anthropological Perspective,* edited by Anthony Oliver-Smith and Susannah Hoffman, 89–112. New York: Routledge.

Chan, Chang-Chuan, Yi-Ping Lin, Hsiu-Hsi Chen, Ta-Yuan Chang, Tsun-Jen Cheng, and Li-Sheng Chen. 2003. "A Population-based Study on the Immediate and Prolonged Effectsofthe1999TaiwanearthquakeonMortality."*AnnalsofEpidemiology*13,502–8.

Chrislip, David. 2002. *The Collaborative Leadership Fieldbook.* New York: Jossey-Bass.

Coles, Eve, and Philip Buckle. 2004. "Developing Community Resilience as a Foundation for Effective Disaster Recovery." *Australian Journal of Emergency Management* 19, no. 4, 6–15.

Cuevas Muñiz, Alicia, and José Luis Seefoo Luján. 2005. "Reubicación y Desarticulación de La Yerbabuena: Entre el Riesgo Volcánico y la Vulnerabilidad Política." *Desacatos* 19, 41–70.

Cutter, Susan. 2005. "The Geography of Social Vulnerability: Race, Class and Catastrophe." Retrieved 15 December 2016 from http://understandingkatrina.ssrc.org/

Davidson, Colin H., Cassidy Johnson, Gonzalo Lizarralde, Nese Dikmen and Alicia Sliwinski. 2007. "Truths and Myths about Community Participation in Post-disaster Housing Projects." *Habitat International* 31, 100–115.

Davis, Elizabeth, Rebecca Hansen, Maria Kett, Jennifer Mincin, and John Twigg. 2013. "Disability." In *Social Vulnerability to Disaster,* edited by D. Thomas et al., 199–234. Boca Raton, FL: CRC Press.

D'Ooge, Charlotte. "Queer Katrina: Gender and Sexual Orientation Matters in the Aftermath of the Disaster." Paper presented at the annual meeting of the National Women's Studies Association, Sheraton Denver Downtown Hotel, Denver, CO, 27 November 2014.

Eads, Marci. 2002. "Marginalized Groups in Times of Crisis: Identity, Needs, and Response." Quick Response Report #152. Boulder: University of Colorado, Natural Hazards Research and Applications Information Center. Retrieved 16 June 2016 from http://www.colorado.edu/hazards/qr/qr162/qr152.html.

Enarson, Elaine. 1999. "Violence against Women in Disasters: A Study of Domestic Violence Programs in the United States and Canada." *Violence against Women* 5, no. 7, 742–68.

———. 2000. "Gender Equality, Work, and Disaster Reduction: Making the Connections." Revision of Gender and Natural Disasters, Working Paper#1 prepared for the ILO InFocus Programme on crisis response and Reconstruction.

———. 2001. "What Women Do: Gendered Labor in the Red River Valley Flood." *Environmental Hazards* 3, 1–18.

Enarson, Elaine, and P. G. Dhar Chakrabarti, eds. 2009. *Women, Gender and Disaster: Global Issues and Initiatives.* New Delhi: Sage.

Enarson, Elaine, Alice Fothergill, and Lori Peek. 2007. "Gender and Disaster: Foundations and Directions." In *Handbook of Disaster Research,* edited by Havidan Rodríguez, E. L. Quarantelli, and Russell Dynes, 130–46. New York: Springer.

Enarson, Elaine, and Betty Hearn Morrow, eds. 1998. *The Gendered Terrain of Disaster: Through Women's Eyes.* Miami: International Hurricane Center.

———. 1998 "Women Will Rebuild Miami: A Case Study of Feminist Response to Disaster." In *The Gendered Terrain of Disaster,* edited by E. Enarson and B. H. Morrow, 185–200. Miami: International Hurricane Center.

Farmer, Paul. 2012. *Haiti after the Earthquake.* Pittsburgh: University of Pittsburgh Press.

Felten-Biermann, Claudia. 2006. "Gender and Natural Disaster: Sexualized Violence and the Tsunami." *Development* 49, no. 3, 82–86.

Fisher, Sarah. 2005. *Gender based violence in Sri Lanka in the After-math of the 2004 Tsunami Crisis.* Master's thesis, University of Leeds, UK.

Fordham, Maureen, William Lovekamp, Deborah Thomas, and Brenda D. Phillips. 2013. "Understanding Social Vulnerability." In *Social Vulnerability to Disaster,* edited by D. Thomas et al., 1–32. Boca Raton, FL: CRC Press.

Galea, Sandro, Melissa Tracy, Fran Norris, and Scott F. Coffey. 2008. "Financial and Social Circumstances and the Incidence and Course of PTSD in Mississippi during the First Two Years after hurricane Katrina." *Journal of Traumatic Stress* 21, no. 4, 357–68.

Gamburd, Michele Ruth, and Dennis B. McGilvray. 2010. "Sri Lanka's Post-tsunami Recovery: Cultural Traditions, Social Structures and Power Struggles." *Anthropology Faculty Publications and Presentations,* Paper 38. Retrieved 2 June 2016 from http://pdxscholar.library.pdx.edu/anth_fac/38.

Gorman-Murray, Andrew, Scott McKinnon and Dale Dominey-Howes. 2015. "Queer Domicide: LGBT Displacement and Home Loss in Natural Disaster Impact, Response, and Recovery." *Home Cultures* 11, no. 2, 237–61.

Grant, Joanne. 1999. *Ella Baker: Freedom Bound.* New York: John Wiley & Sons, Inc.

Handmer, John. 1985. "Local Reaction to Acquisition: An Australian Study." Working Paper #53, Centre for Resource and Environmental Studies, Australian National University, Canberra.

Haynes, Katherine, John Handmer, John McAneney, Amalie Tibbits, and Lucinda Coates. 2010. "Australian Bushfire Fatalities 1900–2008: Exploring Trends in the 'Prepare, Stay and Defend or Leave Early' Policy." *Environmental Science & Policy* 13, no. 3, 185–94.

Hoffman, Susanna M. 2017. "Disasters and Their Impact: A Fundamental Feature of Environment." In *Routledge Handbook of Environmental Anthropology,* edited by H. Kopnina and E. Shoreman-Quimet, 193–205. New York, Routledge.

Horn, Rebecca. 2010. "Exploring the Impact of Displacement and Encampment on Domestic Violence in Kakuma Refugee Camp." *Journal of Refugee Studies* 23, no. 3, 356–76.

Horton, Myles, and Paolo Freire. 1990. *We Make the Road by Walking: Conversations on Education and Social Change.* Philadelphia: Temple University Press.

Hummon, David. 1990. *Commonplaces.* New York: SUNY.

Hyndman, Jennifer. 2008. "Feminism, Conflict and Disasters in Post-tsunami Sri Lanka." *Gender, Technology and Development* 12, no. 1, 101–21.

Ikeda, K. 1995. "Gender Differences in Human Loss and Vulnerability in Natural Disasters: A Case Study from Bangladesh." *Indian Journal of Gender Studies* 2, no. 2, 171–93.

Jenkins, Pam, and Brenda D. Phillips. 2008. "Battered Women, Catastrophe and the Context of Safety." *NWSA Journal* 20, no. 3, 49–68.

Johnson, Cassidy, Gonzalo Lizarralde, and Colin H. Davidson 2006. "A Systems View of Temporary Housing Projects in Post-Disaster Reconstruction." *Construction Management and Economics* 2, no. 4, 367–378.

Jonkman, Sebastian N., and Ilan Kelman. 2005. "An Analysis of the Causes and Circumstances of Flood Disaster Deaths." *Disasters* 29, no. 1, 75–97.

Joshi, Chandhi, and Mihir R. Bhatt. 2009. "Engendering Tsunami Recovery in Sri Lanka: The Role of UNIFEM and its Partners." In *Women, Gender and Disaster: Global Issues and Initiatives*, edited by E. Enarson and P. G. Chakrabarti, 304–20. New Delhi: Sage.

Kottegoda, Sepali. 2007. "In the Aftermath of the Tsunami Disaster: Gender Identities in Sri Lanka." Retrieved 2 June 2016 from http://www.isiswomen.org/index.php?option=com_content&view=article&id=910.

Krajeski, Richard, and Kris Peterson. 1999. "But She's a Woman and This is a Man's Job: Lessons for Participatory Research and Participatory Recovery." *International Journal of Mass Emergencies and Disasters* 17, no. 1, 123–30.

MacDonald, Rhona. 2005. "How Women Were Affected by the Tsunami: A Perspective from Oxfam." *PLOS Med* 2, no. 6, e178. doi:10.1371/journal.pmed.0020178.

Montas-Dominique, Monique. 2011. "Sim Pa Rele (If I Don't Shout)." In *Haiti after the Earthquake*, edited by P. Farmer, 259–72. Pittsburgh: University of Pittsburgh Press.

Lund, Francie, and Tony Vaux. 2009. "Work-focused Responses to Disasters: India's Self-Employed Women's Association." In *Women, Gender and Disaster: Global Issues and Initiatives*, edited by E. Enarson and P. G. Chakrabarti, 212–23. London: Sage.

Mottet, Lisa, and John M. Ohle. 2003. *Transitioning our Shelters*. Washington, DC: National Gay and Lesbian Task Force/National Coalition for the Homeless.

Natural Hazards Center. 2005. *Holistic Disaster Recovery: Ideas for Building Local Sustainability after a Natural Disaster*, 2nd ed. Boulder, Colorado: Natural Hazards Center.

Neal, David M., and Brenda D. Phillips. 1990. "Female-Dominated Local Social Movement Organizations in Disaster Threat Situations." In *Women and Social Protest*, edited by Guida West and Rhoda Lois Blumberg, 243–55. New York: Oxford University Press.

Neumayer, Eric and Thomas Plümper. 2007. "The Gendered Nature of Natural Disasters: The Impact of Catastrophic Events on the Gender Gap in Life Expectancy, 1981–2002." *Annals of the Association of American Geographers* 97, no. 3, 551–66.

Norris, Fran H., Matthew J. Friedman, and Patricia. J. Watson. 2002. "60,000 Disaster Victims Speak: Part II. Summary and Implications of the Disaster Mental Health Research." *Psychiatry* 65, no. 3, 240–60.

Norris, Fran. H., and Matthew. J. Friedman, Patricia J. Watson, Christopher M. Byrne, Eolia Diaz, and Krzysztof Kaniasty. 2002. "60,000 Disaster Victims Speak: Part I. an Empirical Review of the Empirical Literature, 1981–2001." *Psychiatry* 65, no. 3, 207–39.

Oxford Committee for Famine Relief (OXFAM). 2005. *The Tsunami's Impact on Women*. Retrieved 10 July 2016 from http://www.oxfam.org.uk/what_we_do/issues/conflict_disasters/downloads/bn_tsunami_women.pdf.

Peacock, Walter, and Kathleen Ragsdale. 1999. "Social Systems, Ecological Networks, and Disasters: toward a Socio-political Ecology of Disasters." In *Hurricane Andrew: Ethnicity, Gender and the Sociology of Disasters*, edited by Walter. G. Peacock, Betty Hearn Morrow and Hugh Gladwin, 20–35. London: Routledge.

Pearce, Laurie. 2003. "Disaster Management and Community Planning, and Public Participation: How to Achieve Sustainable Hazard Mitigation." *Natural Hazards* 28, 211–28.

Petersilia, Joan. 2000. "Invisible Victims: Violence against Persons with Developmental Disabilities." *Human Rights* 27, no. 1, 9–12.

Phillips, Brenda. 1995. "Creating, Sustaining and Losing Place: Homelessness in the Context of Disaster." *Humanity and Society* 19, 94–101.

Phillips, Brenda D., and Pam Jenkins. 2013. "Violence in Disasters." In *Social Vulnerability to Disaster,* edited by D. Thomas et al., 311–40. Boca Raton, FL: CRC Press.

———. 2016. "Gender-based Violence and Disasters: South Asia in Comparative Perspective." In *Women and Disasters in South Asia: Survival, Security and Development,* edited by Linda Racioppi and Swarna Prajnya. New York: Taylor and Francis Group.

Phillips, Brenda, David M. Neal, Thomas Wikle, Aswin Subanthore, and Shireen Hyrapiet. 2008. "Mass Fatality Management after the Indian Ocean Tsunami." *Disaster Prevention and Management* 17, no. 5, 681–97.

Phillips, Brenda, Patricia Stukes, and Pam Jenkins. 2012. "Freedom Hill Is Not for Sale and Neither Is the Lower Ninth Ward." *Journal of Black Studies* 43, no. 4, 405–26.

Prober, Suzanne, and Michael H. O'Connor, Fiona J. Walsh. 2011. "Australian Aboriginal Peoples' Seasonal Knowledge: A Potential Basis for Shared Understanding in Environmental Management." *Ecology and Society* 16, no. 2, 12.

Rodda, Annabel. 1994. *Women and the Environment.* Zed books. London.

Ruwanpura, Kanchana. 2008. Temporality of Disasters: The Politics of Women's Livelihoods 'after' the 2004 Tsunami in Sri Lanka." *Singapore Journal of Tropical Geography* 29, 325–40.

Sayeed, Azra T. 2009. "Victims of Earthquake and Patriarchy: The 2005 Pakistan Earthquake." In *Women, Gender and Disaster,* edited by E. Enarson and P. G. Chakrabarti, 142–52. London: Sage.

Sharkey, Patrick. 2007. "Survival and Death in New Orleans: An Empirical Look at the Human Impact of Katrina." *Journal of Black Studies* 37, no. 4, 482–501.

Soares, Judith, and Audrey Y. Mullings. 2009. "A We Run tings': Women Rebuilding Montserrat." In Enarson and Chakrabarti, *Women, Gender and Disaster,* 250–60.

Stukes, Patricia. 2014. *A Caravan of Hope-Gay Christian Service: Exploring Social Vulnerability and Capacity-Building of Lesbian, Gay, Bisexual, Transgender, and Intersex Identified Individuals and Organizational Advocacy in Two Post Katrina Disaster Environments.* Ph.D. thesis, Texas Woman's University, Denton.

Stringer, Ernest. 2013. *Action Research,* 4th ed. Thousand Oaks, CA: Sage.

Tanida, Noritoshi. 1996. "What Happened to Elderly People in the Great Hanshin Earthquake." *British Medical Journal* 313, 1133–35.

Thomas, Deborah, Brenda Phillips, William Lovekamp, and Alice Fothergill, eds. 2013. *Social Vulnerability to Disaster.* Boca Raton, FL: CRC Press.

Thomas, Saji. 2018. "Bhopal Disaster Survivors Revive Compensation Demands." Retrieved 16 April 2019 from https://www.ucanews.com/news/bhopal-disaster-survivors-revive-compensation-demands/83966.

Thurnheer, Katharina. 2009. "A House for a Daughter? Constraints and Opportunities in Post-tsunami Eastern Sri Lanka." *Contemporary South Asia* 17, no. 1, 79–91.

United Nations (UN). 1982. *Shelter after Disaster.* New York: United Nations.

Walby, Sylvia, Jo Armstrong, and Sofia Strid. 2012. "Intersectionality: Multiple Inequalities in Social theory." *Sociology* 48, no. 2, 1–17.

Wallace, T. 1994. "Saharawi Women: Between Ambition and Suffering." In *Women and Emergencies,* edited by Bridget Walker, 50–53. Oxford, UK.

Webb, Gary, Kathleen Tierney, and James. Dahlhamer. 2000. "Businesses and Disasters: Empirical Patterns and Unanswered Questions." *Natural Hazards Review* 1, no. 2, 83–90.

Wisner, Ben. 2001. "Development of Vulnerability Analysis." Emmittsburg, MD: Federal Emergency Management Agency (FEMA) Higher Education Project.

Wisner, Ben, Piers Blaikie, Terry Cannon, and Ian Davis. 2003. *At Risk*. London: Routledge.

Yonder, Ayse, and Sengül Akçar, Prema Gopalan. 2009. "Women's Participation in Disaster Relief and Recovery." In *Women, Gender and Disaster: Global Issues and Initiatives*, edited by E. Enarson and P. G. Chakrabarti, 175–88. New Delhi: Sage.

CHAPTER 8

Resettlement for Disaster Risk Reduction

Global Knowledge, Local Application

ANTHONY OLIVER-SMITH

Introduction

One of the consistent problems that confronts not just applied anthropology, but also every discipline that purports to have application to real world human problems and challenges, is the gap that occurs between science, policy, and practice. Discussed by Schenkel (2010) as the tension between scientific information and societal and political priorities, the gap between what we know and what we do continues to elude solution in many fields. In that context, those elements that make for good policy, i.e., policy that mobilizes political support, can often produce policies that are un-implementable within its framework of methodologies and goals (Mosse 2004). This situation is particularly true of the two fields in which I have worked for the past forty plus years: disasters associated with natural hazards, and development forced displacement and resettlement (DFDR). In the 1970s, I worked for about a decade with the survivors of a disaster who tenaciously and successfully resisted resettlement to an area of lower exposure to environmental hazard. Many of the same problems that afflicted those survivors of a disaster who then faced further displacement and resettlement continue to appear regularly today. This chapter discusses the difficult issue of the gap between science and policy in the context of the confluence of these two fields in terms of recent developments and examples of resettlement for disaster risk reduction (DRR), starting with a brief overview of both disaster research and resettlement research.

Disaster Research: Learning More and Losing More

Over the past seventy years a great deal of knowledge on disasters has been acquired through systematic research across multiple disciplines. Since 1970 we have learned to trace the development of causality of disasters in society rather than in nature, which has led to current strategies for DRR rather than focusing solely on emergency response and recovery. Our models and approaches toward disaster risk and causality now focus on the issues of socially generated root causes, risk drivers, and unsafe conditions (Oliver-Smith et al. 2016; Wisner et al 2004). However, the transfer of these perspectives to both policy and practice has been at best uneven, partially due to a prevailing emphasis on response and recovery in both the policy and practice community. Thus, as White, Kates, and Burton put it with regard to the concurrent increase in understanding of disasters and mounting economic and human losses, "knowing better and losing even more" (2001, 81). Weichselgartner and Kasperson (2010) explore the factors that limit a coherent identification and framing of problems in vulnerability and disaster research between the producers of knowledge and the potential users of that knowledge as (1) functional (divergent objectives, needs, scope, and priorities), (2) social (divergent cultural values, communication, understanding, and mistrust), and (3) structural factors (different institutional settings and standards).

Resettlement Research: Understanding Risks, Yet Impoverishing More

Similarly, seventy years of research on development projects has generated a much more fine-grained understanding of the process of displacement and resettlement. DFDR research has provided important insights regarding the risks imposed on displaced and resettled people, facilitating the identification and diagnosis of processes that lead to the impoverishment of people and failed projects (Bennett and McDowell 2012; Cernea and McDowell 2000; Oliver-Smith 2009; Scudder 2005). The stages that involuntarily resettled people pass through and the tasks and costs they face in adapting to the new circumstances have been documented (Scudder 2009; Scudder and Colson 1982). The risks entailed in involuntary resettlement have been identified and used to enable both diagnosis and remedy (Cernea 1997). The problems in project organization and implementation have also been identified and analyzed (de Wet 2006; Downing and Garcia Downing 2009). Both research tools and policy guidelines have been developed to mitigate the effects of displacement and to improve resettlement (World Bank 2004). However, although there have been some improvements, resettlement outcomes continue to be poor, since most projects are the product of an ideology that privileges infrastructural development and relegates the human rights of affected populations to secondary consideration, producing projects that continue to impoverish and disempower resettled people. Recent assessments of resettlement projects suggest that many of the factors that Weichselgartner and Kasperson point to are extremely relevant to the case of DFDR as well (Downing and Garcia-Downing 2009; de Wet 2009). Moreover, the significant advances that have been made in the development of the World Bank guidelines (2004) are now in the process of being dismantled under pressure from borrower countries and competition from private development funders.

Environmental Disruption and the Resettlement Option for Disaster Risk Reduction

The failure of disaster research and resettlement research fields to successfully bring knowledge and practice together to achieve consistent positive outcomes is a source of concern to researchers, policymakers, and practitioners in each. This is particularly so since resettlement is increasingly seen as a viable option for DRR as displacements due to natural hazard–related disasters and climate change correspondingly increase. It is also true because, even though the realities of environmental disturbance and climate change are generally accepted, the impacts of actual and projected effects are still much debated in both scientific and political forums. In some contexts, the projected effects of environmental change, particularly as they pertain to specific human communities, have entered as much into political controversy as they have into academic and scientific debate (Black 2001; Foresight Project on Migration and Environmental Change 2011; Oliver-Smith 2012).

The research and scholarship focusing on the relationship between environment and displacement, often glossed as migration, is equally rife with controversy, centering largely around the issues of predicted numbers, appropriate terminology for people uprooted by environment, and the political implications of both research and policy pertaining to environmentally displaced people. Over the past decade, the issue has generated considerable scientific debate: first, on the role of the environment in driving migration and projecting local manifestations of environmental change; and second, on what adjustments will need to be induced in natural and human systems (Dessai, O'Brien, and Hulme 2007). Nonetheless, a consensus has now coalesced around the need to address the probability of significant environmental displacement in the future.

The United Nations High Commissioner for Refugees (UNHCR) sees five displacement scenarios emerging in the near future: (1) hydro-meteorological disasters, (2) population removal from high-risk areas, (3) environmental degradation, (4) the submergence of small island states, and (5) violent conflict (UNHCR 2009, 4). Although the figures for environmentally displaced people are notoriously unreliable (Gemenne 2011), it has been noted that statistics for people displaced by natural hazard-induced disasters have a better empirical base and are generally more reliable. Probably the most reliable figures at present are produced by the Internal Displacement Monitoring Center (IDMC) and the Norwegian Refugee Council (NRC). Those groups place the number of people displaced by so-called natural disasters in 2017 at 18.8 million, using a baseline of events from the Epidemiology of Disasters-International Disaster Database (EM-DAT) database to produce a core data set for events where more than fifty thousand people were affected (IDMC 2018). Data on the displaced from each event is then sought from organizations involved in relief for those events (Yenotani 2011).

The IDMC recently reported that 184.6 million people were displaced by disasters between 2008 and 2015, a number largely due to the increase in exposure and vulnerability on a global scale (IDMC 2015). Anticipating still more increase, resettlement of both disaster victims and people at risk of hazard impact is being seriously considered as a viable form of DRR. As anthropologists have worked on displacement and resettlement beginning in the mid-twentieth century and on to the present, they have arguably made the single strongest, most tangible, and most internationally documented and recognized contribution to development policy and practice toward such displaced people over the past quarter century (Oliver-Smith 2006). Since displacement and resettlement affect virtually every domain of individual and community life, anthropology's inductive approach has equipped it well to address the inherent complexity of the resettlement process. In the 1970s anthropologists linked the problems experienced by people impacted by development-induced displacement and resettlement to those of people displaced by conflicts and disasters triggered by natural hazards (Hansen and Oliver-Smith 1982). Since the 1980s the resettlement of communities located in high-risk zones for DRR has also attracted greater attention (Correa 2011a, 2011b; Ferris 2011; Oliver-Smith 1991; Perry and Mushkatel 1984).

Geophysical hazard-related disasters—volcanic eruptions, earthquakes, and tsunamis—while not subject to climate change intensification, displaced 1.5 million people in 2009 and more than 4 million in 2010 (Yenotani

2011). As the frequency of and intensity of extreme climate-related events grows, climate change events and processes represent a large potential for even greater population displacement as well. Global climate change has been accepted by both the scientific community and the general public as a reality that must be addressed at both the level of policy and the level of practice.

Although the initial focus in climate change adaptation tended to see resettlement relatively unproblematically, basically as a solution to intensified storms, sea level rise, and desertification, the findings of research on development-induced displacement and resettlement, research largely done by anthropologists, have now brought the complexity of the process into a more realistic perspective. In effect, the lessons learned in development-induced displacement and resettlement research are now being used to expand the array of approaches and methods that address the challenges presented by disasters and climate change displacement and resettlement at the local community and project levels, in national and international political discourse, and in the policy frameworks of multilateral institutions.

Whatever the root causes of displacement, communities that have been displaced and resettled are communities that must be reconstructed, either by themselves or with assistance (Birkmann et al. 2013; Oliver-Smith 2005). In either case, an infrastructure has to be built to replace the one that has been lost and a community, as a social body, has to reconstitute itself. Resettlement project–affected people are confronted with a complex, cascading sequence of events and processes most often involving dislocation, homelessness, unemployment, the dismantling of families and communities, adaptive stresses, loss of privacy, political marginalization, a decrease in mental and physical health status, and the daunting challenge of reconstructing one's ontological status, including reconstituting family and community contexts (Birkmann et al. 2013; Cernea 1997; Oliver-Smith 2005; Scudder and Colson 1982). All suffer the endangerment of structures of meaning and identity, and all must mobilize social and cultural resources in their efforts to reestablish viable social groups and communities and to restore adequate levels of material and cultural life (Bennett and McDowell 2012). Given the disruption and trauma that displacement may generate, the resettlement of either disaster victims or people at risk should be undertaken only in the case of risks that under no circumstances can be mitigated or reduced.

Global Knowledge Construction on Resettlement for Disaster Risk Reduction

Over the past decade the significant amount of attention paid to resettlement for DRR has led to a number of strategies. Recent policy-oriented initiatives include "Guidance on Protecting People from Disaster and Environmental Change through Planned Relocation" (Brookings Institution–Georgetown University's Institute for the Study of International Migration–UNHCR (2015); "Fleeing Floods, Earthquakes, Droughts and Rising Sea Levels: 12 Lessons Learned about Protecting People Displaced by Disasters and the Effects of Climate Change" (Nansen Initiative 2015), "Peninsula Principles on Climate Change Displacement within States" (Displacement Solutions 2013); and "Humanitarian Crises and Migration: Causes, Consequences and Responses" (Martin, Weerasinghe, and Taylor 2014). In general, these and other proposals predominately focus on the protection of the rights of people displaced by disasters and climate change effects and do not delve into the practical matters of designing and implementing actual resettlement projects.

To date, the most comprehensive treatment of DRR and resettlement is contained in the two-volume publication published by the World Bank called *Populations at Risk of Disaster: A Resettlement Guide*, along with an edited companion volume of case studies entitled *Preventive Resettlement of Populations at Risk of Disaster* (Correa 2011a, 2011b). While there is neither the time nor the space here to conduct an extended review and critique, the two volumes, the product of three years of research and practice led by Elena Correa, one of the Bank's senior social specialists on resettlement, attempt to cover the task of DRR resettlement, recognizing the complexity of the process and the heterogeneity of affected populations, laying out a holistic understanding, and providing a step-by-step approach to the challenges of resettlement. The concern for the well-being of both resettled and receiving communities is clearly demonstrated, as is an awareness of the traumatic aspects of the displacement and resettlement process.

Following a logical framework approach the manual (Correa 2011a) identifies clearly the goals, purposes, activities, expected results and monitoring, and evaluation steps that need to be taken. Each section of the guide anticipates many of the problems that may emerge, but there are some problems that may not be predictable. Since the task of resettlement is not entirely a linear process, there may be issues that emerge in the process of implementation that are unanticipated and thus the manual emphasizes the need for flexibility.

The manual (Correa 2011a) and the case study volume (Correa 2011b), in addressing issues like employment and capacity building, housing provision and distribution, and participation, implicitly establish that they are designed to address vulnerability as well as exposure; exposure is the fundamental issue at the heart of resettlement. In its attention to social, cultural, and economic detail and complexity, the manual makes clear that resettlement is not just about physically relocating a community in its social, cultural, and economic entirety in a new place: it also asserts that the project is not about reproducing in another setting the same social, economic, and physical arrangements that constrained the welfare of the affected population in the original site. After all, as the manual points out, the task of diminishing disaster risk should not generate social and economic risk.

Both works (Correa 2011a, 2011b) reveal an important recognition at the national level that disaster management and reduction embrace a wider domain than emergency response and immediate recovery. The realization that DRR entails a broader array of responsibilities, and also includes legislative action and administrative restructuring to deal with long-term vulnerabilities, is indeed an important step. Furthermore, the requirement that resettlement be undertaken only for high-risk contexts in which no effective hazard mitigation is possible, indicates that the only way of assuring a reasonable level of security in the face of severe natural hazard is resettlement. The two volumes, the manual and the case studies, clearly demonstrate that resettlement, if done poorly, without taking local culture into account and without proper consultation and participation of affected people, can constitute a form of secondary disaster, disabling livelihoods, shredding social networks, and generally impoverishing people. Furthermore, the two volumes constitute solid evidence that knowledge derived from DFDR is highly relevant to resettlement for DRR.

Resettlement for Disaster Risk Reduction: Toward Progress?

Despite the lessons learned from a half century of research in both disasters and resettlement studies, we continue to see highly problematic outcomes from resettlement projects, particularly in DRR. I would like to discuss three case studies, each of which illustrates some of the problematic aspects of resettlement that could be effectively addressed by research-informed practice.

Case Study 1: The Villages of the Volcán de Colima

The assessment of risk of residing in the original environment versus the risks of resettlement is obviously central to the question of resettlement to reduce disaster risk. In terms of hazards, the risk evaluation must be scientifically sound and entirely credible in local terms. People living in high-risk areas may be all too aware of risks they face, but they also know that being resettled moves them out of the only home they have. Resettlement entails numerous risks in terms of loss of social, economic and cultural assets, and resources. The post-displacement use of vacated terrain can validate or invalidate the project in the view of resettled people. Indeed, the credibility of the entire project rests on the issue of post-displacement land use. If vacated land is reoccupied or used in inappropriate ways, the credibility of the risk assessment is invalidated and the community can only feel that its land has been usurped.

The Colima volcanic complex consists of two volcanoes—Nevado de Colima (4,320 meters high) on the north and the historically active Volcán de Colima (3,850 meters high) to the south. The Volcán de Colima is one of the most active in North America and ranks among the most dangerous. Historical eruptions since the sixteenth century have been recorded, with more than thirty periods of eruptions since 1585. Its activity has increased since 1998, and includes four periods of eruption. Scientific monitoring of the volcano began twenty years ago (Gavilanes-Ruiz et al. 2009; Reyes-Dávila and De la Cruz-Reyna 2002). Currently, the volcano remains extremely active with frequent emissions of steam and gases through fumaroles, and occasional, but continuous, small explosions. In addition, an enormous explosion occurred in January 2015, fracturing the dome of the crater in three places and emitting a significant ash fall.

Nearly half a million people live within a radius of forty kilometers of the volcano, including those living in Colima City (2010 population 137, 383), the capital of Colima State. According to official hazard maps, several villages, such as La Yerbabuena and La Becerrera, both within ten kilometers of the peak, are at risk for pyroclastic flows and *lahars* (a landslide of wet volcanic debris on the side of a volcano or the deposit left by such a landslide) (Gavilanes-Ruiz et al. 2009). The economy of these communities is based on agriculture (mainly sugar cane, coffee, and corn) and livestock (i.e., cattle, goats, and sheep). Some people also work in tourism at a local resort (see below). Land is either privately held or in *ejidal* form (common property). Connections with regional, national, and international contexts are maintained through access to radio and television. In addition, several meetings with local people about the levels of risk of the volcano and the question of resettlement have been held by the University of Colima. The levels of education among adults range from completed primary to completed secondary school. Most of the population is Catholic, but there are communities with a significant proportion belonging to various Protestant sects.

Since the most recent spate of volcanic activity beginning in 1998, there have been six major evacuations of duration varying between several days to a month, each one with diminishing participation by local people. The sometimes extended stays in the shelters provided, which the villagers considered to be inadequate, took people away from their fields and other tasks and created tensions and social conflict. Indeed, the continued evacuations stimulated significant resistance to both evacuation as well as resettlement plans to reduce risk. Doubting the evidence presented by the scientists and the authorities, villagers see that the volcano is dangerous, but, as one person put it, "The volcano is dangerous, but I prefer to take the risk, because if I don't, I lose my means of making a living" (Cuevas Muñiz and Seifoo-Luján 2005, 59).[1]

Notwithstanding these hesitations, most of the people from La Yerbabuena and several other smaller villages consented in 2002 to be resettled, much to their dismay since the housing they were provided is considered to have been designed poorly and the lots are much smaller (two hundred square meters). Their previous homes averaged fifteen hundred square meters, with space for household gardens and small animals. Moreover, the distance to both their fields, as well as to markets and schools, is greater, creating considerable economic hardship. Some families refused resettlement, and continued residing on their land. Their resistance is partially attributable to the fact that the Hacienda de San Antonio, a luxury resort hotel, located a few kilometers from La Yerbabuena, has not been under pressure to move and, in fact, has used the risk from the volcano as a pretext to relocate people from land it owns in La Yerbabuena. In the eyes of many, both resisters and resettled, volcanic risk has been used as a means to usurp land by more-powerful economic and political interests (Cuevas Muñiz and Gavilanes Ruiz 2013).

Today the resisters still reside in La Yerbabuena and have been joined by new families who have bought land from those who resettled. This situation has become the source of significant discontent among the original resisters who continue to criticize the authorities' official high volcanic risk assessment. The resettled families remain deeply unsatisfied with conditions in the new settlement and many of them move regularly between their new houses and their old homes in La Yerbabuena (Cuevas Muñiz, personal communication 2016).

Case Study 2: The Displaced of the Indian Ocean Tsunami in Sri Lanka

The Indian Ocean tsunami hit the coastal regions of Sri Lanka to the north-east, east, and south on 26 December 2004. In Sri Lanka approximately 35,322 people lost their lives and 516,150 persons were displaced by the tsunami event (United Nations Development Programme [UNDP] 2005). In addition, the tsunami caused massive damage to public and private infrastructure and to personal assets. In view of the mass destruction the tsunami caused in the coastal communities of Sri Lanka, the government declared "no construction zones" of one hundred meters on the western and southern coastal areas, and of two hundred meters on the eastern and northern coastal areas. Those displaced by the tsunami and living within these zones had to resettle in donor-built settlements (Birkmann et al. 2013; Fernando 2011).

More than a third of the people who had lived close to the sea before the tsunami were encroachers engaged as daily paid laborers in the fisheries sector or in other informal sectors (e.g., street vendors, daily paid laborers) in Galle City. Most of the encroachers were facing poverty or chronic poverty due to their land insecurity and income insecurity, in addition to coastal hazards before the tsunami.

The displaced people first were placed in temporary camps, often more than six months before they were moved into temporary wooden shelters. When permanent resettlement began, most people were not involved in the planning and implementation process of resettlements. The Sri Lanka government provided its guidelines for housing to nongovernmental organizations (NGOs) and other donors. However, donors did not follow the guidelines and local authorities did not monitor them. In consequence, there was no coordination between donors and technical officers at the local level. The majority (nearly 62 percent) of those displaced were not informed about the housing structure and quality (i.e., single or two story), or the availability of common facilities and social infrastructure (e.g., schools, community house, etc.) in the new location (Fernando and Punchihewa 2011).

Despite the lack of appropriate information, according to Fernando, the government had a good resettlement plan to resettle people on government land who were forced to leave under the buffer zone regulation. Nonetheless, the resettlement process quickly became politicized. People resisted the buffer zone regulation with the help of some opposition politicians since the eve of presidential election occurred in the middle of the relocation process, leaving the government candidate little choice but to reduce the buffer zone to forty-five meters, unofficially at first, and later officially.

In general, there was huge pressure after the tsunami, both from politicians as well as from displaced people, to speed up the housing construction in resettlements, which increased the likelihood of failures in planning and construction. Most of the respondents in the new settlement site complained that building contractors used poor materials purposely to increase their profits, since there was no proper supervision either from relevant government officials or from donors. The lack of a functioning common social infrastructure (e.g., schools, public transport, etc.) before settling beneficiaries in their new location, also contributed to new risks of livelihood disruption and additional costs for commuting, such as for sending children to school (Birkmann et al. 2013).

In most of the relocation sites a proper drainage system for rainwater or wastewater, neither for individual houses nor for the settlement as a whole, was in place. Some household even reported that they were now exposed to new hazards, such as flooding of the house during monsoons. Health risks were also likely to increase in the new relocation sites due to the lack of public garbage collection or due to the inappropriate treatment of wastes, such as whether to burn it, bury it in the garden, or throw it on abandoned lands or streets corners, as reported by some of the interviewed households (Birkmann et al. 2013).

Due to lack of funding for purchasing privately held land, the majority of relocated settlements were constructed on available government land far from Galle City, despite officials' intention to relocate people close to their former villages. Moreover, some of the sites selected proved to be subject to flooding during the rainy season. Thus, people who were in the relocation settlements suffered great difficulties, especially with regard to living conditions and the distance to the city. In fact, many returned to their previous settlements in the buffer zone. Local government institutions and politicians responsible for providing public services were unable to provide those services due to lack of resources (Fernando 2011).

While the relocation process eight to twelve kilometres from the city reduced the spatial exposure of households to coastal hazards, including tsunamis, it increased the socioeconomic susceptibility due to two main factors: (a) disruption of income-earning activities and (b) new household expenses. The two factors have cumulatively reduced the monthly income of households relocated and increased household expenses. Notably, almost two thirds (nearly 66 percent) depend on public transport (bus) to commute to the city and other places of work. Hence, the reduction in exposure to coastal hazards in the new settlement increased the difficulties in accessing the place of work that is, for a large part of the relocated households, still located in the coastal zone. The difficulties of the daily commute pose a new threat to livelihoods and livelihood recovery.

Commuting to the city for income-earning activities and access to social infrastructures, such as schools or hospitals, is a new monthly expenditure for the majority of households relocated (Birkmann et al. 2013; Fernando and Punchihewa 2011).

All newly built housing units received individual water and electricity connections. While these connections are a clear improvement of the housing situation for the majority of relocated people, they now need to pay for electricity and piped water. Somewhat more than a third of the poor households could not pay, which meant that they were disconnected from these services. Their socioeconomic susceptibility, therefore, increased due to the improved infrastructure services and the respective costs linked to it (Birkmann et al. 2013).

None of the new occupants in the three study locations received formal landownership for their new land until five years after the tsunami disaster. Increases in susceptibility emerged from conflicts between host and guest communities over resources, such as land, access roads and the community center. In addition, conflicts also emerged due to the different cultures the host and guest communities belong to in terms of different ways of communication, attire, employment, and even eating habits.

A small proportion of resettled householders want to move back to their old location close to the sea or to another location near Galle City. A number of new resettlement houses were reported as closed, sold, or rented even prior to this research. Those who moved back to their previous place in the buffer zone, or somewhere close to the sea, run the risk of being exposed to tsunamis or other coastal hazards in the future, but they avoid the various stresses related to forced relocation (Fernando 2011).

Case Study 3: Displaced Indigenous People in Guatemala from Hurricane Stan

A study entitled "Guatemala. La primera ciudad Tz'utujil del siglo XXI" by Eduardo Aguirre Cantero (2011) focuses on the resettlement of people impacted by Hurricane Stan in October 2005 who remained on lands deemed at risk in the aftermath. In the province of Solalá, in the district of Santiago Atitlan, Hurricane Stan unleashed an avalanche from the Toliman Volcán that devastated the indigenous villages of Panabaj and Tz'anchaj. Two hundred eighty-seven families lost members, houses, possessions, and crops. Six hundred people were killed and 205 houses were destroyed.

The resettlement project was undertaken under conditions of certain urgency in the aftermath of the hurricane, thus integrating it into the various initiatives comprising the post-disaster reconstruction process. Reconstruction in Latin America is increasingly seen conceptually as contributing to societal transformation. Explicit links between reconstruction processes and societal transformation are made, emphasizing greater popular participation, social solidarity, institutional strengthening, and advancing democratization. Significant advances from a focus on disaster response and reconstruction toward strategies of risk reduction as part of public policy had been achieved, particularly since the impact of Hurricane Mitch in 1998.

The government reconstruction plan was articulated with the national land planning strategy to promote economic and social development focusing on rebuilding houses, schools, health centers, water and sanitation systems, and other public infrastructure in low-risk areas (Aguirre Cantero 2011). The plan also included an emphasis on reducing economic, social, cultural, and environmental vulnerability through improved land use planning and the establishment of an early warning system. An economic and productive focus was aimed at supporting agriculture, livestock raising, agro-forestry, tourism, and handicraft micro-enterprises. A further dimension included strengthening the social fabric through community participation.

The selection of land for the resettlement of people living in high-risk areas was achieved through consultation with community representatives who insisted that the Tz'utujil indigenous people were born, had lived, and wished to die on their land (Aguirre Cantero 2011). Sites with low hazard exposure were, therefore, selected in the district for housing with easy access to roads, services, urban facilities, and proximity to the municipal administrative center. Moreover, the new land chosen for the project turned out to be the original pre-Columbian settlement, thus discovering land use lessons learned long ago and forgotten, as well as reinforcing the deep cultural ties of the people to the site. Indeed, the ties to an ancestral past facilitated the resettlement process in the face of a population that manifestly refused to relocate outside the immediate region.

Indeed, the Guatemalan case study is noteworthy for the integration of culturally based understandings about the links between human social relations and disasters. Mayan culture believes that disasters are caused by imbalances between humans and humans, or humans and nature. Therefore, reconstruction that would be accepted by the people had to be framed in such a way that addressed those imbalances. This process was furthered by the extensive participation of the affected population in housing and urban design issues. The more-detailed discussion of housing and urban design also details the more-active role played by the affected people in the resettlement project. The first families were resettled from emergency shelters in 2007, more than two year later, after a long wait during which many considered returning to their original high-risk sites.

However, in January 2010 about seven hundred families were resettled in the new site.

The case study is candid about the challenge that reconstruction after Hurricane Stan presented to the Guatemalan government that already faced lack of confidence in government in the aftermath of war, ethnic differences between the national society and the affected population, physical isolation and poor transportation infrastructure, and extreme poverty. In many respects, the success of the resettlement project will have developmental implications for the entire region in terms of national healing and inter-ethnic relations in Guatemala.

The Guatemalan case illustrates well that with appropriate collaborations between project personnel and the affected people, based on respect for local culture and values, successful resettlement can be achieved, despite the challenges presented by geographical isolation, ethnic differences, and extreme poverty. Moreover, a successful project can contribute to the process of national healing in the aftermath of conflict (Aguirre Cantero 2011).

Discussion

The three cases studies illustrate a variety of the challenges that resettlement for DRR presents, progressing from resistance through various gradations of consultation and participation with affected people in the design of projects, particularly regarding the four basic elements that I outlined in my 1991 *Disasters* article (Oliver-Smith 1991), which are articulated in various formats and contexts in the World Bank publications (e.g., World Bank 2004). Principal among these four basic elements is authentic participation. The word "participation" has become a development buzzword, more often honored in the breach than actually nourished, often because it is highly political, is time consuming, and interferes with elite notions of efficiency, accountability, and control. Abandoning these touchstones of bureaucratic culture is very difficult for public sector (and private sector) agencies to do. It is very threatening for most government agencies and also some NGOs to relinquish to beneficiaries any degree of control in the use of funds.

There is a complexity in resettlement that is inherent in "the interrelatedness of a range of factors of different orders: cultural, social, environmental, economic, institutional and political—all of which are taking place in the context of imposed space change and of local level responses and initiatives" (de Wet 2006, 190). Despite this daunting complexity, planners and administrators of projects dealing with uprooted peoples have generally approached their tasks as a straightforward material transfer. Indeed, the goals of such undertakings frequently stress efficiency and cost containment over restoration of community. Such top-down initiatives have a poor record of success because of a lack of regard for local community resources. Planners often perceive the culture of uprooted people as an obstacle to success, rather than as a resource.

However, besides the issues of guidelines and ethics, we also have to address the issue of limits of competence. In the best of cases, adequate resettlement has been only partially achieved, and those cases are miniscule in number in comparison to the truly awful failures. Guidelines, well intentioned though they might be, will not help if we really do not know how to carry out resettlement projects that benefit the people. There is more than a fair amount of hubris involved in the idea of resettlement. "We" think "we" can do it, but resettlement is not just picking up communities and setting them down somewhere else. It really is trying to replace what has been a historical process of community evolution with an administrative process. In point of fact, resettlement (and the reconstruction of community) may not be entirely amenable to administrative process.

Are there ethical constraints against engaging in actions for which the competence to achieve successful outcomes does not exist? Surely there are in medicine, engineering, law, and other professions and trades in which examinations and licenses establish competence. But competence in resettlement is still informally established. Do guidelines point toward what works or what we should avoid? How do we know "best practices" are best if a good track record of success does not exist? Given that there is a scant established canon based on successful cases, can current projects be considered a form of experimentation masked as a development project (Visvinathan 1990)? Are there not ethical guidelines regarding experimentation in cases where we are not sure that such procedures actually work?

Although the need to resettle people facing serious hazards may be based on solid evidence and analysis of risk, a variety of constructions, frequently based on notions of class, race, or ethnicity, are held of those to be resettled by authorities that may influence how the process of displacement and resettlement are planned and implemented. The people may be perceived as underdeveloped, primitive, unsophisticated, incapable of understanding; if they resist, they are deemed ignorant or shortsighted. These condescending and negative constructions of the people to be resettled may justify in the view of the authorities a variety of omissions, deceptions, and false economies made in the guise of efficiency. Indeed, in this regard, the performance of governments may not inspire great confidence in local people, particularly if the people are indigenous or poor. Since the rights and protections that people facing resettlement because of high-risk hazard onset are still only being defined and are not yet protected by formally recognized law, there is often little recourse left to them beyond outright resistance, or a refusal to move (Bronen 2011; Deng and Cohen 1999; Displacement Solutions 2014; Ferris 2012; Inter-Agency Standing Committee [IASC] 2006).

Conclusions

The challenge to resettlement for DRR, as in other causes, thus, becomes the development and application of policy that supports a genuine participatory and open-ended approach to resettlement planning and decision-making (de Wet 2006). The case studies I have briefly outlined demonstrate a broad array of significant intertwined problems in the way resettlement is planned and carried out after disasters, including elite cultural insensitivity and perception of displaced peoples' needs, misguided concepts of economic efficiency and cost containment, corruption, institutional and administrative competence, and lack of political will. However, as one of the case studies I explored (Guatemala) reveals, some advances are being achieved in making efforts toward involving the displaced in the planning and construction of their new settlements. Such a trend may demonstrate that we may be beginning to reduce the gap between what we know and what we do.

Note, however, that the cases I have explored are all quite recent and their relative success or failure is still open to question. Thayer Scudder, one of the pioneers of DFDR asserts that we cannot assess the actual success of a resettlement project until at least two generations have passed (Scudder 2009). As a final thought, we must recognize that resettlement, for whatever reason and despite stated purposes and legal requirements, does not simply pick a community up and reconstitute it intact in another location. Resettlement may or may not reduce risk, improve resource or service access, and provide employment or other means of livelihood. It may or may not reduce hazard exposure or vulnerability, but it does, in the vast majority of cases, entail an additional quotient of risk, particularly of impoverishment, that must be addressed. Given the considerable knowledge that has been accumulated over the past half century or more, the application of the knowledge to improve resettlement project implementation must be paramount.

Anthony Oliver-Smith is Professor Emeritus of anthropology at the University of Florida. In 2013 he received the Bronislaw Malinowski Award of the Society for Applied Anthropology for his work in disasters and involuntary resettlement. He held the Munich Re Foundation Chair on Social Vulnerability at the United Nations University Institute on Environment and Human Security in Bonn, Germany, from 2005 to 2009. He has authored, edited, or coedited eight books and more than one hundred journal articles and book chapters on disasters and involuntary resettlement. He is currently working on issues of climate change, disasters, displacement, and resettlement in coastal environments.

Note

1. "El volcán es peligroso, pero prefiero arriesgarme por si no, pierdo mi sustento económico."

References

Aguirre Cantero, E. 2011. "Guatemala: The First Tz'utujil City of the Twenty-First Century." In *Preventive Resettlement for Populations at Risk of Disaster: Experiences from Latin America*, edited by E. Correa, 85–106. Washington, DC: The World Bank, GFDRR.

Bennett, O., and C. McDowell. 2012. *Displaced: The Human Cost of Development and Resettlement.* London: Palgrave McMillan.

Birkmann, J., M. Garschagen, N. Fernando, V. Tuan, A. Oliver-Smith, and S. Hettige. 2013. "Dynamics of Vulnerability: Relocation in the Context of Natural Hazards and Disasters." *Measuring Vulnerability to Natural Hazards,* 2nd ed., edited by Joern Birkmann, 505–50. Tokyo: United Nations University Press.

Black, R. 2001 "Environmental Refugees: Myth or Reality?" *UNHCR Working Papers 34,* 1–19.

Bronen, R. 2011 "Climate Induced Community Relocations: Creating an Adaptive Governance Framework Based on Human Rights Doctrine." *N.Y.U. Review of Law and Social Change 35,* 356–406.

Brookings Institution–Georgetown University's Institute for the Study of International Migration–UN High Commissioner for Refugees. 2015. "Guidance on Protecting People from Disaster and Environmental Change through Planned Relocation." http://www.brookings.edu/~/media/research/files/papers/2015/10/07-planned-relocation-guidance/guidance_planned-relocation_14-oct-2015.pdf.

Cernea, M. 1997. "The Risks and Reconstruction Model for Resettling Displaced Populations." *World Development* 25, no. 10, 1569–88.

Cernea, M. and C. McDowell. 2000. *Risks and Reconstruction: Experiences of Resettlers and Refugees.* Washington, DC: World Bank.

Correa, E., ed. 2011a. *Populations at Risk of Disaster: A Resettlement Guide.* Washington, DC: World Bank.

————. 2011b. *Preventive Resettlement of Populations at Risk of Disaster*. Washington, DC: World Bank.

Cuevas Muñiz, A., and J. C. Gavilanes Ruiz. 2013 "La Historia Oral de una comunidad reubicada: Estrategias adaptativas en los procesos de riesgo-desastre." *La Historia Oral y la Interdisciplinaridad. co-ords. K. Covarrubias Cuéllar y M. Camarena Ocampo* (coords), 71–102. Colima, Mexico: Universidad de Colima.

Cuevas Muñiz, A., and J. L. Seifoo-Lujan. 2005. "Reubicación y desarticulación de la Yerbabuena. Entre el riesgo volcánico y la vulnerabilidad política." *Desacatos 19: Vulnerabilidad social, riesgo y desastres*, 41–70. Centro de Investigaciones y Estudios en Antropología Social, Mexico City, Mexico.

de Wet, C. 2006. "Risk, Complexity and Local Initiative in Involuntary Resettlement Outcomes." Development-Induced Displacement: Problems, Policies and People. edited by C. de Wet, 180–203. Oxford: Berghahn Books.

Deng, F., and R. Cohen. 1999. "Masses in Flight: The Global Crisis of Internal Displacement, and The Forsaken People: Case Studies of the Internally Displaced." *Human Rights Quarterly* 21, no. 2, 541–44.

Dessai, S., K. O'Brien, and M. Hulme. 2007. "Editorial: On Uncertainty and Climate Change." *Global Environmental Change* 17, 1–3.

Downing, T. E. 1996. "Mitigating Social Impoverishment when People are Involuntarily Displaced." *Understanding Impoverishment: The Consequences of Development-Induced Impoverishment,* edited by C. McDowell, 33–48. Oxford: Berghahn Books.

Downing, T and C. Garcia-Downing. 2009. "Routine and Dissonant Cultures: A Theory about the Psycho-socio-cultural Disruptions of Involuntary Resettlement and Ways to Mitigate Them without Inflicting Even More Damage." In *Development and Dispossession: The Crisis of Development Forced Displacement and Resettlement,* edited by A. Oliver-Smith, 225–54. Santa Fe, NM: SAR Press.

Fernando, N. 2011. "Vulnerability Assessment in the Context of Disasters and Resettlement Processes." Disaster Risk: Integrating Science and Practice (IRDR) conference 31 October to 2 November 2011, Beijing, China.

————. 2012. *Protection and Planned Relocations in the Context of Climate Change.* UNHCR Legal and Protection Policy Research Series. Geneva: UNHCR.

Fernando, N., and A. Punchihewa. 2011. "Lessons Learnt from the 2004 Indian Ocean Tsunami." Published as a special case study under Migration and Environmental Change project by Government Office for Science, United Kingdom. Retrieved May 2019 from https://www.researchgate.net/publication/261635731_Relocating_the Displaced_Strategies_for_sustainable_relocation.

Foresight Project on Migration and Environmental Change. 2011. *Final Project Report.* London, UK: The Government Office for Science.

Gavilanes-Ruiz, J. C., A Cuevas-Muniz, N., G. Varley., G. J. Stevenson, R. Saucedo-Giron, A. Perez-Perez, M. Aboukhalil, and A. Cortes-Cortes. 2009. "Exploring the Factors that Influence the Perception of Risk: The Case of Volcán de Colima, Mexico." *Journal of Volcanology and Geothermal Research* 186, no. 3–4, 238–252.

Gemenne, F. 2011. "Why the Numbers Don't Add Up: A Review of Estimates and Predictions of People Displaced by Environmental Changes." *Global Environmental Change* 21, s41–s49.

Hansen, A., and A. Oliver-Smith, eds. 1982. *Involuntary Migration and Resettlement: The Problems and Responses of Dislocated People.* Boulder, CO: Westview Press.

Inter-Agency Standing Committee (IASC). 2006. *IASC Operational Guidelines on the Protection of Persons in Situations of Natural Disasters.* Washington DC: The Brookings-Bern Project on Internal Displacement.

Intergovernmental Panel on Climate Change (IPCC), 2014. *Climate Change 2014: Synthesis Report.* Contribution of Working Groups I, II and III to the Fifth Assessment Report of the Intergovernmental Panel on Climate Change [Core Writing Team, R.K. Pachauri and L.A. Meyer (eds.)]. IPCC, Geneva, Switzerland, 151. Retrieved May 2019 from http://www.ipcc.ch/report/ar5/syr/.

Internal Displacement Monitoring Centre (IDMC). 2015. *Global Estimates 2015.* Retrieved May 2019 from http://www.internal-displacement.org/publications/global-estimates-2015-people-displaced-by-disasters.

Internal Displacement Monitoring Centre/Norwegian Refugee Council. 2018. Global Report on Internal Displacement 2018. Geneva. Retrieved May 2019 from http://www.internal-displacement.org/global-report/grid2018/.

Macías Medrano, J., ed. 2000. *Reubicación de comunidades humanas: entre la producción y la reducción de desastres.* Colima, Mexico: Universidad de Colima.

Martin, S., S. Weerasinghe, and A. Taylor eds. 2014. *Humanitarian Crises and Migration: Causes, Consequences and Responses.* New York: Routledge.

Mosse, D. 2004 "Is Good Policy Unimplementable? Reflections on the Ethnography of Aid Policy and Practice." *Development and Change* 35, no. 4, 639–71.

Nansen Initiative. 2015. *Fleeing Floods, Earthquakes, Droughts and Rising Sea Levels 12 Lessons Learned About Protecting People Displaced by Disasters and the Effects of Climate Change.* Geneva, Switzerland. Retrieved May 2019 from www.nanseninitiative.org/staff-member/fleeing-floods-earthquakes-droughts-and-rising-sea-levels-12-lessons-learned-about-protecting-people-displaced-by-disasters-and-the-effects-of-climate-change.

Oliver-Smith, A. 1991 "Success and Failures in Post-Disaster Resettlement." *Disasters* 15, no. 1, 12–24.

———. 2005 "Communities after Catastrophe: Reconstructing the Material, Reconstituting the Social." In *Community Building in the 21st Century,* edited by Stanley Hyland, 45–70. Santa Fe, NM: School of American Research Press.

———. 2006. "Displacement, Resistance and the Critique of Development: From the Grass Roots to the Global." In *Development-Induced Displacement: Problems, Policies and People,* edited by Chris De Wet, 141–56. Oxford: Berghahn Books.

———, ed. 2009. *Development and Dispossession: The Crisis of Development Forced Displacement and Resettlement.* Santa Fe, NM: SAR Press.

———. 2012 "Debating Environmental Migration: Society, Nature and Population Displacement in Climate Change." *Journal of International Development* 24, no. 8, 1058–70.

———. 2014. "Environmental Migration: Nature, Society, And Population Movement." In *Routledge International Handbook of Social and Environmental Change,* edited by Stewart Lockie, David A. Sonnenfeld, and Dana R. Fisher, 142–53. London: Routledge.

Oliver-Smith, A., I. Alcántara-Ayala, I. Burton, and A. Lavell. 2016. *Forensic Investigation of Disasters: A Conceptual Framework and Guide to Research.* Beijing: Integrated Research on Disaster Risk (IRDR).

Perry, R. W and A. H. Mushkatel. 1984. *Disaster Management: Warning, Response and Community Relocation.* Westport, CT: Quorum Books.

Reyes-Dávila, G. A., S. De la Cruz-Reyna. 2002. "Experience in the Short-Term Eruption Forecasting at Volcán de Colima, Mexico, and Public Response to Forecasts." *Journal of Volcanology and Geothermal Research*, 117, 1–2, 121–27.

Schenkel, R. 2010. "The Challenge of Feeding Scientific Advice into Policy-Making." *Science* 330, 24 December, 1749–51.

Scudder, T. 2005. *The Future of Large Dams: Dealing with Social, Environmental, Institutional and Political Costs.* London: Earthscan.

———. 2009. "Resettlement Theory and the Kariba Case: An Anthropology of Resettlement." In *Development and Dispossession: The Crisis of Development Forced Displacement and Resettlement*, edited by A. Oliver-Smith, 25–48. Santa Fe, NM: SAR Press.

Scudder, T., and E. Colson. 1982. "From Welfare to Development: A Conceptual Framework for the Analysis of Dislocated People." In *Involuntary Migration and Resettlement: The Problems and Responses of Dislocated People*, edited by A.Hansen and A. Oliver-Smith, 267–88. Boulder, CO: Westview Press.

Turton, D. 2006. "Who Is a Forced Migrant?" *Development-Induced Displacement: Problems, Policies, and People,* edited by C. de Wet, 13–37. Oxford: Berghahn Books.

United Nations Development Programme (UNDP). 2005. "Survivors of the Tsunami: One Year Later. UNDP Assisting Communities to Build Back Better." New York: United Nations Development Programme.

United Nations High Commissioner for Refugees (UNHCR). 2009. "Climate Change, Natural Disasters, and Human Displacement: A UNHCR Perspective." United Nations High Commissioner for Refugees, Geneva. Retrieved May 2019 from http://www.unhcr.org/cgi-bin/texis/vtx/home/opendocPDFViewer.html?docid=49 01e81a4&query=displacement%20scenarios.

Visvinathan, S. 1990. "On the Annals of the Laboratory State." In *Science, Hegemony and Violence,* edited by A. Nandy, 257–88. Oxford: Oxford University Press.

Weichselgartner, J., and R. Kasperson. 2010. "Barriers in the Science-Policy-Practice Interface: Toward a Knowledge-Action-System in Global Environmental Change Research." *Global Environmental Change* 20, no. 2, 266–77.

White, G. F., R. W. Kates, and I. Burton. 2001. "Knowing Better and Losing Even More: The Use of Knowledge in Hazards Management." *Global Environmental Change Part B: Environmental Hazards* 3, 3/4, 81–92.

Wisner, B., T. Cannon, P. Blakie, and I. Davis. 2004. *At Risk: Natural Hazards, Peoples' Vulnerability and Disasters,* 2nd ed. London: Routledge.

World Bank. 2004. *Involuntary Resettlement Sourcebook.* Washington DC: World Bank.

Yenotani, M. 2011. *Displacement due to Natural Hazard-Induced Disasters. Global Estimates for 2009 and 2010.* Oslo: International Displacement Monitoring Centre (IDMC), and Norwegian Refugee Council (NRC).

CHAPTER 9

From Nuclear Things to Things Nuclear

Minding the Gap at the Knowledge-Policy-Practice Nexus in Post-Fallout Fukushima

RYO MORIMOTO

> Since radioactive clouds do not bother about milestones, national boundaries or curtains, distances are abolished. . . . Any distinction between near and far, neighbors or foreigners, has become invalid; today we are all "proximi."
>
> —Günther Anders, *Theses for the Atomic Age*, 1962

Introduction

Let me begin by describing the consensus shared among policymakers, scientists, decontamination workers, locals, and the general public alike regarding the past four years of the multi-billion-dollar decontamination efforts in Fukushima: decontamination cannot, realistically, decontaminate this once-contaminated land. However, five years after the 2011 nuclear disaster many evacuees who relocated to areas away from the plant in Fukushima (164,865 total in May 2012 and 92,154 in May 2016) have been slowly returning to their original places of residence. Looking at the population statistics alone, it seems that decontamination has done *something*, but this something is yet to be fully understood. The lack of understanding perhaps is justifiable since decontamination in Japan is "missing expertise" (Rajan 2002) that people might draw on in order to make sense of or perhaps to accept "a new species of trouble" (Erikson 1994) that has become the inevitable since the 2011 nuclear catastrophe.

For the country that has historically suffered from the traumatic fear of nuclear contamination and exposure after the bombing of Hiroshima and Nagasaki in 1945 and The Lucky Dragon Incident in 1954, coexisting with the presence of ionizing radiation—living or adjusting to life in the shadow of the nuclear meltdowns in Fukushima—seems to violate a cultural code or else the negation of the lessons learned from these past experiences. From such a cultural historical perspective, decontamination appears to be a necessary act, one that is aimed at preserving the continuity with the past. However, I argue that decontamination is more cultural performance than it is scientific procedure, because (and contrary to Anders's above observation of the Atomic Age) decontamination ideologically aims to reestablish both real and imagined distances between the energy-generating rural peripheries and energy-consuming urban centers that the hydrogen explosions of nuclear reactors in March 2011 abolished.

Decontamination in Fukushima is carried out with rapidly constructed assemblages of expert knowledge, policy, and practice (i.e., the implementation of expert knowledge and policy) with the common goal of reducing the radiation level of a given space as much as possible. As such and just like any collaborative, cross-organizational and cross-disciplinary work, the Gap is unavoidable. Here I use the term "the Gap" to indicate the surprising and unintended outcomes, consequences, or reverberations that the application of policy, in practice, produces. It is not enough to say that the Gap is produced out of the unsuccessful coordination of knowledge, policy, and practice, nor is it enough to say that this Gap arises because there is no common language or clear communication between experts, policymakers, and those implementing policy on the ground. This is precisely because the Gap resides beyond the reach of the knowledge-policy-practice collaborations that have produced it. In the case of decontamination in Fukushima, I argue that the Gap appears in a form of anthropologically significant experiences that any knowledge-policy-practice overshadows. As I will show, this is the ethnographically ascertainable, lived experience of what decontamination means to and impacts the locals who identify Fukushima their home, despite the ongoing catastrophe. The Gap presents disaster researchers with an opportunity for making use of ethnographic analysis in order to illuminate the asymmetrical distribution of risk and vulnerability in society (Oliver-Smith 2004).

In this chapter I aim at exposing the working of the Gap in the context of decontamination after the 2011 nuclear disaster in Fukushima. By analyzing the unexpected divergence between the ideology implicit in government-implemented, disaster recovery decontamination policy (which is aimed at reducing the measurable radiation levels in a given space) and its actual practice in Fukushima as the selective visualization of a particular image of containment of the (omni)presence of nuclear things in the environment, I examine how, in its undertaking, decontamination generates material objects that the locals confront in their everyday lives and that are not clearly defined in policy or by expert, technoscientific knowledge.[1] In Fukushima, the waste that decontamination work produces is contained in large, opaque, and polyethylene bags, which many

Figure 9.1. *Frecon baggu* in a temporary storage facility in Minamisōma City, Fukushima. Photograph by author, 2014.

locals refer to as *frecon baggu* (flexible container bag) (figure 9.1). I will illustrate what the accumulative presence of frecon baggu does to people's ongoing experience of the nuclear disaster and discuss how such experience percolates as a meaningful resource for understating the ideology of decontamination.

In making sense of the Gap and the kind of experiences that work within it that exceeds the reach of the knowledge-practice-policy nexus, I argue that anthropology can provide an explanatory framework that legitimizes locals' inchoate knowledge and the lay expertise (Wynne 1996) that they have cultivated through their everyday encounters with the disaster. The contextually grounded, cultural knowledge (ethnographic expertise [Rajan 2002, 245–48]) of the outcomes of policy implementation is formulated by and at the same time can make an important contribution to the understanding of the Gap that the knowledge-policy-practice nexus generates and does not have the kind of analytic tools to explore.

This chapter engages with decontamination as the ongoing struggle in Fukushima between nuclear things as the spectacle of potentially controllable risk at a far distance and as the everyday experience of uncontainable uncertainty in close proximity. What influence does decontamination as a policy that purports to transform the materiality of nuclear things—from the agent of contamination to the object of containment—have on the local people's and the general public's imagination of nuclear things? How does actual decontamination of locals' places unexpectedly reveal a Gap transpired by the knowledge-policy-practice nexus? In order to illustrate the making of the Gap, and to evaluate the anthropological value of and in the Gap, I will cover the following three dilemmas of decontamination that have emerged in the past five years of the decontamination policy: (1) the lack of proper method and knowledge; (2) the eradication of the local residents' familiar territories; and (3) the production of nuclear waste.

First Dilemma: Decontamination as the Act of Re-moving Nuclear Things

It is important to point out that both local residents in Fukushima and the national and local governments seem to share the general understanding that decontamination is a critical part of the recovery and reconstruction of Fukushima from the nuclear disaster. Decontamination appears to be a mutually accommodating procedure. It appears to fulfill both the local residents' desire (though not all of them desire this) to make their contaminated land feel safer and livable so that they might reclaim their familiar territories, as well as the government's interests in making a gesture of respect toward its citizens' fundamental rights to live in their own property. Decontamination is also significant since the policy-level projection of the reopening of parts of the evacuation zone depends on the successful execution of what the government means by "decontamination." But how, then, does the government define decontamination?

Since the enforcement of the Act on Special Measures Concerning the Handling of Radioactive Pollution (hereafter the Act on Decontamination) on 1 January 2012, the national government has claimed its responsibility only within the Special Decontamination Area (eleven municipalities in Fukushima) for the collection, transfer, monitoring, temporary storage, and final disposal of radioactive materials discharged by the accident in 2011.[2] The national government (the Ministry of the Environment is the responsible for the operation) defines the Special Decontamination Area as the regions that are located either within the 20 kilometers (km) distance from the Fukushima Daiichi Nuclear Plant or within highly contaminated areas where the average radiation level is expected to exceed the annual cumulative exposure rate of 20 mSv/y (e.g., Iitate Village).[3] The rest of the areas, designated as the Intensive Contamination Survey Area where the radiation exposure might exceed the 1 mSv/y standard but is significantly lower than 20 mSv/y (there are one hundred such municipalities in eight prefectures), falls under the responsibility of the local governments to decontaminate. However, the national government has pledged to cover any costs incurred on behalf of Tokyo Electric Power Company Holdings Inc. (TEPCO) (Article 43) if and only if decontamination work follows the stipulated standard procedures.[4] Article 40.2 of the Act on Decontamination specifies this: "A person or entity implementing measures for decontamination of the soil, etc., in a decontamination zone shall comply with the standards set forth in the relevant Ordinance of the Ministry of the Environment, when such measures are entrusted to another party." This meant, according to a decontamination worker, that government regulates the method of decontamination and favors certain general construction companies as their contractors.[5] Therefore, decontamination benefits the economy of a certain group of organizations and individuals. This political economy is one of the reasons why, the above-mentioned worker asserted, that the government presents decontamination to be a necessary national project. He questioned why construction companies such as Kajima Corporation, Shimizu Corporation, Taisei Corporation, and so on that were responsible for building nuclear plants have also been responsible for the decontamination project, pointing out that whether building or cleaning, the same set of companies has been benefiting from the national government's nuclear cleanup policy.[6] However, it would be misleading not to point out that local construction companies have also been benefiting from decontamination work since many have been subcontracted by the above-mentioned large construction companies, and thus the local residents cannot openly criticize the decontamination project even if they think it is questionable for the reasons mentioned above or others. Importantly, regardless of the cost-effectiveness of efforts to reduce the radioisotopes in the living environment, decontamination stimulates both the national and local economies of a selected few.

This is illustrative of what Naomi Klein (2008) calls disaster capitalism, where political leaders use a crisis as an opportunity to maximize their profits while exploiting the needs and wants of vulnerable populations. At the local level, what is more visible is the sudden growth of a nonlocal, temporary decontamination labor force. Many locals have been concerned about the growing presence of strangers and the potential impact that these strangers might have on public security to the extent that many evacuated families perceive the presence of strangers in their hometown as a greater security threat than the potential effect of chronic low-dose radiation exposure.

Despite this disaster economic bubble, however, no one, including TEPCO or the Ministry of the Environment, knows the effective way to

Figure 9.2. The title page of the Ministry of the Environment's pamphlet (2013b) released in February 2013. This twenty-six-page pamphlet describes their experimental decontamination in various regions in coastal Fukushima prior to starting the actual decontamination project. The pamphlet is titled, "What We Have Tested before Decontamination—We Are Decontaminating Based on the Test Results." The title page depicts a series of techniques associated with the decontamination work. For example, toward the left-hand corner of the image two workers are gathering leaves on the ground, and on the top part of the image, the worker is using a machine to strip soil from the ground. Public domain.

remove nuclear contamination other than to collect, contain, and isolate an indefinite amount of radioactive contaminants. Kogure (2013), a geo-environment technology specialist, argues that there is no silver bullet for decontamination, and currently the only known way to decontaminate is through "washing, scraping and diluting" (179) radioisotopes in an environment (figure 9.2). Although there has not been much significant progress made with the decontamination process (the government has announced the completion of decontamination in only three out of eleven municipalities as of August 2015), between 2012 and April 2016 the national government has spent more than $200 billion on decommissioning of the crippled plant, decontamination and on the construction and maintenance of storage sites (Nippon Hōsō Kyōkai [NHK] 2016). The decontamination-related expenditure is still growing, and this national project to initiate the reconstruction process in Fukushima prefecture is covered currently by tax money.

Due to the lack of scientifically validated effective procedures, the government describes its *jyosen* (decontamination; *jyo* = eradicate, *sen* = stain] policy vaguely in the Act on Decontamination as the removal of as many radioisotopes as possible in the environment: "In this Act, 'measures for decontamination of the soil, etc.' means measures taken for soil, vegetation, structure, etc., contaminated with radioactive materials discharged by the accident including the removal of the contaminated soil, fallen leaves and twigs, sludge accumulated in ditches, etc., preventive measures to keep the pollution from spreading, and other measures" (Article 2.3). This general uncertainty about the method and its targets is reflected in the focus of decontamination efforts. Decontamination is not about the actual removal of radioisotopes from the environment but about the reduction of technoscientifically detectable levels of background radiation measured with the Sievert (Sv) scale in a given area. In other words, the government's decontamination policy emphasizes that the measurement of background radiation levels is a means through which to ascertain the effectiveness of decontamination, because it indicates the reduction of technoscientifically ascertainable potential risks to human health. The government measures this by comparing the pre- and post-decontamination levels of background radiation of a given area. According to the Ministry of the Environment, the goal of decontamination is to try to meet the 1mSv per year or 0.23μSv per hour standard within a specific section of the zone (figure 9.3). The International Commission on Radiological Protection (ICRP) recommends the additional 1mSv exposure standard for the general public.[7]

Critically, decontamination as the government defines it does not take internal radiation exposure (i.e., exposure through ingestion, inhalation, and/or injection of radioactive material) into account, which has been a growing concern for the local residents. Moreover, it is important to note that the reduction of the potential for external exposure can occur naturally, since radioisotopes decay over time. The three most prevalent contaminants in many regions of Fukushima are known to be iodine-131 (with a half-life of eight days), cesium-134 (two years) and cesium-137 (thirty years). Thus, in theory, the level of radiation would be reduced in thirty years by more than a half of the original amount without any human intervention. This is one of the things that has created a mismatch between the intended outcome of policy and its actual implementation in practice: many areas within the designated decontamination zone had already achieved the government proposed goal of 1 mSv/y standard prior to decontamination by 2013, when the half-life of cesium-134 was reached. A few local residents in Fukushima I interviewed between 2013 and 2015 questioned the need of any decontamination work, especially when the

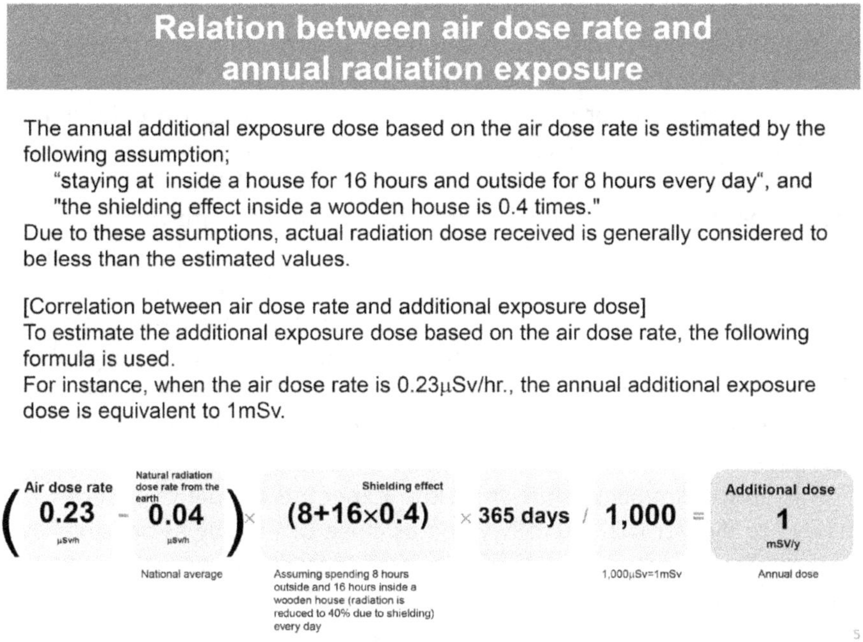

Relation between air dose rate and annual radiation exposure

The annual additional exposure dose based on the air dose rate is estimated by the following assumption;
   "staying at inside a house for 16 hours and outside for 8 hours every day", and
   "the shielding effect inside a wooden house is 0.4 times."
Due to these assumptions, actual radiation dose received is generally considered to be less than the estimated values.

[Correlation between air dose rate and additional exposure dose]
To estimate the additional exposure dose based on the air dose rate, the following formula is used.
For instance, when the air dose rate is 0.23μSv/hr., the annual additional exposure dose is equivalent to 1mSv.

$$\left(\underset{\substack{\text{Air dose rate} \\ \mu Svh}}{0.23} - \underset{\substack{\text{Natural radiation} \\ \text{dose rate from the} \\ \text{earth} \\ \mu Svh}}{0.04} \right) \times \underset{\text{Shielding effect}}{(8+16\times0.4)} \times 365\ \text{days} / 1,000 = \underset{\substack{\text{Additional dose} \\ mSV/y}}{1}$$

National average   Assuming spending 8 hours outside and 16 hours inside a wooden house (radiation is reduced to 40% due to shielding) every day   1,000μSv=1mSv   Annual dose

Figure 9.3. The Ministry of the Environment's proposed equation for the 1 mSv/y calculation. Following this equation, the Ministry of the Environment has determined that the average hourly airborne radiation (external) exposure should be less than 0.23 μSv to achieve the 1 mSv per year standard. Taken from the Ministry of the Environment (2013a) PowerPoint presentation at ICRP in 2013. Public domain.

radiation level is already as low as any other place in the country. I will elaborate on local understandings of the need for decontamination later.

   The issue regarding the assessment of the effectiveness of decontamination is that it focuses primarily on a fixed spatiotemporal delimitation of nuclear things. In other words, the government defines the effectiveness of decontamination according to the reduction of the background radiation in a given area only during the time of decontamination but does not monitor the same site after the decontamination work has ceased. The Act on Decontamination only briefly discusses but does not provide specific details about the ongoing cleanup activities happening inside the nuclear plant, nor does it discuss the potential atmospheric and ecological transfer of radioisotopes across areas in the present and future that decontamination or other human activities might initiate. For example, a heavy rain could transfer concentrated radioisotopes in the mountain into the residential areas through a river, or traffic along the coastline along the nuclear plant or winds could carry radioisotopes from a more highly irradiated areas to less irradiated areas. The precarious movement of the source of radiation, which locals often discover as they consistently measure the radiation levels present at a single spot over time, would make it almost impossible for the government to be able to declare definitively that a decontaminated site has been definitively free of nuclear things and is therefore safe for habitation. In fact, in 2014 the Ministry of the Environment shifted the language of its goal from "1 mSv/y" to "lower than 20 mSv/y and as close to 1 mSv/y as possible" after finding out that such reduction is very difficult to achieve in certain situations, such as in areas with high vegetation and forestation. The implementation of decontamination policy in practice revealed that the policy's previous goal was unachievable. Nevertheless, the government has insisted on maintaining its plan to reopen the evacuation zones by changing the aim of radiation reduction level.

This subtle linguistic shift from the expected annual exposure of "1 mSv/y" to "lower than 20 mSv/y and as close to 1 mSv/y as possible" has caused a few negative reactions from both local residents and the general public. However, many evacuees in and from Fukushima I spoke with did not seem to question the specifics of the decontamination policy. They seemed unconcerned, for example, about the particularities of *how* to decontaminate and *whose* standards of decontamination (the national government, the local government, which ministry, etc.) needed to be met. Instead they questioned the practicality of the timeline of *when* the decontamination of their previous living environment would finally happen. This is because locals, regardless of whether they decide to return to their homes, tend to interpret decontamination and all that it entails as signs of the government's intention to reopen the evacuation zone in the near future (figure 9.4).

Therefore, for the locals, decontamination is not only a technoscientifically sound measure to remedy the fallout, but also a symbolic process, something that must and will happen prior to deciding whether they return to their homes. Put differently, decontamination is a trope of containment that has been providing a sense of structure—a plot—for the evacuees who have been trying to grapple with their present and future uncertainties. The governmentally defined progress of decontamination, measurable by the reduction of the background radiation level, meant there was a semblance of order for the evacuees who are still interested in returning to their homes. Many locals I interviewed voiced that they did not understand the nitty-gritty details of the science of radiation or how decontamination worked, beyond a basic understanding that it was aimed at cleaning up materials that had been contaminated. Yet upon witnessing that decontamination was taking place by accompanying decontamination

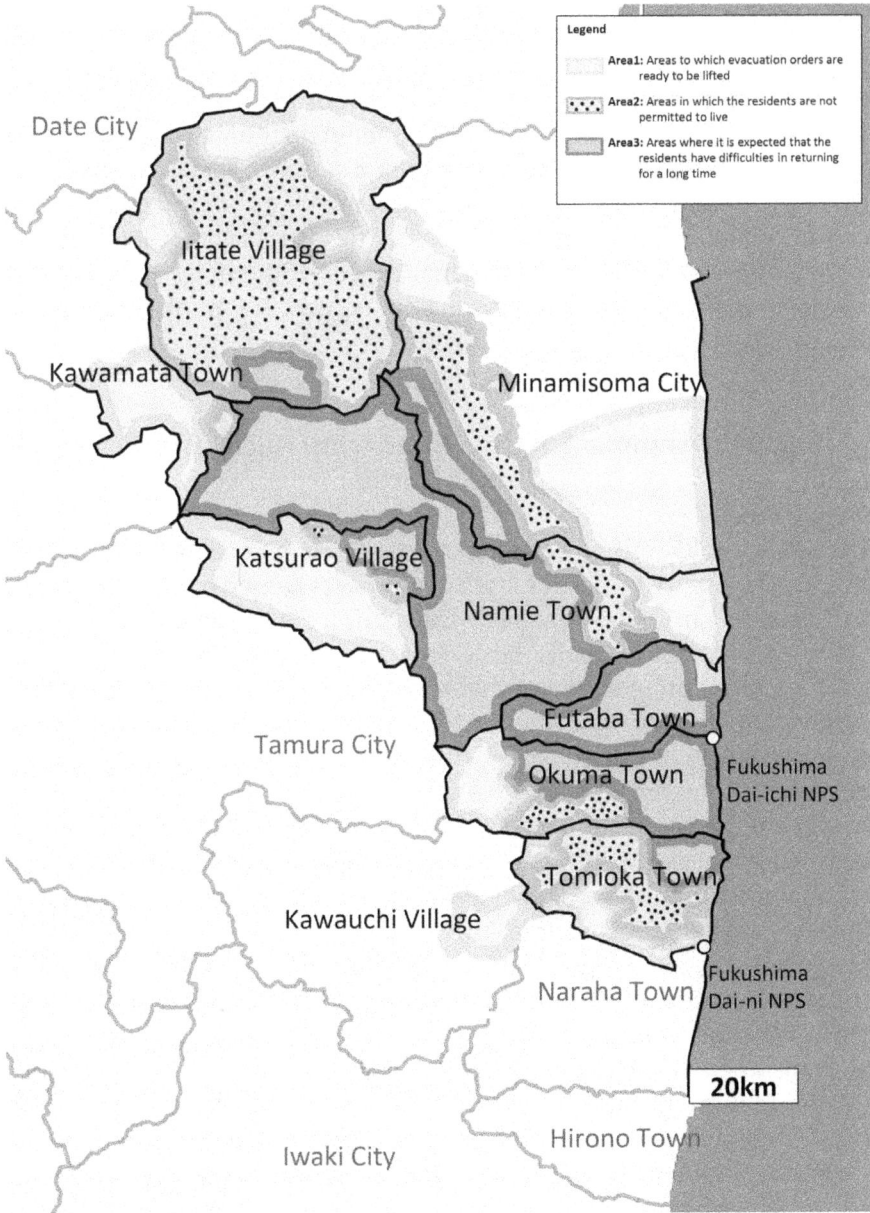

Areas to which evacuation orders have been issued
(September 5, 2015)

Legend

Area1: Areas to which evacuation orders are ready to be lifted

Area2: Areas in which the residents are not permitted to live

Area3: Areas where it is expected that the residents have difficulties in returning for a long time

Date City

Iitate Village

Kawamata Town

Minamisoma City

Katsurao Village

Namie Town

Futaba Town

Tamura City

Okuma Town

Fukushima Dai-ichi NPS

Tomioka Town

Kawauchi Village

Naraha Town

Fukushima Dai-ni NPS

20km

Hirono Town

Iwaki City

Figure 9.4. Areas to which evacuation orders have been issued (5 September 2015) by the Ministry of Economy, Trade and Industry. Public domain. Map's colors modified by the author.

workers' initial inspection, watching the accumulation of frecon baggu bags on their property, and going through the final inspection process to confirm the quality of a work, they had to think realistically about where they would be in the near future.

However, there is one complication. For those who have already decided that they will not return, the same decontamination process is experienced as a threat to the continuation of monetary compensation since the length of compensation is dependent on the length of the legally regimented evacuation order on their pre-disaster residences. Therefore, those people are more likely to be critical of and question decontamination efforts and their effectiveness. All of this is to say that local people understand and define decontamination very differently than experts from within the nuclear science, policy, and practice arenas understand and define it.

Second Dilemma: The Transition from Nuclear Things to Things Nuclear

The second dilemma of decontamination is its emphasis on ends over means. The proposed decontamination process (based on the goal of the "as close to 1 mSv/y as possible" standard) justifies the potential eradication of the local residents' living territories by assuring the public that they are eradicating anything that *might be* contaminated. This process demoralizes the locals who would like to maintain the pre-fallout landscape and ways of living as much as possible. Decontamination, in principle, treats all material objects in a given area as belonging to a relatively homogeneous category named "contaminants," or what the government defines as "pollutants."[8] This is because the endgame of decontamination is the overall reduction of radiation levels via the removal of radioisotopes within a designated area. The removal of supposedly contaminated materials from their original location to another location transforms one conceptual category of nuclear things (invisible pollutants) into things nuclear (visible waste).

Under the Act on Decontamination, the national government defines waste produced through the act of decontamination accordingly: "In this Act, 'waste' means refuse, bulky refuse, burnt residue, sludge, excreta, waste oil, waste acid, waste alkali, carcasses, and *other filthy and unnecessary matter,* which are in a solid or liquid state (excluding soil)" (Article 2.2, emphasis added). Notice that the government's definition of waste does not specify who determines what is or is not "filthy and unnecessary matter." This lack of specification is critical, since it means that the government and decontamination workers have the authority to determine what fits these criteria. The government could argue, contrary to a local person's opinion otherwise, that a given material object is unnecessary by appealing to a shared understanding that the reduction of the background radiation level is important for the safety of the individual and that any contaminated item is unnecessary. In other words, according to the decontamination policy, the more objects in a given space become waste, the better the outcome of decontamination.

Nihei-san, a local resident whose property went through decontamination in the summer of 2014, complained that the decontamination workers removed her beloved tree despite the fact that she marked it as something that needed to be kept.[9] Prior to a decontamination work, residents are given the opportunity to mark which objects on their property are to be saved from the decontamination process. Nihei-san told me that a supervisor of the work apologized for the mistake but tried to convince her that trees in general tend to gather airborne contaminants, and so it was better long term for her and her family's own safety to have it removed. Although in theory the supervisor's explanation made sense to her, in reality the average background radiation of her place prior to decontamination, as she measured, was under 1mSv/y, at the level, which already fulfilled the policy's safety criteria. The reason she was upset with the removal of the particular tree was because her great grandfather planted it; it had been on the family's property for more than seventy years. Even though the removal of the tree, in theory, ensured the safety of her property, Nihei-san lamented to me that "without the tree . . . my house no longer feels like the house I know." In fact, the tree's removal made her feel less certain about returning home, if the opportunity were to arise, to a place that no longer felt familiar.

The logic behind decontamination and its policy and practice, therefore, promotes the ignorance of the significance invested in an individual object and the role the object's meaning might play in enabling the locals to articulate their personal territories and their sense of the self in the world (Goffman 1971; Goodenough 1997). By focusing purely on the scientifically measurable reduction of radiation level in a particular space, made possible by the nonexclusive removal principle, decontamination determines a specific historicity of material objects. That is: Decontamination renders the biography of an object (Kopytoff 1986) prior to the fallout insignificant. Beck (1992) points out that modern risks like radiation exposure are egalitarian. This idea of egalitarianism coincides with how decontamination frames material objects in Fukushima: under the regime of the science of radiation protection and the decontamination policy, all objects in a given space-time were and are contaminated equally from the fallout and thus are subject to removal.

Relatedly, decontamination may disturb a particular organization of individual objects in a bounded space, an organization that might articulate some dimension of social and cultural order (Millar 2002; Sahlins 1976, 180; cf. Bourdieu 1984). The selection of disaster debris to be saved or wasted in times of crisis and transition reflects the contested process of what people do with memory in the present in order to project the continuity of the past life in the future, and from which aspect(s) of the past they are parting (e.g., Gardner 2011; Greenspan 2003; Hastrup 2010, 2011; Morimoto 2012). For Nihei-san, the family tree served as an important marker of her family lineage as well as a mark of her family having made their home in the same location across generations. In losing this symbolic marker that helped her to identify the space as her *place,* she felt less certain about her willingness to go back to her home. In this regard, Nihei-san's experience of loss from decontamination further supports previous findings, which indicate that the selective preservation of material culture is an important dimension of dealing with a sudden change.

When I asked Nihei-san why she asked her property to be decontaminated if it was already considered to be safe according to the science of radiation protection, she replied,

> Getting your place decontaminated is kind of like a ritual. I mean, people feel better to know that their house has been decontaminated even if decontamination cannot, as we all know, do much. At least from cleaning around the house by gathering fallen leaves, weeding the overgrown vegetation, getting rid of old tools and items, and replacing gravel, decontamination makes the place looks nicer and more habitable. I initially feared radiation because it is insensible, but having seen the decontamination work, now I really fear it because decontamination made me imagine that radiation is coming from my personal items. Before I thought radiation came from something foreign.

Her observation is illuminating because it shows how the decontamination process is interpreted as a ritualistic practice. This ritual transforms nuclear things into things nuclear: something unfamiliar (radiation) is made visible through the throwing away of familiar things. This is something neither radiation expertise, nor decontamination policy, nor practice of that policy warrants; decontamination as it is experienced by the local is an experience of estrangement that prevails when decontamination decontextualizes everything in a given area as post-disaster object or contaminant to be removed regardless of people's subjective attachments to or the historicity of a particular object. Local people's experience of the decontamination policy begins when decontamination practice uses some tangible bodies (things nuclear) for the removal of invisible radio-active particles. This experience can be demoralizing in that although only through the removal of things nuclear as nuclear things can people's living environment become once again safe and livable; for the locals such decontaminated space feels less familiar and habitable. Locals experience decontamination as a kind of ritual in a sense that it is an act of cleaning up their past (symbolized by familiar objects) in exchange of achieving government-defined safety. This personal ritual of cleaning up, however, not only demands that the locals experience a sense of loss but also leave a material scar, so to speak. The decontamination policy and its practice do not end with re-moving potential contaminants from one location in the act of modifying one's familiar environment. What is produced in the mixture of nuclear things and things nuclear are decontaminated waste in bags.

Last Dilemma: Living with Nuclear Waste and Its Image

The last dilemma of decontamination is the production of (temporary) waste and its associated negative image. A critical step in the govern-ment-dictated decontamination policy-practice is the assemblage and con-densation of radioisotopes in black bags or frecon baggu .[10] By December 2015 the total number of frecon baggu in Fukushima had reached more than 9.15 million (Mainichi 2015). These bags have accumulated in more than eight hundred temporary outdoor storage facilities across Fukushima prefecture. The number of bags increases each day and as of this writing in June 2016 is still increasing. Moreover, in larger cities like Fukushima City and Koriyama City, decontamination bags have been stored not in a special temporary facility but buried underground, oftentimes some-where in the backyards of individual properties or schools. A Fukushima local newspaper, *Fukushima Minpo*, reported on 23 August 2015 that the total number of frecon baggu stored at the site of decontamination in Fukushima prefecture had increased from 1,809 in May 2012 to 102,093 in March 2015.

The purpose of moving contaminants freely floated in the environment into frecon baggu is to make it possible to transfer the contaminants to elsewhere. The role of frecon baggu is to transform the haphazard spread of radioisotopes in local residents' living environment into a controllable, bounded, and movable entity (figure 9.5). What decontamination intends through the modification of radioisotopes from contaminants to bounded waste therefore is to artificially give a different configuration of space-time (Munn 1977, 1986) to nuclear things.

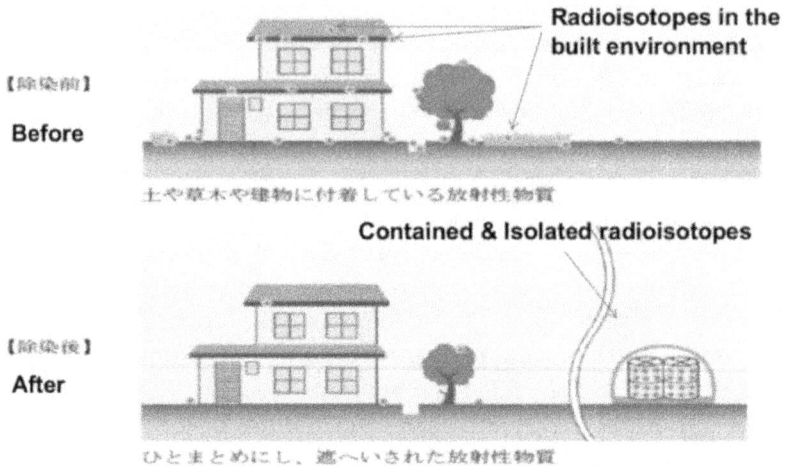

Radioisotopes in the built environment

【除染前】

Before

土や草木や建物に付着している放射性物質

Contained & Isolated radioisotopes

【除染後】

After

ひとまとめにし、遮へいされた放射性物質

Figure 9.5. A rendering of ideal environmental decontamination showing both before (top) and after (bottom) conditions in an affected site (Ministry of the Environment, 2012, 11). Public domain.

Inside each bag there might be a combination of soil, leaves, weeds, muck, gravel, and any other material substances that decontamination workers decide is waste and that therefore is the object of isolation by virtue of it being identified as a source of contamination. Importantly, enclosing contaminants in a container homogenizes what goes inside the bag as decontaminated waste since containers, according to Klose (2015), "decompose [material objects] into at least two parts, of which one part means nothing, apart from containing something . . . and the other part bears a plethora of meanings but cannot contain them for long" (341). For example, Nihei-san's beloved family tree, once inside a bag, becomes just another piece of decontaminated waste.

Importantly, frecon baggu is a generic container used for wide variety of purposes. They have been used to construct temporary dams, to raise the ground, and for the disposal of waste at construction or demolition sites. This is to say that frecon baggu are neither designed to shield radiation or to assert any meaning of its own, nor are they designed to last longer than a few years. Frecon baggu as the intermediary holding space of radio-isotopes/contaminants discharged by the disaster are, by definition, both temporary and inconspicuous. Not unlike Mary Douglas's conceptualization for understanding that objects come to be counted as dirt or dirty, the frecon baggu are a means of controlling and ordering disorder (2002 [1966], 2). Nonetheless, frecon baggu are particularly interesting not only because they are perceived to be socially and culturally indefinable and dangerous (96–97), but also because they materially represent a chasm between technoscientific knowledge, policy, and practice regarding the reduction of radiation on the one hand and local people's experience of loss on the other. Yet frecon baggu also help to transform contaminants into nuclear waste stored in an isolated facility. Once again, frecon baggu are structurally and procedurally not meant to remain present in one location for a long time. The visibility and fragility of frecon baggu always anticipates their eventual removal, though where they are going is still under discussion; as a result, more than 9.15 million bags are still present in Fukushima prefecture. The persistent and looming presence of frecon baggu casts a shadow on the government-led recovery of Fukushima and Japan from the nuclear disaster.

Here I want to emphasize the short-term implications of the decontaminated waste that are pertinent to the everyday lives of the local residents. Although in theory decontamination lessens the presence of nuclear things from the living environment and thus lessens the risk of exposure, the removal of contaminants through decontamination has been producing tangible objects—frecon baggu filled with radioactive materials—that enable the locals and outsiders alike to visualize for the first time the otherwise invisible presence of nuclear things from the disaster in the living environment.

As a result, there is an emerging association between the image of frecon baggu and nuclear contamination in Fukushima prefecture. For instance, many people outside Fukushima have asked me to share pictures of frecon baggu in Fukushima with them so that they can use these images in presentations or workshops about Fukushima. Moreover, a group of artists put together an exhibit, entitled Fukushima Art in 2016, using frecon baggu as a representation of Fukushima. The use of frecon baggu as a way of thinking about energy issues, nature, and culture suggest frecon baggu's capacity to evoke public imagination of the nuclear disaster, although many of the locals with whom I spoke find these artworks to be bewilderingly tasteless.[11] The association between frecon baggu and contaminated Fukushima is an unexpected outcome of the knowledge, policy, and its practice; this association is a potent image that has emerged only after the nuclear disaster in 2011.

Ironically, it appears as if decontamination has made the fact of the nuclear disaster more sensible than ever before. Nevertheless, and even though the bags seem to have come to stand in for or be treated as a sign of the nuclear disaster, not everyone interprets the bags in the same way. While the general public seems to interpret the image and presence of frecon baggu as representing the real and imagined danger in Fukushima at distance, some local residents interpret the same bags as representing their hope for returning to the alienated homes soon.

Regardless of how different parties interpret the presence of frecon baggu, in confronting the emerging image of the nuclear disaster, the problem is no longer the sociopolitical concern over the invisibility of the nuclear disaster as was the case in Chernobyl (Kuchinskaya 2014; Petryna 2002), or the technoscientific concern over how to alleviate contaminated environment, but the material-semiotic concern over how decontamination makes certain things visible, while at the same time making other aspects invisible. Refocusing the locus of analysis on the Gap that has appeared as a result of knowledge, policy, and practice of decontamination—namely, the visualization of nuclear things as a result of its containment—it becomes clear that decontamination demands radical and perhaps irreversible changes to locals' familiar place and landscape and challenges the continuity of material objects in which they might anchor their remembrance of their pre-fallout lives. Moreover, the majority of people who live outside of Fukushima interpret the locals' experience of losses, which increasingly become concealed in frecon baggu, to be a sign of danger.

The Gap therefore helps to illuminate the fact that the government policy selectively imagines nuclear things; nuclear things can be controlled and isolated at distance far from most people. The decontamination policy has demonstrated this in practice by making what was invisible in Fukushima visible as nuclear waste. In this sense, decontamination has been effective in making the contamination of Fukushima visible and localizing the extent of contamination in Fukushima, thereby allowing nonlocals to imagine their place to be in less risk of radiation contamination and exposure. This collective imagination of nuclear things with the asymmetrical distribution of bags of radioactive waste has been undermining the local people, who must live with the ongoing sense of uncertainty regarding whether the nuclear disaster is really over, and if the state of things will worsen at any given minute. Decontamination policy only promises the fixed spatiotemporal removal of nuclear things measured by technoscientific calculations. Decontamination does not promise to remove its by-product—the image of contamination that the bags of radioactive waste has made visible to the general public and what those bags should mean for Fukushima. Nonetheless, decontamination is a spectacle through which Japanese society at large might reimagine the sense of nuclear containment, as the risk of radiation contamination is redistributed to Fukushima.

Conclusions

In this chapter I have argued that decontamination is effective in sustaining the collective imagination surrounding containment—a kind of false consciousness, if you will—and the persistent belief that nuclear things can be and will be contained and isolated at a remote distance far from the majority. The belief in the efficacy of decontamination and perhaps even the work and labor involved in decontamination itself might be an imaginative (performative) act, since in reality the work that falls under the umbrella of government-operated decontamination is a mere act of deferral that only moves radioisotopes from one location to another location.

In the case of the effort to decontaminate Fukushima, policy decisions seem to be enacted as a way to reconstruct a spatial and ideological division between the land of nuclear containment—"waste-wilderness" (Galison 2011)—and the land unwieldy of nuclear contamination. The division between these binary categories of space is significant and even exaggerated through the decontamination work. The nuclear disaster in 2011 did cause a catastrophe, and frecon baggu in Fukushima represent tangible evidence of the catastrophe. The visible presence of nuclear things in Fukushima and only in Fukushima, I argue, helps nonlocals to mitigate the flood of uncertainties about their own safety regarding the potential, though invisible, threat of radiation in the air.[12] Decontamination has enabled this selective visualization of the risk of contamination in the society that has been localized far from the urban centers in Fukushima.

If we shift our focus on the experience of locals who live around or who have once lived near the site of the disaster, the problem of decontamination appears to stem from the government's limited attention to the reduction of the background radiation. This technoscientifically informed policy creates a difficult situation whereby the evacuees from a decontaminated area might be forced to return to their homes only after their familiar landscape has been transformed into a temporary waste-wilderness of frecon baggu. In other words, when nuclear things were invisible but only technoscientifically observable, the evacuees were not allowed to live, but when nuclear things are contained in theory but visible as nuclear waste, the evacuees must return. With the continued presence of so-called temporary frecon baggu throughout Fukushima, the government demands that locals confront a future in which they must live with the visible presence of risks in proximity, despite the fact that the government has framed the bags' radioactivity and their impact on surroundings to be scientifically safe.

In closing, it is critical to reiterate that locals' experience of decontamination as a symbolic process or a ritual is precisely what resides in Gap of decontamination, which was formulated by the knowledge-policy-practice nexus. Although decontamination was intended to achieve the goal of the reduction of accumulative radiation exposure to "as close to 1 mSv/y as possible," its surprising consequence was the production of the particular image of contamination in Fukushima—namely, tangible bags of nuclear waste, many of which contained local people's experiences of loss. The task of anthropology is to locate the Gap when expertise and policy cannot fully explain the on-the-ground experiences it produces once implemented in practice.

Ryo Morimoto is assistant professor of anthropology at Princeton University. Since 2013 he has been studying the aftermath of the nuclear disaster in coastal Fukushima, Japan. He has published numerous articles on Japan's triple disasters, including "Message without Coda: On Rhetoric of the Photographic Records" in *Signs and Society* and "Shaking Grounds, Unearthing Palimpsests: Semiotic Anthropology of Disaster" in *Semiotica*. Before joining Princeton, he was a postdoctoral fellow at the Reischauer Institute of Japan Studies at Harvard University, where he also served as the project manager of the Japan Disasters Digital Archive (jdarchive.org).

Notes

1. I use the expression "nuclear things" after Gabriel Hecht (2007, 2010) to signify those materials, whether natural or artificial, that emit ionizing radiation. In the case of nuclear things in Fukushima, the most common sources of radiation are Cesium-134 and 137 (Imanaka, Hayashi, and Endo 2015).
2. A tentative English translation of the Act is available at Government of Japan (2012). I use this English version for the chapter. For more information on radioactive materials discharged by the accident, see the Ministry of the Environment website (Government of Japan n.d.).
3. The Sievert (Sv) is a unit of ionizing radiation dose, representing the stochastic health risk of cancer induction and molecular damage by the deposit of a joule of radiation energy per kilogram (kg) of human tissue where 1Sv (or 100 rem) = 1,000mSv = 100,000μSv. For example, radiation of 0.12μSv/h means a person would be irradiated with 0.12μSv in an hour. Thus, in two hours the total, annual cumulative dose becomes 0.24μSv/y.
4. The former Tokyo Electric Power Company changed its name to Tokyo Electric Power Company Holdings Inc. (TEPCO), and also changed its organizational structure in April 2016.

5. My conversations with the decontamination worker cited here, and with other informants in this chapter, happened between 2013 and 2015 in Minamisoma city, Fukushima. With regard to government's preference for certain construction companies, in Minamisoma between July 2014 and June 2015 six companies dominated the decontamination work within the Special Decontamination Area with a project budget of more than $566 million.

6. See Samuels (2013) and Aldrich (2014) for a discussion of the resiliency of many pre-disaster policies such as national security, energy, and local governance in the post-3.11 Japan.

7. The International Atomic Energy Agency (IAEA) has suggested against adhering strictly to the 1 mSv/y standard and instead encourages spending time and money on building consensus across local communities (IAEA 2013). Scientists are divided on their opinions regarding the health effects of low-dose radiation, and the question is highly politicized. For examples see Nakagawa (2011) for the history of radiation exposure and Kodama, Shimizu, and Noguchi (2014) as well as Hosoda et al. (2016) for technoscientific debates about Fukushima.

8. The language of pollutant reflects the government's view of the nuclear disaster: it is an environmental disaster like the one in Minamata (e.g., George 2001) but not a technological disaster. When I talked to the former minister of reconstruction, Takumi Nemoto in May 2015, he implied that ministries other than the Ministry of the Environment should have been responsible for the cleanup. Moreover, he mentioned that the Ministry of the Environment did not know much at all about nuclear science and had to learn more about it after it was assigned. Thus, it took them some time to take action.

9. I use a pseudonym to protect my informant's identity. I have translated all my commentaries.

10. There are also blue and beige colored *frecon baggu* used to contain the decontaminated waste as well. For this chapter, I focus on the black bags, since these are most prevalent type in coastal Fukushima.

11. The website link of Fukushima Art (in Japanese) curated by Mitsuhiro Okamoto is http://www.kunstarzt.com/VvK/180FukushimaBijyutsu/FukushimaBijyutsu.htm.

12. There are other prefectures in Japan that have been going through decontamination but the scale and extent of decontamination in Fukushima is incomparable.

References

Aldrich, P. Daniel. 2014. "Postcrisis Japanese Nuclear Policy: From Top-down Directives to Bottom-up Activism." In *Japan at Nature's Edge: The Environmental Context of a Global Power,* edited by Ian Miller, Julia Thomas, and Brett Walker, 280–92. Honolulu: University of Hawaii Press.

Anders, Günther. 1962. "Theses for the Atomic Age." *Massachusetts Review* 3, no. 3, 106–25.

Beck, Ulrich. 1992. *Risk Society: Towards a New Modernity.* London: Sage.

Bourdieu, Pierre. 1984. *Distinction: A Social Critique of the Judgment of Taste,* trans. Richard Nice. Cambridge: Cambridge University Press.

Douglas, Mary. 2002[1966]. *Purity and Danger: An Analysis of the Concepts of Pollution and Taboo.* London: Routledge.

Erikson 1994. *A New Species of Trouble: The Human Experience of Modern Disasters.* New York: W. W. Norton & Company

Galison, Peter. 2011. "Waste-Wilderness: A Conversation with Peter L. Galison." Friends of the Pleistocene. Retrieved 10 July 2015 from https://fopnews.wordpress.com/2011/03/31/galison/.

Gardner, B. James. 2011. "September 11: Museums, Spontaneous Memorials, and History." In *Grassroots Memorials: The Politics of Memorializing Traumatic Death,* edited by Peter Jan Margry and Cristina Sánchez-Carretero, 285–303. New York: Berghahn Books.

George, S. Timothy. 2001. *Minamata: Pollution and the Struggle for Democracy in Postwar Japan.* Cambridge. Harvard University Press.

Goffman, Erving. 1971. "The Territories of the Self." In *Relations in Public: Microstudies of the Public Order,* 28–61. New York: Basic Books.

Goodenough, H. Ward. 1997. "Moral Outrage: Territoriality in Human Guise." Zygon 32, no. 1, 5–27.

Government of Japan. 2012. *Act on Special Measures Concerning the Handling of Environment Pollution by Radioactive Materials Discharged by the NPS Accident Associated with the Tohoku District—Off the Pacific Ocean Earthquake That Occurred on March 11, 2011.* Retrieved from http://josen.env.go.jp/en/framework/pdf/special_act.pdf?20130118.

Greenspan, L. Elizabeth. 2003. "Spontaneous Memorials, Museums, and Public History: Memorialization of September 11, 2001 at the Pentagon." *Public Historian* 25, no. 2, 129–32.

Hastrup, Frida. 2010. "Materializations of Disaster: Recovering Lost Plots in a Tsunami-Affected Village in South India." In *An Anthropology of Absence: Materializations of Transcendence and Loss,* edited by Mikkel Bille, Frida Hastrup, and Tim Flohr Sorensen, 99–112. New York: Springer.

———. 2011. *Weathering the World: Recovery in the Wake of the Tsunami in a Tamil Fishing Village.* New York: Berghahn Books.

Hetcht, Gabriel. 2007. "A Cosmogram for Nuclear Things." *Isis* 98, 100–8.

———. 2010. "The Power of Nuclear Things." *Technology and Culture* 51, 1–30.

Hosoda, Masahiro, Shinji Tokonami, Yasutaka Omori, Testsuo Ishikawa, and Kazuki Iwaoka. 2016. "A Comparison of the Dose from Natural Radionuclides and Artificial Radionuclides after the Fukushima Nuclear Accident." *Journal of Radiation Research* 1–9. Retrieved 1 August 2019 from https://www.ncbi.nlm.nih.gov/pmc/articles/PMC4973637/.

Imanaka, Tetsuji, Gohei Hayashi, and Satoru Endo. 2015. "Comparison of the Accident Process, Radioactivity Release and Ground Contamination between Chernobyl and Fukushima-1." *Journal of Radiation Research* 56, S1, i56–i61.

International Atomic Energy Agency. 2013. *Final Report: The Follow-up IAEA International Mission on Remediation of Large Contaminated Areas Off-Site the Fukushima Daiichi Nuclear Power Plant.* Tokyo and Fukushima Prefecture, Japan, 14–21 October 2013. Retrieved 1 August 2019 from https://www.iaea.org/sites/default/files/final_report230114.pdf.

———. 2015. *The Fukushima Daiichi Accident.* Vienna: IAEA Books. Retrieved 10 July 2015 from https://www-pub.iaea.org/MTCD/Publications/PDF/Pub1710-ReportBy TheDG-Web.pdf.

Klein, Naomi. 2008. *The Shock Doctrine: The Rise of Disaster Capitalism.* New York: Picador.

Klose, Alexander. 2015. *The Container Principle: How a Box Changes the Way We Think*, trans. Charles Marcrum III. Cambridge: The MIT Press.

Kodama, Kazuya, Shuji Shimizu, and Kunikazu Noguchi, eds. 2014. *Science and Social Studies of Radiation Exposure: The Truth of Fukushima in the Fourth Year*. [Houshasen Hibaku no Rika-Shakai: Yonenme no 'Fukushima no Shinjitsu']. Kyoto, Japan: Kamogawa Shuppan.

Kogure, Keiji. 2013. *Decontamination and Waste Management*. [Houshanou Jyosen to Haikibutsu Shori]. Tokyo: Gihoudou Shuppan.

Kopytoff, Igor. 1986. "The Cultural Biography of Things: Commoditization as Process." In *The Social Life of Things: Commodities in Cultural Perspective*, edited by Arjun Appadurai, 64–91. Cambridge: Cambridge University Press.

Kuchinskaya, Olga. 2014. *The Politics of Invisibility: Public Knowledge about Radiation Health Effects after Chernobyl*. Cambridge: The MIT Press.

Mainichi, Shimbun. 2015. "Decontaminated Waste Piling Up." [Tsumiagaru Jyosenhaikibutsu]. (in Japanese). All translations by author. Retrieved 30 May 2016 from http://mainichi.jp/articles/20151210/k00/00e/040/160000c.

Millar, Daniel. 2002. "Artefacts and the Meaning of Things." In *Companion Encyclopedia of Anthropology: Humanity, Culture and Social Life*, edited by Tim Ingold, 397–419. London: Routledge.

Ministry of Economy, Trade, and Industry. 2015. "Areas to which Evacuation Orders Have Been Issued" (5 September 2015). Retrieved 18 April 2019 from https://www.meti.go.jp/english/earthquake/nuclear/roadmap/pdf/150905MapOfAreas.pdf.

Ministry of the Environment. 2012. "Decontamination Guideline." Ministry of the Environment, Government of Japan, Tokyo. Retrieved from http://josen.env.go.jp/kousyuukai/pdf/rule120416.pdf (in Japanese). All translations by author.

———. 2013a. "Progress on Off-site Cleanup Efforts in Japan." Ministry of the Environment, Government of Japan, Tokyo. Retrieved from http://www.icrp.org/docs/Tsutomu%20Sato%20Progress%20on%20Off-site%20Cleanup%20Efforts%20in%20Japan.pdf.

———. 2013b. "What We Have Tested before Decontamination—We Are Decontaminating Based on the Test Results." Ministry of the Environment, Government of Japan, Tokyo. Retrieved 18 April 2019 from https://web.archive.org/web/20160429191254/https://josen.env.go.jp/material/pdf/jissyou_jikkenn.pdf.

———. n.d. "Environmental Remediation." Ministry of the Environment, Government of Japan, Tokyo. Retrieved from http://josen.env.go.jp/en/.

Morimoto, Ryo. 2012. "Shaking Grounds, Unearthing Palimpsests: Semiotic Anthropology of Disaster." *Semiotica* 192, 263–74.

Munn, D. Nancy. 1977. "Spatiotemporal Transformations of Gawa Canoes." *Journal de la Société de Océanistes* 54–55(33), 39–55.

Nakagawa, Yasuo. 2011. *The History of Radiation Exposure: From the U.S. Development of A-bombs to The Fukushima Nuclear Accident*. [Houshasen Hibaku no Rekishi: America Genbaku Kaihatsu kara Fukushima Genpatsu Jiko made]. Tokyo: Akashi Shoten.

Nippon National Kyōkai (NHK). 2016. "An Estimated Decontamination-related Expenses by 2016 is over 30 billion Dollars." [Jyosen Nadono Hiyou Kotoshidakede 3Choen Ijyouno Shisan]. (in Japanese). All translations by author. Retrieved 30 May 2016 from http://www3.nhk.or.jp/news/genpatsufukushima/20160406/1717_3chou_yen.html.

Oliver-Smith, Anthony. 2004. "Theorizing Vulnerability in a Globalized World: A Political Ecological Perspective." In *Mapping Vulnerability: Disasters, Development and People*, edited by G. Bankoff, G. Frerks, and D. Hillhorst, 10–24. London: Earthscan.

Petryna, Andrea. 2002. *Life Exposed: Biological Citizens after Chernobyl*. Princeton, NJ: Princeton University Press.

Rajan, S. Ravi. 2002. "Missing Expertise, Categorical Politics and Chronic Disasters: The Case of Bhopal." In *Catastrophe and Culture: The Anthropology of Disaster*, edited by S. Hoffman and A. Oliver-Smith, 237–59. Santa Fe, NM: School of American Research Press.

Sahlins, Marshall. 1976. *Culture and Practical Reason*. Chicago: University of Chicago Press.

Samuels, J. Richard. 2013. *3.11: Disaster and Change in Japan*. Ithaca, NY: Cornell University Press.

Wynne, Brian. 1996. "May the Sheep Safely Graze? A Reflexive View of the Expert-Lay Knowledge Divide." In *Risk, Environment and Modernity: Towards a New Ecology*, edited by S. Lash, B. Szerszynski and B. Wynne, 44–83. London: Sage.

CHAPTER 10

"Haitians Need to Be Patient"

Notes on Policy Advocacy in Washington Following Haiti's Earthquake

MARK SCHULLER

Introduction

The earthquake that struck Haiti on 12 January 2010 is one of the most iconic disasters in recent history. The exact death toll is a matter of debate—estimates range from 65,000 to 316,000—but there is no question that it was far deadlier than a quake five hundred times more powerful in Chile weeks later, and certainly a seismic event of similar magnitude and similar proximity to a metropolitan area in New Zealand September 2010, in which only one person died.

Haiti's earthquake was also a mass media event, creating what could be called a global imagined community (Anderson 1987), shaping a collective consciousness and forging a global citizen. The iconic disaster narrative led to one of the most generous foreign aid responses to date: $13 billion in pledges from governments and international organizations and just over $3 billion contributed by individuals. As I have argued elsewhere (Schuller 2016), the narrative also normalized and justified foreign control of the response. Social movements and nongovernmental organizations (NGOs) in Haiti organized a challenge not only to this foreign-led aid effort—what some activists have called a humanitarian occupation—but also to take the opportunity to denounce foreign-led neoliberal economic policies that augmented Haiti's vulnerability to disasters in the first place. At the same time, a group of NGOs, most of them in Washington, organized an advocacy effort as the Haiti Advocacy Working Group (HAWG).

This chapter identifies the gaps in post-disaster advocacy: how principled advocacy in Washington, DC, works, how it does not work, why, and what it means for communities impacted by disasters, particularly foreign communities. While my analysis is necessarily a particular and singular perspective, a partial truth (Crapanzano 1986), I am writing in the spirit of opening the discussion to invoke self-analysis and critique in hopes that the gaps of NGOs and advocacy I identify might thus be addressed. HAWG eventually succeeded in convincing Congress to pass the Assessing Progress in Haiti Act, which was signed into law by President Barack Obama on 8 August 2014. The legislation offers useful tools to call the U.S. government, particularly the U.S. Agency for International Development (USAID), to task for the lackluster outcomes of the highly visible relief effort. Despite HAWG's victory, however, implementation has been uneven at best. Overall, this chapter argues that the advocacy effort was limited by a range of factors, including the political process in Washington marked by increasing partisanship, the reward structure of documenting process over results, the lack of a clear constituency, and structural inequalities and racism that restrict who can speak for Haiti. These political processes limit the perspectives of Haitian interlocutors, in effect speaking for the entire country.

Haiti as Political Casualty in Washington

This chapter focuses on one international relationship, privileging the United States. The reason for this focus is clear: not only am I a U.S. citizen, but also the role played by the United States as regional hegemon is unmatched. Traditionally the United States is by far the largest single donor to Haiti. And diplomatically and geopolitically, United States influence is felt heavily in Haiti and throughout the Caribbean region.

The Haitian earthquake struck not quite a year into the presidency of Barack Obama, the first African American president, who was elected using slogans of hope, change, and optimism. The most enduring slogan was borrowed from the United Farmworkers' *sí se puede*, translated into English as, "Yes we can." Commentators across the Caribbean and Latin America saw particular justification or opportunity in Obama's election to make claims for rights at home, drawing on shared histories of the slave trade (e.g., Anderson 2015; Maddox 2015; Main 2015; Page 2015). Nowhere was this Diasporic connection more apparent than in Haiti, the world's first free black modern nation-state to arise, and the only one born of a slave revolt. Obama himself drew on what Kamari Maxine Clarke (2010) termed "humanitarian diasporas," particular nodes of connection across the "Black Atlantic" (Gilroy 1993) based on solidarity activism. Published on 14 January online and 21 January in print, Obama authored the cover story for *Newsweek* entitled "Why Haiti Matters" (Obama 2010). He reported, "I have instructed the leaders of all agencies to make our response a top priority across the federal government." Explaining those actions, the first reason given cited the lives of U.S. citizens and the third the ties between the United States to its close neighbor. Obama also explained that the United States acted "for the sake of the Haitian people who have been stricken with a tragic history, even as they have shown great resilience." Concluding this section of the article, Obama explained, "But above all, we act for a very simple reason: in times of tragedy, the United States of America steps forward and helps. That is who we are. That is what we do."

Haiti's earthquake allowed the president to define what might be called the Obama Doctrine—a direct challenge to his predecessor: "When we show not just our power, but also our compassion, the world looks to us with a mixture of awe and admiration." His presidency already an assertion of a narrative of racial equality and progress, Obama twice invoked U.S. leadership in its role in protecting common humanity. In his concluding paragraph, Obama noted, "But it is also in these moments, when we are brought face to face with our own fragility, that we rediscover our common humanity. We look into the eyes of another and see ourselves." *Newsweek*'s decision to print the article might have had the intention of saving the publication's print version, which was already losing money and became defunct two years later, but the visibility of providing a U.S. president the opportunity to write had the discursive impact of underscoring the significance of the president's directive and hearkened back to President Kennedy's American Century and Franklin Delano Roosevelt's response to Pearl Harbor and his fireside chats. These Democratic presidents perfected the role of the bully pulpit—sharing their perspective with the public through the media.

Whether avenging George W. Bush, who had not been given—or did not take—a similar space to publish his message about 9/11, as an open challenge to the racial uplift message, the role of the government, or a bristling at the assertion of an idea of common humanity, Republicans correctly identified Haiti, and the *Newsweek* article, as a target for the U.S. administration, as it had been for Clinton in 1994. Obama's message—or the United States voting public's fatigue with the Bush Doctrine and the seemingly permanent War on Terror—carried fifty-seven Democratic senators (and two independents, including 2016 presidential hopeful Bernie Sanders), one vote shy of having a filibuster-proof majority in this more cautious, and structurally powerful house of Congress. The year 2010 was marked by increasing partisan rancor in the United States as an Astroturf social movement bankrolled by the billionaire Koch Brothers, the Tea Party, aimed to take Congress back. The 2010 elections brought the U.S. House of Representatives back to Republican control. With only a third of Senators running for election every two years, the U.S. Senate was to switch hands following the 2014 midterm election. From 1–16 October

2013, congressional Republicans led a successful shut down of the federal government in a dispute over the Patient Protection and Affordable Care Act, usually shorthanded as "Obamacare," which was signed into law that fateful year of 2010.

This was the context surrounding the NGO advocacy effort about Haiti. Obama's article put Haiti under the crosshairs. It also exposed what was to become a central theme in the partisan brinksmanship, precisely defining the role of the executive and that of the legislative branch in the U.S. model of divided government. In addition to Obamacare, other highly visible issues that exposed this constitutional conflict were immigration reform and closing the military detention center in Guantanamo Bay, Cuba. Since Secretary of State Condoleezza Rice's 2006 restructuring of the State Department, USAID had become even more closely interwoven institutionally and financially with the rest of the State Department. But Congress still has budgeting authority over federal agencies including USAID. And while the State Department, and the secretary of state, have the power to negotiate treaties, Congress is vested with the authority to ratify them. A particularly embarrassing moment in U.S. diplomatic history resulted from this fact: the United States was never a member of the League of Nations, brainchild of U.S. President Woodrow Wilson following the end of World War I.

At the 31 March 2010 United Nations Donors Conference on Haiti, Secretary of State Hillary Clinton pledged $1.15 billion in aid to the country, upping the ante after Venezuelan president Hugo Chavez' pledge of 1.18 billion U.S. dollars. Like any such commitment, and certainly because it involved appropriations, Congress had to ratify Clinton's pledge. On 29 July 2010, almost seven months after the earthquake and four months after the pledge, Congress passed the Supplemental Appropriations Act of 2010. Actually, the $1.15 billion statistic is a bit misleading, since $588 million, or 53 percent, went to the U.S. government, mostly to reimburse the military and other agencies involved in Haiti (Ramachandran and Walz 2012, 7). The U.S. Department of Defense received $483 million (12). A significant portion of the planned aid, $339 million, was earmarked to go to U.S.-based firms, NGOs, and for-profit firms. These for-profit firms are often called Beltway Bandits; former State Department official Robert E. Maguire (2014) argued that these firms wielded great influence in Washington policymaking. One such Beltway Bandit, Chemonics, received several contracts to work in Haiti, including no-bid contracts, totaling $175 million by October 2012, despite a bad performance review (Center for Economic and Policy Research [CEPR] 2012). HAWG pressed for the immediate passage of aid to Haiti; it was an early point of unity in an otherwise diverse coalition. There was a lot of activity in Congress concerning Haiti at the time: 102 bills were introduced during the 111th Congress (2009–10), and fifteen became law.

Despite the appropriation being made in Congress, there were several steps that had to be accomplished before funds were released and sent to Haiti, or more precisely to the agency acting on behalf of the Haiti relief and reconstruction effort (Ramachandran and Walz 2012). Senator Tom Coburn (R-OK) set a condition on the release of the funds of a successful election in Haiti, setting the stage for a high-stakes game of political brinkmanship in which Secretary of State Clinton played a heavy role (Peck 2013; Seitenfus 2015).

Haiti's Representatives in Washington

My interviews with USAID and other development officials since 2004 outlined the need for a constituency in the United States—an organized group of concerned citizens, or a voting bloc—to support their efforts. In Washington, particularly after the 2010 United States Supreme Court *Citizens United* ruling removed barriers to corporate financing of elections, there developed a perception of pay to play, with high-paid lobbyists wielding enormous power to shape the country's legislative agenda. Before the earthquake there were only two consistent constituencies for Haiti issues: first was the Haitian Diaspora, that had flexed its muscles in the 1980s to protest the Centers for Disease Control (since renamed Centers for Disease Control and Prevention) naming Haitians as one of the four high-risk groups for HIV/AIDS, and again in 1991 to protest the coup against Haiti's first democratically elected president. Second was the Congressional Black Caucus (CBC). The historical ties between the U.S. African American community and Haiti have been strong since Haiti's independence became a beacon of black freedom in 1804 (see, e.g., Horne 2015; Plummer 1988). Intellectuals in Haiti were in dialogue with leaders of the Harlem Renaissance, and black Civil Rights organizations helped end the nineteen-year U.S. occupation of Haiti in 1934 at the same time as others, such as the Tuskegee Institute, collaborated with the occupiers in their *mission civilisatrice* (civilizing mission; Polyné 2010). Only after the generation of the 1950s and 1960s Civil Rights struggle ended the Jim Crow voting laws was the way paved for a critical mass of Black legislators and the establishment of the CBC.

Founded in 1971, the CBC has forty-four voting U.S. Representatives, a representative of the U.S. Virgin Islands, and as of 2016, a U.S. Senator. In 2016 when the chapter was written, all but one member belonged to the Democratic Party. Since its inception, the CBC has led U.S. legislation on international and diasporic issues, such as the effort to end apartheid in South Africa, attempts to prevent genocide in Rwanda, cutting off aid to Jean-Claude Duvalier, and restoring Haiti's constitutional democracy in 1994. The members have an annual weeklong conference in February during Black History Month that brings together activists, scholars, and legislators to debate the black legislative agenda. Given the CBC's traditional leadership role, other legislators in the U.S. Congress defer to the group on Haiti issues.

Haiti Advocacy after the Earthquake

Although HAWG was formed shortly after the earthquake, plans for it were initially hatched in 2009 when it looked as though there would be significant investment in Haiti. In 2009 noted British economist Paul Collier published a report (Collier 2009) outlining solutions to Haiti's problems: tourism, offshore apparel factories, and high-value agricultural exports, especially mangos (see Chancy 2013; Schuller 2009). Key to the solutions was a public-private partnership that has come to define the Clinton Global Initiative. Indeed, the Collier Report suggested that none other than former president Bill Clinton get involved as special envoy. There was also a showdown in 2009 for a minimum wage increase, from under $2 to $5 per day, in which Clinton and Secretary of State Hillary Rodham Clinton suggested that Haiti not upset its so-called comparative advantage of low wages. During the mobilization, United Nations (UN) troops shot tear gas into demonstrations of people demanding an increase in Haiti's minimum wage.

HAWG is a voluntary organization comprising NGOs. At its peak, HAWG counted more than forty member organizations, ranging from single-person, single-issue voluntary associations to large international NGOs (INGOs) that have dozens of staff in Washington and for whom Haiti is one of many countries of focus. Since 2011, when it launched a formal website, HAWG has listed a membership of thirty organizations. No formal membership structure exists, and no dues, so the assemblage relies on member organizations' willingness to donate personnel time to staff the organization. Immediately following the earthquake, HAWG held a weekly conference call with all involved, which gradually reduced to a monthly call. As needed, committees formed to address particular topics, such as violence against women, housing, agriculture, aid accountability, and elections. In recent years, climate change and mining have been added as topical foci.

Since HAWG has no dedicated, paid staff, involvement and activity within these subgroups was dependent on individual NGO staff taking the lead on organizing conference calls, writing white papers, and planning for advocacy efforts, including "Days on the Hill" or visits from Haitian activists, particularly visiting delegations. Given the ad hoc nature of leadership, activity ebbed and flowed as particular staff moved in and out of involvement with the issue or the NGO. For obvious reasons, organizing travel from Haiti was the most time- and resource-intensive task, and as such it required a member NGO committing resources.

One of the primary activities the HAWG engaged in was organizing briefings in the U.S. legislative branch to keep Haiti visible in the eyes of Washington-based policymakers. HAWG made special efforts whenever a member organization was able to pay for colleagues in Haiti—from what they called civil society organizations—to present at briefings where congressional staffers attended, and to a lesser extent at formal committee hearings, which comprise part of the official congressional record and have members of Congress present. Member organizations of HAWG had pre-established relationships with certain members of Congress. A natural ally was the CBC. Especially following the 2010 midterm elections, and later the 2014 elections, when the Democrats lost one then both houses of Congress, HAWG's ability to pass legislation was diluted. As such, a gap widened between those who were selected to represent the most affected, Haitian beneficiaries, and the decision makers able to address Haitian people's concerns. Given this political gap the gap between knowledge, policy, and practice also widened.

Citizen Lobbying: An Ethnographic Encounter with NGO Practice

As an individual, I am not and cannot be a member of HAWG, because the prerequisite for membership is NGO or nonprofit status. Several groups are little more than a single passionate individual volunteering personal time. But I decided not to create an NGO as my platform. My work is as a scholar conducting research. Nonetheless, I was invited to Washington six times following the earthquake, usually by HAWG or in coordination with HAWG. The following discussion offers an analysis of one visit in the fall of 2012, wherein my role was simply as translator and accompanier for a colleague of mine, whom I will call Kato, who was one of the founders of a social movement for housing, LoJistis. Since my role was primarily to translate, I had more of an opportunity to observe and precisely identify the conversation, as well as time to write more-detailed fieldnotes. This experience also typified when Haitian activists were invited, and highlights HAWG activities, and therefore brings up some unresolved questions about the transnational advocacy process and the gap it reveals in disaster response and advocacy. As noted in the introduction to the chapter, this description represents only a partial truth, and is meant to engender further reflection and dialogue, and not to represent the totality of HAWG activities or even an overall assessment.

Kato Goes to Washington

Kato and I were each battling a headache, not helped by the frenetic pace of events over the past twenty hours since he arrived—two hours late—at Ronald Reagan National Airport the day before. We had just passed through a metal detector, had our photos taken, and were granted a temporary badge to enter the World Bank building. We had crisscrossed the Federal Triangle, going from the House of Representatives to an NGO two blocks south of the White House, back to the U.S. Senate, and were now again in the executive turf—now the World Bank, this after our initial meeting at the U.S. Department of State.

Thankfully, the man who worked for the World Bank—but not really, since his paycheck came from the Haiti Reconstruction Fund (HRF), still sort of since this had been his office as a desk officer for several years, with awards from previous engagements strategically at eye level—spoke French.[1] So I did not have to translate, since my friend Kato who typically and vociferously refuses to speak the colonizer's tongue, replied in French, being of the minority of well-educated who could speak it. So, if I wanted to, I could partake in the dialogue. But then the World Bank desk officer said something in English, ostensibly to Kato but really to the two of us U.S. citizens, myself and the HAWG staff. So I had to translate it: "Haitians have to be patient."

Kato's visit as a representative of the housing rights coalition called LoJistis was anomalous. LoJistis had received funding from a United States–based NGO the previous year, to participate in an advocacy week sponsored by the HAWG. But the U.S. government had denied him a visa, so funds remained unspent. Kato was upset, not only at our government's policies that discriminated against lower-income Haitian people or less-established groups, but also because the sponsoring NGO seemed not to care that the money was not spent. It appeared to Kato—and to many others in Haiti—that it was more important for NGOs to spend money than it was that they do their job. So a colleague and I worked with Kato to contact this sponsoring U.S. NGO and arrange a U.S. tour, not only including Washington but also cities with a large Haitian Diaspora, such as New York, Miami, and New Orleans.

Given that few development and policy staff speak Haitian Creole, Kato needed someone to translate back and forth. Since I had invited him (unsuccessfully) with me to Washington the previous year to participate in a panel sponsored by the CBC and I had known him since 2008, I was an obvious choice. Up to that point, I had taken three trips to Washington to share my own analyses of the earthquake recovery effort, particularly, based on research that I had been encouraged to do by LoJistis and others, about the internally displaced persons (IDP) camps. But this time I was just a translator, not a participant, and could slip back into a behind-the-scenes role more comfortable for this anthropologist. Since I had to translate the English to Haitian Creole and back, I have a much greater recollection of the talking points than if I had been an advocate. It is these experiences— and my reflection—that form the basis for this chapter.

Once Kato was in Washington, the fact that he was not initially invited by the HAWG was ignored. He was welcomed to a meeting of a large women's NGO and a conference call for the HAWG. He was also identified as a HAWG collaborator in the meetings with policymakers. We were accompanied by four people at different times, one a colleague I had known and three others I had just met. Of the two co-conveners of the HAWG, one was out on family leave and the other had left her NGO. Someone new stepped in for this role. The new convener was self-effacing, humble, and a consensus builder. Since Kato was delayed by U.S. Customs in Miami as a first-time traveler from Haiti, being given additional questioning, he had missed his original flight to Washington. I met this convener at his office, dropped off my bags, and we headed out to the airport together.

Because of Kato's delay, we had to rush on the way back with him to the city, and took a cab to our first appointment at State. The intervals in the cabs were the basis of an interrupted conversation about the state of the HAWG, of Haiti, and of the precise talking points to be raised with each particular office we would be visiting. It could have been the lack of precision in Haitian Creole or my own limited ability, or Kato's "weapons of the weak" (Scott 1985) or inexperience, but the talking points I translated into English did not vary much at all. Neither, as it turned out, were those I translated from the English.

The two other handlers whom I had just met both spoke French, so they talked directly to Kato in the back of a cab. One of them brought issues like corruption into the conversation despite Kato's insistence that this issue was not germane, since we also had to talk about corruption within the very USAID, World Bank, and NGOs institutions we were meeting, in addition to the Haitian government. With one exception, the policymakers turned the discussion to what is wrong with the Haitian government and attempted to enlist Kato to support their efforts. As a Haitian, it was natural that Kato focused on the Haitian government and what it should do, so the opportunity was lost to keep pressure on donors like USAID who had previously pledged $1.15 billion in reconstruction.

Kato is neither a lobbyist, nor an architect, nor a policymaker, economist, or structural engineer. Yet these apparently were the topics of expertise our Washington interlocutors demanded of him when they finally did ask questions. When he could only respond to what he knew—the horrible living conditions in the IDP camps for the almost four hundred thousand people still living in them at the time, the increasing violence and insecurity as tools of forced evictions from the camps, the urgent necessity for public consultation in what little housing is being constructed, and that cooperative models from Latin America might be adopted—our busy bureaucrats frustratingly retreated.

The talking points from Washington policymakers that I dutifully translated into Haitian Creole were that the Haitian government should take the lead but they were not, that there was a capacity issue, that Haiti's current government of Michel Martelly—supported by foreign donors in an election that could hardly be called free, transparent, and fair—was doing better than others, that neoliberal macroeconomic policy frameworks are the necessary first steps to build growth that can support development, and that the public sector should not be involved in the "business" of housing. This information was useful to me as a solidarity activist, the public transcripts. More useful would have been to get at what James Scott (1990) called "hidden transcripts," the behind-the-scenes understandings of how decisions are made—that is, the architecture of Washington policymaking and a social mapping of power. This information, in fact, was specifically asked by Haitian activists of the HAWG. How, where, and with whom, do we intervene in order to shift directions?

I posed these questions at numerous occasions to the HAWG advocates. For example, since 2011, with the Republican Party in control of the U.S. House of Representatives, the chair of the Foreign Affairs committee at the time, Ilana Ros-Lehtinen, represented an area of greater Miami with a large Haitian constituency, with an even greater Haitian constituency residing nearby. Why then did the NGOs only talk with House Democrats? The diplomatic answer I was given was that they were afraid of what the Republicans, emboldened at the time by the Tea Party surge, would do to the USAID budget if they owned the issue, lexicon for taking the lead on setting the agenda. I responded to the few who were listening that exactly, threatening to work with the Republicans who would eviscerate its budget was the best leverage we could have with USAID, suggesting a good cop/bad cop approach that had worked when I was a community organizer in Minneapolis and St. Paul. Later I was told by my colleague who, in addition to HAWG, was a member of the Haiti Response Coalition (HRC) that positioned itself as a more grassroots, radical, left-leaning group, that some members of the HAWG had received funding from USAID.

This is not to diminish the HAWG's legislative successes. It was able to get a commitment for appropriation for the Haiti earthquake effort when the process dragged on. HAWG also got the House of Representatives to pass the Assessing Progress in Haiti Act, sponsored by CBC member Barbara Lee (D-CA), on 26 April 2012, months later and after the establishment of a panel (described below). The Assessing Progress in Haiti Act was reintroduced during the next Congress (since it did not pass the previous session) and was signed into law in 2014.[2] However, with USAID not only controlling the purse strings but also setting the terms of engagement with the United States, NGOs, and in theory also the Haitian government, and with USAID being a branch of the U.S. government, USAID was our best target for changing the system. My first monograph, *Killing with Kindness* (Schuller 2012), details just how powerful USAID is in influencing recipient NGOs. Had USAID changed the rules or priorities to require genuine local participation and collaboration with the Haitian government, there might have been real, tangible, and immediate progress. The importance of needing new approaches was underscored following the outbreak of cholera in October 2010, which UN troops had introduced to Haiti (Hendriksen, Price, and Shupp 2011; Piarroux et al. 2011); this disease had killed almost 9,500 people as of 2017.[3] Donors like USAID could have demanded full coverage of water and sanitation services, contracting with Haitian providers, or supporting local government.

Such a transformation of the reward structure, one that might have made a real difference, did not happen, in part because HAWG did not include the request in the discussion of aid accountability. The issue of interrogating NGOs and their relationships with groups in Haiti, unlike focusing upward on USAID and its for-profit contractors, had the potential of being divisive. In addition to the potential conflict of interest, wherein some NGOs were hesitant to bite the hand that feeds them, was a series of structural barriers. Kato's saga of getting his visa suggests another reason for the limited progress in legislation, to say the least about on-the-ground impacts. The upshot was, once again, a problem that poststructuralist scholars have long discussed: the issue of representation. Who speaks for Haiti?

Politics of Representation

Knowledge about Haiti—as anywhere else—is controlled by people who represent it. As such the issue of representation is a central concern for addressing the knowledge gap.

Haitian American anthropologist, activist, and performance artist Gina Athena Ulysse (2010, 2015) focused the attention of scholars, activists, and policymakers on the ways in which Haiti is represented by foreign media agencies, NGOs, and scholars. The politics of representation has been a core concern within feminist scholarship, especially feminists of color and transnational feminism. Chandra Mohanty (1988) critiqued the ways in which women of color get essentialized "under Western eyes." In the United States, a range of activist scholars such as Cherrie Moraga and Gloria Anzaldua (1983), Patricia Hill Collins (1990), and bell hooks (1994) all added specific analyses about the politics of representation. Gayatri Spivak (1988) asked, "Can the subaltern speak?"

To answer Spivak's question, it is useful to note the speakers at these Haiti events. The main events were those organized by the CBC in which members of Congress themselves would be speaking. A HAWG colleague told me that two members from California "own" Haiti in terms of legislative advocacy. Barbara Lee of Oakland and Maxine Waters of Los Angeles do not represent districts with the most Haitian people but are among the longest involved members. Both have led numerous delegations to Haiti. During the 2004 political crisis in Haiti that saw the forced removal of President Aristide, for the second time, both had defended him. Waters had attended the bicentennial celebration in 2004. These briefings, and certainly hearings, also included United States–based solidarity organizations who shared the reputation of being pro-Aristide. One group's Haiti partner, a human rights activist and attorney, often presented at these legislative events, and was usually the only Haitian on the list of speakers. Sometimes the legislators used their clout to interrogate USAID and other donors, such as the World Bank and Inter-American Development Bank. As a reflection of the partisanship in Washington, the discussions could be somewhat heated. As hearings, the discussion become part of the official public record; C-Span sometimes televises the hearings.

Representation of people from Haiti was greater in briefings, the purpose of which is to brief legislative staff on a given situation; briefings are far more numerous and common than hearings. Whereas hearings have to be called for by standing committee chairs, briefings can be organized by a single member of Congress. In my experience, often as not, a single member of Congress attended the briefing. Attendance at these issue-specific panels was usually smaller than the main events—the hearings—

and usually the tone was less combative and more technical. It was these specific events that I was invited to participate in, since I was considered to be an expert on the IDP camps. One exception to the usual collegiality was a panel I was also part of called "Haiti: Where Did the Money Go?," which incidentally was also the title of a documentary that aired on the U.S. Public Broadcasting System for the second anniversary of the quake in January 2012. The documentary, which included an interview with me, was highly critical of the lack of progress in the IDP camps, while juxtaposing images of young foreign workers living in luxury. The film had struck a raw nerve, and the Red Cross went on the attack during the meeting, spreading one-page leaflets criticizing the film, which I was seeing for the first time. One of the panelists from CEPR had debunked the Red Cross's rebuttal. Another panelist, from the Disaster Accountability Project, was also similarly critical of the lack of accountability in NGOs. The Red Cross representative interrupted the moderator, the director of TransAfrica Forum, to attack her for what the Red Cross took to be TransAfrica's insincere gestures at contacting them. Yvette Clarke, the representative from Brooklyn, one of the two districts with the largest Haitian communities, was in the front row during the briefing.

According to emails I received about hearings or briefings, five individuals predominated in representing Haiti, especially in hearings and particularly during the first couple years after the earthquake. With the exception noted above, most of the speakers were professional members of Haiti's NGO class (Schuller 2009). In Haiti, NGOs provide the most stable sources of employment and to some, constitute a social class, in the Weberian if not Marxian sense (see also Louis 2018). These NGO professionals regularly attended regional conferences and events like the World Social Forum. They were already established public figures, heads of NGOs (most of which, incidentally, were technically networks of other NGOs), by 2004. With one exception, all were also vocal opponents of Aristide and all had received aid from either USAID directly or through a contracting NGO. One in particular, a human rights organization, received a human rights prize from the U.S. embassy following Aristide's ouster, which raised many questions in Haiti and in solidarity circles about this NGO's legitimacy and independence. The established leaders shared the status with their pro-Aristide colleague of having a U.S. passport and international backers, as well as being a member of the so-called 1986 generation (referring to the year that Haitian social movements forced dictator Duvalier out).

I had asked a colleague about the limited participation by Haitian individuals in these critical events, and he said that while there were genuine democratic tendencies within HAWG, two NGOs dominated it. Those

NGOs provided the funding for travel from Haiti, so they were able to choose their partners, creating a smaller version of pay to play. Whether there was behind-the-scenes chatter about the two NGOs' domination, no one dared bring it up during conference calls I participated in, since it was understood that these NGOs were shouldering the financial burden of inviting the Haitian partners. As a result the list of those who spoke for Haiti was relatively small and remained consistent throughout the legislative action days. Over time, as Haiti moved from center stage, more spaces opened up for additional spokespersons for the country.

One of the first groups to break this particular class ceiling and become one of the go-to groups in Washington was a women's organization called Commission of Women Victims for Victims (KOFAVIV). KOFAVIV was a grassroots organization that spun off an established women's organization in 2004, out of concern for the latter's militant anti-Aristide activism and with the support of a young U.S. activist. KOFAVIV maintained strong ties to their grassroots, which consisted of dozens of volunteers trained to fight gender-based violence. At least until the earthquake, they operated under a shoestring budget. The earthquake thrust them into a position of leadership and high visibility, since the two leaders and most of their volunteer field agents lived in the IDP camps, where the issue of rape and sexual harassment became much more rampant (d'Adesky and Poto Fanm+Fi 2012; Institute for Justice and Democracy in Haiti et al. 2010; Nolan 2011; Schuller 2015). Their visibility also jeopardized the leaders' personal lives: they were threatened with kidnapping. A few of us were working to help them with media coverage and what little grassroots support we could access. Because of their high visibility they attracted a large USAID grant administered by one of Haiti's largest and most established NGOs, and a media advocacy organization headquartered in New York. CNN named one of their leaders a hero in a series of biographical sketches culminating in a big Hero of the Year award in 2012. This award, however, was linked to their downfall since it came with a large sum of cash and the CNN filming was very invasive. A crew of nine people went inside the camps to document KOFAVIV's heroism. Leaders of one of the camps believed that, since the camp residents had been used as props, they were owed part of the prize money. One woman in particular waited outside the airport following a ceremony involving the Haitian government. Immediately thereafter, death threats to the two KOFAVIV leaders began. KOFAVIV had to move offices in the middle of the night and leaders went into hiding. As of this writing, one leader has been in the United States under asylum for two years. Despite their quick rise in media profile, at least initially, KOFAVIV leaders also had difficulties obtaining their U.S. visas, just as Kato did. Like Kato, their lower socioeconomic status meant that they had a much more difficult time convincing the U.S. government that they were not flight risks to overstay their tourist visa and become undocumented, and so could be granted a visa. Still, even with the KOFAVIV addition, Haitian people still did not comprise a majority of panelists in legislative events about Haiti in Washington.

The silencing of Haitian voices infuriated many in the Haitian Diaspora, whose national networks in the United States spanned a portion of the Haitian political spectrum. In U.S. advocacy and legislative circles the organized Haitian Diaspora groups tended to overrepresent a professional constituency. Two issues were high on the Diaspora's radar screen: temporary protected status (TPS) and dual citizenship. The Obama administration finally granted TPS on 21 January 2010 and extended it because of Hurricane Sandy in November 2012.[4] The issue of dual citizenship was addressed in a round of constitutional reforms presented in 2009 and published in May 2011.[5] Many Haitian people living in the United States choose not to give up their Haitian citizenship—since it is historically a symbol of Black freedom—and as such do not represent a voting bloc proportionate to their numbers. Aside from this, there is very little formal engagement with the diaspora in the congressional briefings of HAWG or the HRC that had a more grassroots or radical self-identity.

Not engaging the Haitian diaspora was a critical error, not only because of reasons of justice and inclusion, but also in terms of effectiveness at obtaining legislative commitments, which was a stated goal of HAWG. One of the briefings I attended centered on the perspective of five Haitian American students who had joined me in researching the IDP camps, taking advantage of Washington's proximity to Baltimore where the Society for Applied Anthropology (SfAA) Conference took place in March 2012, and where we presented our work. All five of the students lived in Brooklyn, all but one in Representative Clarke's district. Clarke explained that she was moved by students' testimony and had pledged her help, but she expressed her impatience with the slow progress, unlike the World Bank official mentioned earlier. Her advice to the students and me was direct: You need to flex your muscles and mobilize the population. You cannot imagine how many people come into my office every day passionate about this or that issue, she said. But I am only one vote. You need to go out there and rattle other people's cages, not as advocates but as citizens, as taxpayers, as constituents. A young woman during the struggles of the Civil Rights movement, Clarke pointed to the successful efforts when rights were advanced. One thing the efforts had in common was citizen pressure. Haiti's earthquake represented an opportunity to mobilize, she analyzed. Sixty percent of U.S. households contributed to the earthquake response. That is a huge potential citizen base, she said.

Eighty percent of African American households contributed, suggesting a coordinated approach through churches. Absent this direct engagement, and painstaking community organization, Clarke pointed to social media. Not without irony (see Cole 2012, for an early critical perspective), she discussed the Kony 2012 campaign that had gone viral.[6] She told my students that they had to take their show on the road: organize the community and create a Facebook page. She was visibly frustrated when sharing that a citizen constituent engagement had been lacking from Haiti earthquake advocacy efforts in the United States Absent this citizen pressure, the power that advocacy coalitions like HAWG or HRC wield is limited to persuasion. You see where this has gotten us, she intoned. This was in direct contrast to the successful campaign to cancel Haiti's debt organized by the Jubilee coalition from 2006 to 2009. Subsequently, the students did create a Facebook page but they lost momentum due to several factors, among them that I failed to secure a space for the students to activate the community before I moved out of New York. And, when I was out of state, I had even less contact (Schuller 2014).

This discussion highlights the critical importance of who gets to speak for, to represent, Haiti, and the importance of participation in when and how knowledge influences policy and practice, in effect knowledge's ability to close the gap. In the following concluding sections of the chapter I assess the general advocacy effort and attempt to bring up general theoretical lessons.

Conclusions

Assessing Progress in Haiti Advocacy

The Washington encounters described above represent moments where knowledge is being created, translated, and transmitted. As political events, legislative progress has been frustratingly slow for many in the HAWG. I have suggested that this may be because of structural limitations such as gaps in language barriers, U.S. visa policies, limited funds for travel, limited public attention to foreign countries, and competing issues for legislative action. As pedagogical events, however, these serve to show constituents of the groups involved—HAWG and its member NGOs as well as the CBC and staff—that people still care about Haiti and are trying to do something about the ongoing conditions there. For the first three anniversaries of the earthquake, and then for the fifth, Congress issued resolutions "Recognizing the anniversary of the tragic earthquake in Haiti on January 12, 2010, honoring those who lost their lives, and expressing continued solidarity with the Haitian people."

This chapter has documented that the circuits of knowledge production have remained relatively closed, as a few Washington-based NGOs have traditionally controlled the access to Haitians. With few exceptions speakers invited from Haiti tend to be middle-class NGO professionals. In Haiti, NGOs provide the most stable sources of employment, representing an NGO class. Whether this closed circle is because these happen to be the groups that Washington NGOs support, or because of a shared NGO habitus (Bourdieu 1980), the perspectives presented for congressional staffers and the occasional CBC member are but a fraction of available opinions, realities, and issues that groups within Haiti are confronting. Interestingly, to my knowledge, none of my homologues at the State University of Haiti or for that matter any scholar within Haiti, was invited to Washington until 2014, long after Haiti had lost its visibility. Consequently, the HAWG NGOs tend to reproduce their perspectives in a feedback loop. The closed loop was augmented when one of the NGOs decided to create a social movement campaign within Haiti, inaugurated by a conference at one of Haiti's high-end hotels where the majority of speakers were white U.S. professionals, some of whom had spent no longer than a few days in Haiti.

One reason why these advocacy efforts matter is that the U.S. voting public and government were caught off guard by a scandal following a report published by NPR/ProPublica in June 2015 (Elliot and Sullivan 2015), to say the least about a sex-ring by Oxfam GB that the *Times* of London uncovered in 2018 (Porter 2018).[7] Opportunities to uncover the structural reasons for limited, incomplete, or lackluster NGO performance did not materialize in any meaningful investigation, to say the least about policy. While gender-based violence received some traction and some new voices were brought to Washington, solutions did not address the intersectionality of the experiences of low-income women living in the camps (Crenshaw 1991; Purkayastha 2012; Yuval-Davis, Anthias, and Kofman 2005). Therefore, official government and humanitarian agencies did not implement policy and practice that would have had a transformative effect. Furthermore, the aid that was given in general did not match the priorities of either the Haitian government or local communities. More grassroots voices and new social movements challenging the inequalities in the system, denouncing it as imperialist, seldom made their way to even a briefing. As a result, the creative solutions that Haitian activists came up with on their own lacked support.

Limitations to NGO Praxis

As the above discussion suggests, HAWG has an uneasy relationship with the U.S. government. HAWG is a diverse coalition, including large INGOs as well as one-person outfits whose only focus is Haiti, some of whom have a very specific mission and sector of intervention. As a voluntary coalition, the HAWG, like many coalitions, has a least-common-denominator approach—sometimes the word "consensus" is applied—that means that the most cautious, moderate, and even conservative approaches to the issue are those that are tried. A few member NGOs had been or were actively receiving or seeking USAID or contracting NGO funding. Pointing to this fact, one individual employed by the U.S. government accused HAWG of being self-serving during a meeting. Even if the prospect of federal funding did not cause many NGOs to self-censor or moderate their advocacy as structures, NGOs mirror the institutions they lobby. The less dependent an NGO is on federal funding, the freer it is to openly challenge the government. The HAWG experience has also demonstrated that smaller NGOs have more flexibility and are able to move more quickly into new arenas. But they lack the resources to bring "their Haitians" (using that language) to Washington.

Structural Racism

The language citing "my Haitian" is troubling as it echoes the real ownership that the Haitian Revolution did away with. Bristling at the colonialism inherent to this relationship, some colleagues in Haiti even call their U.S. partners their "masters." In particular, these are NGOs that have a U.S. NGO for support, not those who are more marginalized. Kato and even leaders of KOFAVIV, which became a media product, were initially denied visas. Even if the pay to play of big NGOs to direct which Haitian NGO professionals were given a platform in Washington did not exist, the racism inherent in U.S. immigration policy regarding Haiti limited the choices to all but the members of an international NGO class. As a result of these processes, United States–based NGOs and the rules of the game set up a feedback loop wherein the priorities of the large U.S. NGO became represented in the Haitian NGOs that are brought to Washington and other, more grassroots, marginalized perspectives were silenced as a result.

There are counter-currents from within HAWG, as a diverse coalition. Furthermore, HAWG's long-term engagement is noteworthy: as the U.S. media and voting public turned away from Haiti, HAWG continued their advocacy, albeit with fewer people and resources. The HRC petered out sometime in 2013, with only random emails sent to the Listserv from peripheral members such as me once in a great while. Either because of or despite now lower visibility, beginning in late 2013 and 2014 a greater number of Haitian people have been invited to Washington from Haiti. They have also tended to be younger, more directly involved in social movements, include scholars, and are not only directors of organizations but also activist members of collectives. To be sure, at the bigger events, such as earthquake anniversaries in January, the traditional leaders are also present, but the conversational practice and events have been more inclusive and they are occurring at different locations, including academic spaces such as George Washington University's Elliott School for International Affairs, with ties to the U.S. State Department.

This story is—along with its author—aging. In between writing and publication, another major disaster struck Haiti, while I was chair of the board of a United States–Haiti organization. My own perspectives have continued to evolve. The passage of Hurricane Matthew in October 2016 also shifted the contours of advocacy, opening new spaces and opportunities for applying lessons learned from the earthquake. In terms of the official response, however, Matthew was destined to be in the shadows: pledges for aid, for example, were 1 percent of what they had been for the earthquake. Along with colleagues I am attempting to untangle the meanings and lessons from this shadow disaster. I hope that by formalizing my own limited understandings through my own participation that this conversation can continue to open spaces for greater participation and inclusion.

Acknowledgments

First I would like to thank LoJistis and KOFAVIV leaders Malya Villard and Eramithe Delva, and other comrades in Haiti's grassroots movement for justice for their inspiration and fierce devotion to the cause, and their support and at times needed critique. I would also like to thank colleagues at the HRC and the HAWG for their time, dedication, and support. Gina Athena Ulysse helped facilitate my entrée into the *Huffington Post,* and the National Science Foundation (# 1122704) and City University of New York provided funds to make the research—and the student presentations at the 2012 SfAA meeting and in Washington—possible. I would also like to thank the SfAA's Risk and Disasters Topical Interest Group for their support and feedback on the paper, first given at the 2013 meeting. A hearty thank you to research assistants Marie Laviode Alexis, Sabine Bernard, Théagène Dauphin, Jean Dider Deslorges, Mackenzy Dor, Jean Rony Emile, Junior Jean François, Robenson Jn-Julien, Sandy Nelzy, Adlin Noël, Rose-Mercie Saintilmot, Stephanie Semé, Roody Jean Therlius, Jimmy Toussaint, Tracey Ulcena, Castelot Val, and Jude Wesh. All statements and analyses—especially any errors—are the responsibility of the author at the time, and are not to be attributed to my funders, HAWG, or HRC.

Mark Schuller is associate professor at Northern Illinois University. Supported by the National Science Foundation Senior and CAREER Grant, Bellagio Center, and others, Schuller has more than three dozen peer-reviewed publications, and even more public media. He published two monographs, including *Humanitarian Aftershocks in Haiti* (2016) and coedited five volumes, including *Tectonic Shifts: Haiti since the Earthquake* (2012). He is codirector/coproducer of documentary *Poto Mitan: Haitian Women, Pillars of the Global Economy* (2009). Recipient of the Margaret Mead Award and the Anthropology in Media Award, he is active in several solidarity efforts.

Notes

1. This Fund was created by the Interim Haiti Reconstruction Commission, run by the World Bank.
2. Once passed, however, there was little engagement with assessing progress until late 2016, two and a half years after its passage.
3. The UN formally apologized for cholera in late November 2016, six years after the outbreak and with almost ten thousand deaths, and no concrete plan to eradicate the disease or to compensate victims' families.
4. The Trump administration has publicly declared its intention to end TPS, with the commander in chief allegedly calling Haiti and African nations "shithole countries" (Vitali, Hunt, and Thorp 2018)
5. There remain several uncertainties about the constitutional amendment. Some people question its validity given that the Creole language version was never officially published. There are also several inconsistencies among the various versions.
6. The campaign involved a video made by an NGO called Invisible Children about Ugandan general Joseph Kony, that the video called a war criminal. The video had more than 100 million views and played a role in a U.S. Senate resolution and sending troops to the African Union.
7. https://www.thetimes.co.uk/edition/news/oxfam-in-haiti-it-was-like-a-caligula-orgy-with-prostitutes-in-oxfam-t-shirts-p32wlk0rp (last accessed 15 May 2019).

References

Anderson, Benedict R. O'G. 1987. *Imagined Communities: Reflections on the Origin And Spread of Nationalism*. London: Verso.

Anderson, Judith M. 2015. "Will the Real Negros Please Stand Up? Understanding Black Identity Politics in Buenos Aires, Argentina." *Transforming Anthropology* 2, no. 23, 69–78.

Bourdieu, Pierre. 1980. *Le Sens Commun*. Paris: Aux Editions de Minuit.

Center for Economic and Policy Research (CEPR). 2013. *Inspector General Finds Lack of Oversight of Chemonics . . . Again* Center for Economic and Policy Research,

1 October 2012 [cited 20 January 2013]. Available from http://cepr.net/blogs/haiti-relief-and-reconstruction-watch/inspector-general-finds-lack-of-oversight-of-chemonicsagain.

Chancy, Myriam. 2013. "A Marshall Plan for a Haiti at Peace: To Continue or End the Legacy of the Revolution." In *Haiti and the Americas*, edited by Carla Calargé, Raphael Dalleo, Luis Duno-Gottberg, and Clevis Headley, 199–218. Jackson: University Press of Mississippi.

Clarke, Kamari Maxime. 2010. "New Spheres of Transnational Formations: Mobilizations of Humanitarian Diasporas." *Transforming Anthropology* 1, no. 18, 48–65.

Cole, Teju. 2012. The White Savior Industrial Complex. *The Atlantic Monthly,* 21 March. Retrieved 2 April 2012 from http://www.theatlantic.com/international/archive/2012/03/the-white-savior-industrial-complex/254843/.

Collier, Paul. 2009. *Haiti: From Natural Catastrophe to Economic Security—A Report for the Secretary General.* New York: United Nations Secretary General.

Collins, Patricia Hill. 1990. *Black Feminist Thought : Knowledge, Consciousness, and the Politics of Empowerment,* 2nd rev., 10th anniversary ed. New York: Routledge.

Crapanzano, Vincent. 1986. "Hermes' Dilemma: The Masking of Subversion in Ethnographic Description." In *Writing Culture*: The *Poetics and Politics of Ethnography,* edited by James Clifford and George Marcus, 51–76. Berkeley: University of California Press.

Crenshaw, Kimberle Williams. 1991. "Mapping the Margins: Intersectionality, Identity Politics, and Violence against Women of Color." *Stanford Law Review* 6, no. 43, 1241–99.

d'Adesky, Anne-Christine, and Poto Fanm+Fi. Beyond Shock—Charting the Landscape of Sexual Violence in Post-quake Haiti: Progress, Challenges, and Emerging Trends, 2010–2012. Foreword by Edwidge Danticat, and photo essay by Nadia Todres. Port-au-Prince: Poto Fanm+Fi Initiative.

Elliot, Justin, and Laura Sullivan. 2015. How the Red Cross Raised Half a Billion Dollars for Haiti and Built Six Homes. New York: ProPublica and NPR.

Gilroy, Paul. 1993. *The Black Atlantic: Modernity and Double Consciousness.* Cambridge: Harvard University Press.

Hendriksen, Rene, Lance B. Price, and James M. Shupp. 2011. "Population Genetics of *Vibrio cholerae* from Nepal in 2010: Evidence on the Origin of the Haitian Outbreak." *mBIO* 4, no. 2, 1–6.

hooks, bell. 1994. *Teaching to Transgress.* New York: Routledge.

Horne, Gerald. 2015. *Confronting Black Jacobins: The United States, the Haitian Revolution, and the Origins of the Dominican Republic.* New York: Monthly Review Books.

Institute for Justice and Democracy in Haiti, MADRE, TransAfrica Forum, University of Minnesota Law School Human Rights Litigation and Advocacy Clinic, and University of Virginia School of Law International Human Rights Law Clinic and Human Rights Program. 2010. *Our Bodies Are Still Trembling: Haitian Women's Fight against Rape.* Boston: IJDH; MADRE; TAF; UMN; UVA.

Louis, Ilionor. 2018. "Classe ONG et Distance Sociale." *Chantiers* 1, no. 4.

Maddox, Camee. 2015. "'Yes We Can! Down with Colonization!' Race, Gender, and the 2009 General Strike in Martinique." *Transforming Anthropology* 2, no. 23, 90–103.

Maguire, Robert E. 2014. "Priorities, Alignment and Leadership: Improving United States' Aid Effectiveness in Haiti." *Cahiers des Amériques latines* no. 75, 59–78.

Main, Meredith. 2015. "Situating the Obama Effect in Ecuador." *Transforming Anthropology* 2, no. 23, 104–15.

Mohanty, Chandra Talpade. 1988. "Under Western Eyes: Feminist Scholarship and Colonial Discourses." *Feminist Review* no. 30, 61–88.

Moraga, Cherrie L., and Gloria E. Anzaldua. 1983. *This Bridge Called My Back: Writings by Radical Women of Color*. New York: Kitchen Table/Women of Color Press.

Nolan, Clancy. 2011. "Haiti, Violated." *World Policy Journal* no. 128, 93–102.

Obama, Barack. 2010. "Why Haiti Matters." *Newsweek*, 21 January.

Page, Sarah E. 2015. "'Your President Is Not Black!': Jamaican Reflections On Identity, Race, Class, and (Global) Politics." *Transforming Anthropology* 2, no. 23, 79–89.

Peck, Raoul. 2013. *Assistance Mortelle*. Paris: Arte France.

Piarroux, Renaud, Robert Barrais, Benoît Faucher, Rachel Haus, Martine Piarroux, Jean Gaudart, Roc Magloire, and Didier Raoult. 2011. "Understanding the Cholera Epidemic, Haiti." *Emerging Infectious Diseases* 7, no. 17, 1161–67.

Plummer, Brenda Gayle. 1988. *Haiti and the Great Powers, 1912–1915*. Baton Rouge: Louisiana State University Press.

Polyné, Millery. 2010. *From Douglass to Duvalier: U.S. African Americans, Haiti, and Pan Africanism, 1870–1964*. Gainesville: University Press of Florida.

Porter, Catherine. 2018. "Haiti Suspends Oxfam Great Britain After Sex Scandal." *New York Times*, February 22. Retrieved from https://www.nytimes.com/2018/02/22/world/americas/haiti-suspends-oxfam.html.

Purkayastha, Bandana. 2012. "Intersectionality in a Transnational World." *Gender & Society* 1, no. 26, 55–66. doi:10.1177/0891243211426725

Ramachandran, Vijaya, and Julie Walz. 2012. *Haiti: Where Has All the Money Gone?* Washington, DC: Center for Global Development.

Schuller, Mark. 2009. "Gluing Globalization: NGOs as Intermediaries in Haiti, 2007 APLA Student Paper Competition Winner." *Political and Legal Anthropology Review* 1, no. 32, 84–104.

———. 2012. *Killing with Kindness: Haiti, International Aid and NGOs*. Foreword by Paul Farmer. New Brunswick, NJ: Rutgers University Press.

———. 2014. "Being an Insider Without: Activist Anthropological Engagement in Haiti after the Earthquake." *American Anthropologist* 2, no. 116, 409–12.

———. 2015. "Pa Manyen Fanm Nan Konsa: Intersectionality, Structural Violence, and Vulnerability Before and After Haiti's Earthquake." *Feminist Studies* 1, no. 41, 184–210.

———. 2016. "The Tremors Felt Round the World: Haiti's Earthquake as Global Imagined Community." In *Contextualizing Disaster*, edited by Gregory V. Button and Mark Schuller, 66–88. New York: Berghahn Books.

Scott, James C. 1985. *Weapons of the Weak: Everyday Forms of Peasant Resistance*. New Haven, CT: Yale University Press.

———. 1990. *Domination and the Arts of Resistance: Hidden Transcripts*. New Haven, CT: Yale University Press.

Seitenfus, Ricardo. 2015. *L'échec de l'aide internationale à Haïti: Dilemmes et égarements*. Port-au-Prince: Éditions Université d'État d'Haïti.

Spivak, Giyatri. 1988. "Can the Subaltern Speak?" In *Marxism and the Interpretation of Culture*, edited by Cary Nelson and Lawrence Grossberg, 271–313. Urbana: University of Illinois Press.

Ulysse, Gina Athena. 2010. "Why Representations of Haiti Matter Now More Than Ever." *NACLA Report on the Americas* 5, no. 43, 37–41.

————. 2015. *Why Haiti Needs New Narratives: A Post-Quake Chronicle. With a preface from Robin Kelley.* Middletown, CT: Wesleyan University Press.

Vitali, Ali, Kasie Hunt, and Frank Thorp V. 2018. "Trump Referred to Haiti and African Nations as 'Shithole' Countries." NBC News online, January 11. Retrieved March 2019 from https://www.nbcnews.com/politics/white-house/trump-referred-haiti-african-countries-shithole-nations-n836946.

Yuval-Davis, Nira, Floya Anthias, and Eleonore Kofman. 2005. "Secure Borders and Safe Haven and the Gendered Politics of Belonging: Beyond Social Cohesion." *Ethnic & Racial Studies* 3, no. 28, 513–35. doi:10.1080/0141987042000337867.

Part III

REVAMPING APPARATUS
AND OUTCOME

In the final part of this book, several prominent scholars and practitioners in the field of disaster studies and disaster risk reduction (DRR) examine the key challenges of overcoming the divides between expert and local knowledge, the priorities of politically and sociopolitically powerful actors, and those who are impacted by a disaster. First, Susanna Hoffman examines the importance of taking culture—as it is understood and recognized by anthropologists—into consideration in disaster mitigation. Although the culture concept, the idea that people's varying behavior, values, and material objects are reflections of local cultures and that culture is something every human group has, is widely accepted by the general public today, this was not always the case. In the early nineteenth century, for example, people used the term "culture" to refer to the music and etiquette of European elites and did not recognize the practices and cosmologies of populations around the world as such. It was not until the latter part of the nineteenth century that the founding figures of anthropology began to make the case that the beliefs, lifeways, mores, material artifacts, and kinship systems of all human beings should be recognized as culture, and that such cultures should be evaluated on their own merits and not in comparison to an arbitrarily imposed elite Eurocentric standard.

During the past 150 years or so, anthropologists and scholars in related disciplines have been quite successful at changing public opinion on what the term "culture" means and who is deemed to have it and who is not. Despite this success, there is much work that remains to be done. For one, Hoffman explains, many governmental, inter-governmental, and nongovernmental organizations (NGOs) have defanged the term "culture" as they have adopted it in their programs, assessments, and evaluations. The watering down of the term has taken the form of governmental officials and some disaster recovery experts conceptualizing it as a superficial collection of differences (often in the form of material culture or ritualized performance) that is unrelated to other dimensions of human experience like kinship, economics, livelihoods, and knowledge making. Hoffman, in contrast, calls on readers to recognize that everything human beings do is in one way or another cultural and that to ignore or disregard such differences in disaster reduction and recovery is to set projects up for failure or, worse, to be culturally disruptive and possibly harmful. At the same time, Hoffman challenges us to recognize that no culture is by any means simple and unchanging. People living within a specific cultural horizon may have different perspectives on life and disaster risk, perspectives that are influenced by a variety of social positionalities (e.g., gender, race, class). Cultures also change over time and their boundaries are not impermeable, but are in a constant process of exchange, interpretation, and growth.

Building on Hoffman's insights, the second chapter of this part is written by members of an internationally recognized group of researchers, practitioners, and scholars (Katherine E. Browne, Elizabeth Marino, Heather Lazrus, and Keely Maxwell) known as the Culture and Disaster Action Network (CADAN). A central part of CADAN's mission is to enhance the ways governmental, intergovernmental, and NGOs that deal with DRR recognize, respect, and integrate local culture in their projects. Because CADAN not only is composed of academics and researchers but also incorporates members who work for U.S. federal agencies and intergovernmental organizations, Browne et al. are able to identify what can actually be done to enhance cultural sensitivity and integration in DRR projects in order to distinguish what are reasonable from what are unrealistic expectations of institutional change. Additionally, Browne et al. provide a concise review of advances that have emerged during the past half century in what is known about disaster and its root causes as well as what is understood about the global rise in disaster risk.

One of the many critical issues raised by Browne et al. is that of discrepancies between the timelines of disaster-affected people and communities and those of the institutions charged with the task of assisting them in DRR and recovery. They write, "We know that, whether in anticipation of disaster or in response to disaster, the timelines of management agencies and personnel and those of communities do not always align. ... Thus, part of the misalignments among disaster responders and communities are expressed via misaligned needs given different senses of time and timing." This concern with temporality sets the tone for the book's final chapter by Ann Bergman in which she explores the importance of the imagination, voice, and futurity in DRR. Bergman is right in calling our attention to issues of time and power in disasters. It is well established in the anthropological literature that establishing when a disaster begins and ends is an act of power that is often contested between liable parties and disaster survivors. Take the case of the Bhopal technological disaster in 1984: For many people of the region, the disaster continues to this day in the form of long-lasting injuries and disabilities and the continued birth of infants with congenital malformations. For liable parties like Union Carbide, however, it is in their best interest that the end of the disaster be determined as close to the catastrophic event as possible in order to avoid increasing reparation costs.

But Bergman's chapter is not so much about the determination of timelines as it is about imagining a less-disaster-prone future than the present, especially when considering Browne et al.'s note that disaster risk continues to grow worldwide. Imagining a less-disaster-prone future, however, is as contested as determining the beginning and end of a disaster. As Susanna Hoffman also tells us, risk is never a self-evident objective reality. Who determines what constitutes a risk is subject to contestation since elite members of any given society may attempt to impose their social norms and priorities on the less fortunate. In addition, those familiar with an environment may not consider an embedded hazard particularly perilous. In the same way, imagining the future is inevitably a polyvocal and contested process, where people from a variety of positionalities in terms of gender, sexuality, ethnicity, class, and race may weigh in with very different perspectives on what is a risk and how to mitigate it. Imagining a less-disaster-prone future will also require us to devise a means of civic engagement and participation that does not merely co-opt the voices of many in favor of the interests of a few, but enables us to collectively and equitably address the challenges of a warming and ever-more disaster-prone planet.

CHAPTER 11

The Scope and Importance of Anthropology and Its Core Concept of Culture in Closing the Disaster Knowledge to Policy and Practice Gap

SUSANNA M. HOFFMAN

If I can't solve certain matters, at least I can expose them and with exposure, I hope the community will rise to the occasion of fixing them.
—Edward Bates, U.S. Attorney General under Abraham Lincoln

Introduction

The forerunning chapters in this volume cover many factors that contribute to the disparity between what is known about risk and disasters and what enters risk and disaster policy and practice. The issues brought up range from the philosophical background behind approaches, to loopholes within international frameworks, power plays of scenario participant parties, situational priorities, managerial discord, and quandaries within the realm of expertise. The chapters also detail the fact that the gap is growing wider due to recent demographic trends, advancing climate change, expanding displacement, the menace of nuclear contamination, and clamorous gender and other remonstrations. All considered, it is evident that the elements contributing to the gap between what disaster experts know and what has—or largely not has—seeped into governing policy and applied procedures are many and have amassed for years.

There would seem to prevail no handy platter of solutions. However, each of the chapters mentions one iterative item in relation to bridging the gap between reducing risk and ameliorating disaster: the need to incorporate local people along with their knowledge and culture into the equation. The questions that arise then are, "What exactly does inputting local people, their ken, and ethos mean?" and "How can integrating locals and their understandings be accomplished?"

In answer to those queries, many agencies in recent years have turned to anthropology. In fact, in some circles anthropology is considered the cutting edge in risk and disaster understanding. This ratification stems from the realization that a people's own cognizance of their situation is neither superficial nor narrow. It goes far beyond simple agreement to or resistance toward environmental alterations, developmental projects, and other readjustments. Rather, it involves taking into account a people's life mastery in the most comprehensive manner, something anthropologists in the disaster field have been promoting for a long time.

Since the inception of the field of anthropology, the idea that groups of people share a particular system of living, termed their "culture," has been its main innovation and core explanatory tool. It emerged as academics and others began to recognize that the many people around the globe functioned, thought, and behaved in diverse manners, and the concept came to refer to these distinctions. In its most holistic framework, culture embraces subsistence, economics, kinship, politics, exploitation of place, scheduling of time, arrangement of space, classification of humans, social roles, language and verbal style, etiquette, child raising, laws, rules, and more. It entails a people's cosmology, how they see their gods, ancestors, the universe, earth, events, and fate, along with their ability to act. It speaks to aesthetics, beliefs, spirituality, explanations, and how various groups envision their identity, rights, and property. Very significantly, culture also operates as a guide to a people's perception—that is, how they see colors, smell odors, hear sounds, feel touch, and judge beauty—thus, how they filter and format information, which, while seemingly unconnected, is in fact highly significant in how people measure their peril and assess their well-being. Every human from childhood on absorbs the culture of the society in which they are raised and by and large lives their entire lives according to its grid (Boas 1940; Kroeber 1948, 1952; Tyler 1871). It is this taking into account the breadth and depth of a people's way of life that risk and disaster anthropologists have been advocating.

In endorsing culture, I largely adhere to the concept in the original deep, traditional anthropological way. That does not mean that I do not recognize that that there are many definitions of the term both historically and in modern times. Nor does it mean that I do not acknowledge that many anthropologists interpret the concept differently, and that to some its use has become somewhat controversial. Some, in fact, refrain from using the term at all. I have further heard the argument that anthropology is the only social science that has lost control of its core concept. Dealing as I do with the realm of risk and disasters, I see it another way. I contend rather that, as opposed to losing control of its core concept, anthropology is the only social science whose core precept has disseminated to and is grasped and used by a widespread audience. That bears great import. The general populace may not use the term "culture" in the deep sense an anthropologist does, but people understand that it refers to different ideas and behaviors of different groups, ethnicities, and even corporations. That widespread discernment of what not only what culture is but also its importance gives those of us in the risk and disaster endeavor cheer. It means that people's knowledge of their own circumstance and their own version of how they wish to live has realized broad cognizance.

Still, even a broad and traditional definition of culture does not imply that culture is unchanging in the face of threats, altered circumstance, or simply through time. It does change. However, even though the dissemination of technological advances and the grasp of global reach have seemingly caused culture to change rapidly in recent times, by and large culture remains sluggish and conservative. Nor does it mean that all the people of a society live exactly according to their culture's dictates. They do not. Still, the extent of their variance is rather like the stretch of a bungee cord. It goes only so far. It also does not mean that at times people do not contest their culture. They do. There further sometimes exist differing definitions of what the culture is and power struggles over which definition dominates (Ortner 1996). As well, subaltern voices can express modifying ideas and influences to the point that some scholars see culture altogether as a sundry process rather than a relatively congruous entity (Brightman 1995; Fischer 2007; Gupta and Ferguson 1992; Zhang and Barrios 2017). Nevertheless, the contestations that arise largely emerge from people's reflective comprehension of their own life ways. Ironically, as recent vocal indictment on many topics including hazards and disasters has shown, it is the very contestation that often leads to cultural change (Firth 1959; Frake 1962; Harris 1968; Hoffman 2014; Steward 1955).

Outsiders to a people, including hazard and disasters experts, officials, and advisors, have all too often merely acknowledged a surface lamina to a people's culture. They have harkened solely to language, dress, food, religion, and a smattering of outward customs. In terms of vulnerabilities and exposure, they have also noted land usage, subsistence economy, and building structures. In addition, the concept of culture itself has been watered down due to its appropriation by other disciplines and especially by corporations that have taken to using the term as a facile way to describe their idiosyncratic styles. But it is culture in its most discerning and anthropological sense that is key to understanding why and how various people worldwide deal with threat and undergo calamity (Hoffman 2014; Hoffman and Oliver-Smith 2002). And, it is culture in that same deep sense that many believe can provide experts, officials, and agents guidance more successfully to reduce exposure and enable a people's rehabilitation after an event.

It is through cultural intricacies that people determine how they calculate peril, undergo and experience catastrophes, recover or do not, protect, or—particularly from an outsider's point of view—fail to safeguard themselves. It is the complex intertwining of numerous background threads that makes for the underpinnings of vulnerability and further mediate how that vulnerability is manifested. Yet, instilling the knowledge of culture in its most profound connotation into much disaster prevention and aid, and most crucially into the policies and practices of international, national, and nongovernmental organizations (NGOs) dealing with disaster, has proven recalcitrant. So have the efforts at altering people's habits in any of these regards, although it is that very knowledge that can help fill the lackings in the gap.

Along this line, it is worth noting that policymakers and agents possess culture, too, and quite regularly use their own ingrained cultural assumptions to legitimize the programs and practices they enact, thereby corroborating the very topic that this chapter is about.

Environment(s), Perception(s), and the Intimate Linkage between Them

Generally the first—and sometimes only—subject that risk and disaster researchers, policymakers, and practitioners ask about when addressing a people's exposure to hazard or their reconstruction from an event is how a people deal with their basic geophysical platform. They then possibly turn attention to what the people have erected on that platform. In short, they inquire about a people's awareness of the faults and foibles under and surrounding them, such as flood propensities, earthquake occurrence, storm and eruption histories, and subsequently ask about the sorts of edifices the people erect and clusters of habitation they occupy. But people do not live just on their landscapes and within their weather zones and buildings. Their ecologies are far more complex. People of disparate furthermore societies do not necessarily perceive their surroundings in the same way as outsiders might. They see them through their own culturally imbued kaleidoscope.

Humans actually dwell in four environments and, to a varying degree, are cognizant of them all. All humans, as well, generally follow long-standing dictums.

The first environment is the basal terrain in which a people perch. This is what is usually referred to by the word "environment," although it might be better termed the "physical plane." By and large, people, unless they are newcomers, know about what lies under, around, or is liable to burst on them. The second environment a people inhabit is their modified circumstance. Humans almost never live in a place without altering it. They sculpt their surroundings. They terrace hillsides, channel streams, lob off mountain tops, purloin sea beds, take down timber, and revise vegetation, often importing it from elsewhere. In all these maneuvers, wittingly or not, they potentially make their domain less stable and allow for, or create, fresh hazards. In addition, wherever humans reside, they occupy a third environment, a built one. Upon their basic orbit, humans erect houses, barns, skyscrapers, temples, memorials, and marinas. Along rustic, riparian, hilly, or flat stretches they assemble villages and cities. To compile their structures, they use mud, straw, brick, wood, concrete, steel, glass, and stucco; they sink them in the soil, lay them on top of loam, or put them on stilts. In short, they superimpose on a physical stage a contrived milieu in which they live, eat, sleep, and work (Hoffman 2017).

All three of these environments—physical, modified, and built—can give rise to peril and destructive events. But, on top of these, there is yet a fourth environment in which people reside, the milieu of their life ways: in other words, again their culture. It is a people's patterns of customs and beliefs that informs them about their physical sphere, how it should be utilized and what they should build on it. Specifically, a people's culture, itself however informed to a certain extent by the other environments, formulates yet another encompassing context around them, one of thinking and doing. Human engagement with a place is neither completely a biological nor a direct response to physical or material conditions, prevailing or constructed. It is filtered (Oliver-Smith and Hoffman 1999).

Key to that filter, and essential ken to any who may dictate policy over or attempt to aid them, is people's own assessment of their circumstance. It is a people's culture that bespeaks what they among themselves calculate as the perils enveloping them, and that estimation may be quite unlike what outsiders appraise. The upshot is that what constitutes actual danger and what comprises risk are not the same thing. Danger refers to an extant, factual hazard that exists in a people's surroundings. It could be a floodplain, cyclone path, fault line, or leaky factory. Risk sits between people and danger much like a water faucet (Ingold 1992; Paine 2002). It relies on assessment of jeopardy, not fact. The regulating faucet can be completely open so a people see all the dangers around them, be they windstorms, eruptions, or tsunamis—that is, the totality of all their perils. Or, and most commonly, the faucet can be only partly open so that people acknowledge some of the encircling hazards but are bedimmed to others. The residents of California are a good example. Californians perceive that earthquakes are a threat and prepare for them. Almost every house has a gas main turn-off key and almost every building is earthquake resistant. Yet, the state's denizens remain in almost total denial—and fail to prepare for—firestorms, floods, and avalanches, all of which are far more common, kill more people, and destroy more property yearly.

A culture's perceptual faucet may also be completely turned off so that the people recognize no danger at all. My favorite example comes from the Greeks who live on the island of Santorini. Santorini is an active volcano seven times larger than Krakatoa. Despite the fact that the volcano has erupted twice in the past century, caused a tremendous earthquake in 1956 and five small ones since, the people refute any exposure. When I replied to a young woman there who asked what I did, she declared, "Thank goodness we don't have problems like that here." As Cosgrove and Daniels (1988, 1) say, "A landscape is a cultural image, a pictorial way of representing or symbolizing surrounding." That image does not necessarily match how outsiders see it.

Occasionally, a people see the risk of exploiting a particular physical place as too high. Some of the inhabitants of the Mississippi River floodplain have moved to higher ground. Others assess that a piece of build environment as too dangerous. The actual residents of Bhopal, India, protested the expansion of the Union Carbide facility to no avail (Rajan 1999, 2002). What the entire set of environments combined with the assessment of danger means, however, is that convincing a people of an outside policymaker's or practitioner's idea of what constitutes a foreboding peril, and then asking for deterrents to it, may comprise a two-part, and often thwarted, struggle. First, it requires convincing people that the hazard exists. Then it requires inducing people to take steps to counteract it. "The environment in which people are active should not be confused with the actual physical world of 'nature'" (Gibson, 1979, 8).

In following, it is also beneficial to realize that not all people share the Westerners distinction of what is, or comes from, nature, and what is domestic, or civilized, space. To many, this separation does not exist. There is, instead, a single continuum. In terms of contravening peril and aiding reconstruction, the distinction matters. Those who separate nature as distinct often advocate an attempt to control it, generally with some physical means. They erect levees, fire walls, and hurricane straps, or redirect rivers. To those who view the physical world as all one unbroken span, often what hazards exist and what events occur are simply a dovetailed sequence and little can be done but to accommodate to it. Hence, frequently indigenous people have a means of dealing with what their arena brings forth. They have houses that they can pick up during floods and then bring back after the event. They cover their abodes with thatch roofs that do no harm should they fall during earthquakes. The erect structures that do not resist wind but comply with it. For these people such events are not disasters. They need neither intervention nor outside rehabilitation. Their means of handling the untoward, which has often evolved over millennia, might just be preferable.

Another cultural factor that plays a role in determining hazard exposure is how a people arrange the space they populate—that is, their density of habitat (Feld and Basso 1996; Low and Lawrence-Zuniga 2003). Some people consider any and all space around them available for occupation, whether safe or not. Take, for instance, Houston or Istanbul where people paved over runoffs and fault lines to allow for habitation. While some communities call for almost wall-to-wall closeness, others strictly define what can be lived on and what must be left unoccupied. They mandate spaced out elbowroom and privacy. Layouts may follow a culturally fancied sense of distance, the demarcation of sacred areas, or, in some laudable cases, proscribed safety codes. For others, no rule abides. People also hold cultural rules on what direction houses should face, what room arrangements should be, whether cooking, bathing, and sleeping take place inside or out, whether walls should surround a home and shelter the activities within it or whether houses and yards should be openly accessible and visible, and whether people's private abode or the community accommodate restive niches. All these cultural norms constitute important factors on which either old homes can be made secure or new ones can be made acceptable. All are part of a culture's accustomed and ingrained topography.

Few outsiders, coming to a strange place, realize that different cultures also configure perceptions differently, and the perceptions contribute to how the people identify danger and judge what constitutes a calamity. For example, many interceding emissaries assume that everyone sees color, hears sounds, smells odor, and feels touch as they do. But again, this is not the case, and these subtle distinctions affect how a people accept protective revisions and post-disaster reconstruction or resettlement. While all humans can physically detect the entire range of observable hues, people from one culture to another merge the colors into different groupings that they regard as essentially the same. Some people, for instance, divide the entire color chromatic into only essentially two shades, basically a distinction between light and dark. Others organize colors into only three merged pigments, something like black, white, and red. Others might distinguish four or five tints, adding yellow, green, and blue. With more uniformity of products, urban people tend to distinguish items more in terms of color, and, therefore, resort to more categories. Those living in more isolate circumstance often differentiate objects by size and shape, and so those people see fewer chromatic divisions (Berlin and Kay 1969). Color in most cultures also involves concomitant meaning. In consequence, what is brown to a re-builder may be red to the client with attendant alarm. What a painter finds soothing may appear gruesome to others. Different people also evidence different preferences in pigment tone. Some adore brights over muted, others pale over vivid. People in some cultures disdain hue almost entirely, while others relish any and all pigmentation (Taussig 2010).

People hear sounds, see sights, feel touch, appraise smell, and regard taste differently, too. To some, the clamor of motors, loud speech, and social hubbub constitutes a familiar ambience, while to others it is an untenable horror. Some are attuned to quite delicate noise, twitters, rustles, and whispers, and for them these sounds make a sought after context. Some people's sight is schooled to notice things at a distance, such as figures in a faraway field or the darkening of waves on the horizon. Others are culturally imbued to direct their vision to matters close by, such as a page, a screen, an intersection. Their long-distance vision becomes stymied just as others find close objects inscrutable. The visual matter of accustomed lives also bears distinction. Some are attuned to blades of grass, not bumper stickers; a bird's flight, not a hundred strangers on the sidewalk; one basket's design, not a hundred boxes on a shelf. And, while some people are socialized to discern and shoo away one fly or only shake arm's-length hands with even close fellows, others are unbothered by the touch of a hundred flies and flock altogether happily on a single chair. Food smells, waste odor, musky perfume, no aromas at all—divergent cultures inculcate their members to variant acceptances. To some people fetid animal fat is sweet and sugar is foul. Some shun chili peppers while others are perplexed by anything bland; the lack of understanding, much less acknowledging of these diverse perception, can lead to significant blunders.

People also do not calculate time in the same way either, yet the treatment of time is of the essence to risk understanding and disaster recovery. Various people think of time, along with their understanding of events, in a linear fashion. They see life unfolding in a direct, straight-ahead progression that happens to be occasionally punctuated with episodes. They further ascribe those episodes to a ticked-off timetable. Others, such as the Tibetans, appraise time in a cyclical fashion, meaning all manner of things, including disasters, reoccur on a revolving calendar no matter what intervenes. The implication for practitioners is that linear thought and time generally imply a better chance of mitigation. Intervention merely tweaks the beeline. With cyclical thinking, alleviation may be dismissed as useless because time and fate are intertwined with all catastrophes preordained (Friedel, Schele, and Parker 1993; Hoffman 2002; Nicholson 1967).

Some people, contrary to either linear or cyclical version, contemplate time in a rather comprehensive manner, something like a surround sound of happenings. To them, all things occur more or less at once, with a piece here and a piece there. As illustration, while many lineal-thinking folks thought the Greek Olympic and the Brazilian World Cup facilities would never be completed before the games ensued because the venues were being incomprehensibly built a little bit on this stadium and a little bit on that. They predicted the events would end up taking place in unfinished debacles. Yet, proceeding in an aggregate but nonetheless overall fashion, all the structures were finished on time.

Styles of calculating time are, however, are not always as simple might appear. Sometimes a mix occurs. Within their culture's computation, some people, such as those in Western societies who generally reckon time in a strict lineal fashion, indulge a covert cyclical forecast in relation to disasters. They, even scientists, claim a calamity to be "the hundred-year flood" or "the fifty-year earthquake." While the basis for such a docket may emanate from a scientific standard of assessment, nonetheless the proclamation instills the notion that a similar calamity is alleviated for a specified time cycle, though, in truth, the same sort of disaster may recur and recur over and over (Hoffman 2002).

The concept of beauty also comes into play and should not be discounted. Most people find their native landscapes, village, town, city streets, and surrounding, whether flat cornfields or towering mountains, beautiful. Beauty provides a sense of indefinable pleasure (Sircell 1975). The accustomed panorama provides a psychological sense of winsomeness. The surroundings bespeak harmony and evoke a feeling that all is well. Disruption or removal from place, therefore, entails lasting disquiet. In addition, people almost everywhere enhance their settings with objects they find congenial. They decorate structures with furnishings, graphics, icons, and objects from which they derive satisfaction. Neither the beauty of place nor the loss of delights are aspects much thought of by those allaying peril or reformulating habitat. They do not consider such matters relevant. Yet, as Scarry (1989) finds, beauty is related not only to pleasure but also to the sense of justice; Kawabata and Zeki (2004) find lack of beauty associated with depression. Still, that "beauty is in the eye of the beholder" proves a truism, and one that counts.

The cultural inculcation of all manner of perception begins in infancy. Like language, the delineated set of perceptual formats are what a child is taught to hear, see, smell, savor, and feel. The acquisition of perceptual canons closes down by about the age of fourteen (Piaget 1926). After that, instilled arrangements are difficult to alter (Berlin and Kay 1969). Importantly, however, all the sensory perceptions—sight, smell, sound, touch, taste, space, and time—play a major part in what has often been called place attachment, and which is perhaps the biggest hobgoblin of risk reduction and disaster recovery programs. Place attachment refers to a people's desire to remain where they are no matter the risk, not to leave despite the extant crisis, and to return to their original circumstance no matter the redundant peril and ruin; it directly affects both displacement and reformulation. Place attachment entails identity, independence, and prosperity. Some researchers have suggested that place attachment and all the precepts it involves may be left over from early human evolution and that it evokes a basis survival instinct for familiarity. Whatever it comes from, it is strong and it is global (Altman and Low 1992) as many officials and program agents have no doubt come to know.

To describe place attachment, I have come to borrow a term from epicurean vocabulary "*terroir.*" *Terroir* describes the loam of the earth, the texture of climate, and waft of air that engenders a special nature to what grows there, like wines. The marrow of a people's life also seems to be a product of the ground they occupy, the atmosphere they breathe, and the way they live. Denizens of a place feel it, smell it, taste it, as much as the maneuver it and manipulate it, and in some sense they themselves— their accents, dress, gestures—are a product of *terroir.* Despite the risks, they prefer to remain where they are. Along the Gulf Coast of Mississippi after Hurricanes Katrina, Rita, Ike and Gustav, where new codes prohibit rebuilding, inhabitants have turned their old house foundations into picnic platforms on which they congregate every weekend. Various survivors of the Oakland-Berkeley firestorm rebuilt not only on the same house lot but also rebuilt the same house. Place attachment also seems to call out to people whose entire community moved to ensure safety. Slowly but surely, many drift back to the former location (Hoffman 2017; 1999b). "A place," says Ingold, "owes its character to the experiences if affords to those who spend time there—to the sights, sounds and indeed smells that constitute its specific ambience. And these, in turn, depend on the kinds of activities in which its inhabitants engage. It is from this relational context of people's engagement with the world, in the business of dwelling, that each place draws its unique significance. Meanings also are not just attached to a certain landscape, they are 'gathered from it'" (Ingold 2011, 192). Separation from a specific place, therefore, not only bears implication of people's engagement but also holds a deeper significance akin to spirituality that can seem irreplaceable. In Japan, the idea of *furusato* (home or ancestral town) carries potent emblematic identity even for those not born there and who may never have seen it (Gill 2013). Place attachment additionally carries critical implications not only for those facing risk and catastrophe currently, but also for those facing the looming matter of climate change and population displacement (Hoffman 2016).

Still, in support of many disaster policymakers and practitioners faced with today's increasing circumstances, there exists a decided exception to affirming local knowledge and acceding to divergent perceptions. Due to the demographic processes that are drawing massive numbers anew to sprawling cities and fragile coastlines, many of those at risk or suffering impact may not be aware of the characteristics of the physical plane surrounding them. They have no cognizance of how it has been altered or what has been constructed on it. Citing Houston again, due to the city's rapid expansion, great numbers of residents apparently did not know that their houses sat on paved-over wetlands, water runoffs that had been filled in, and delta rivulets ignored until the advent of Hurricane Harvey. They, as with many elsewhere, remained blithely unaware of the hidden costs of development in which they unwittingly participated. The citizens of Izmet, Turkey, were not cognizant of the subverted building codes that led to unstable high-rise dwellings. Those rushing to California's golden sparkle have rarely been told of the erosion, waves, and potential fires. And worse, the people in Sichuan, China, were ignorant of the defective building materials used in their schools. For many in today's shifting world, local knowledge is missing and original perceptions are topsy-turvy.

History, Authority, Connectivity, and Other Critical Determinants to Consider

No researcher, policymaker, or aid agent when officiating over or entering a site in peril or ruin enters a place that is without a long, lead-up history of how the culture developed. Virtually all places and people are heirs to a chronicle that has been assembled over years, sometimes centuries, often millennia. Anthropology takes these histories and the effects on the people living there into account; indeed, to do so is also paramount to every intervening party bearing a policy or program. The tales tell of surrounding hazards and saga of disasters. They reveal decisions made and adaptations undertaken. The tales describe who came, bearing what, doing what, and they illuminate the weave of interplaying customs. Even when the policymakers and program administrators come from the same social background as those in peril or in crisis and believe they know their region's history, too, they need to become proficient in the local archive, for the chronicle of every particular spot is unique.

Sometimes, the stories tell of beneficial adjustments, of fruitful sustenance and successful continuity. Other times they speak of miscalculations, misperceptions, lack of foresight, and greed (Oliver-Smith 1999). They show management of risk, its creation, and the long winding path of forgetting or of people simply not understanding what they were doing that in seeming innocence eventually led to a crisis. The eucalyptus trees that so fiercely burned in the Oakland-Berkeley firestorm of 1991 were not newly planted. They had been imported by Spanish land grantees centuries before to replace original fire-resistant forest that they had logged (Hoffman 1998, 1999a, 2013). The Tennessee Valley ash spill in 2008 resulted not from a new mishap, but from a long record of gross actions enacted by a flagrant industry (Button 2010, 135–59.) The severity of the Kobe, Japan, earthquake stemmed not from recent cheap housing, but from shelters reconstructed after World War II to serve the rebirth of the industrial hub. Those most injured by Hurricane Katrina had been pushed into the low-lying Ninth Ward over decades by nefarious economics, politics, and racial machinations long compiled (Browne 2015; Hartman and Squires 2006).

Along with a people's factual—or quasi factual, for history is also chiseled—the tale of a threats and past happenings can also be found in a people's expressive culture, something anthropology also delves into. Policymakers and program staff may not consider a people's art very important, yet in every culture the character of the surroundings and incidents that once took place appear in mythology, legends, art, and writings. Told around a fire, read in books, hanging on a wall—the portrayals function as oracle and speak of consequence. They set forth an illustration of potential and, most importantly, foretell what a catastrophe will be like. Since a people's portrayals also often repeat, possibly in a number of forms, their narrative becomes further fixed in the culture's proficiency. Stories range from Noah's and Manu's floods, to Vesuvius and Gomorrah. They tell of unnamed incidents of shaking earth, monumental eruptions, and gigantic waves. At times the tales impel action, and at other times they prompt dismissal. Accounts spinning the insignificance of prior hurricanes led people in Hurricanes Sandy and Katrina to deny weather predictions and remain in their houses (Hoffman 2002, 2005). As sea levels rise around Chesapeake Bay, the inhabitants invoke long-told anecdotes of erosion rather than acknowledge climate change (Fiske et al., 2014). Northwest Coast songs and art tell of big waves that destroyed whole settlements and forced displacement, yet today's policies speak of tsunamis as if unheralded (Howell 2016). Artistic depictions, including contemporary factual and fictional television and motion pictures, likewise illustrate images of menaces and banes. They, too, convey a pervasive idea of what adversity exists, what mischance might occur, and what action to take enough to influence people. Japanese cartoons and modern painting focused on nuclear meltdown as theme well before Fukushima.

Anthropologist George Lakoff and Nobel Laureate and psychologist Daniel Kahneman both tell us that preformed concepts tend to become the formats by which people filter happenings and determine action, whether or not those formats take into account factual reality. A culture's anecdotes, experiences, and images do just that. They tender a template by which people recognize, dismiss, or readjust actuality. Moreover, when a compelling impression clashes with fact, it is the impression, not the fact, that usually prevails. Kahneman (2011) calls it the "illusion of validity." Memory, he says, automatically provides the narrative about what is going on, true or not, and suppresses alternatives (Kahneman 2011; Lakoff and Johnson 1980). To put it another way, surmises gleaned from story or experience create a pivot differentiating what a culture provides as a model *of* a peril or calamity into what becomes a model *for* a danger or happening. That is, whatever the culture has told people about a risk or disaster contained in their physical, modified, or constructed surrounds is what they will expect and what they will act on. Granting that the model can at times revise, perhaps due to a contradictory event, still this mechanism of mind as it affects any people is most assuredly good to know for anyone in the disaster business, although the impact of prior models has little entered the consideration of officials calling for mitigation, rehabilitation, or evacuation (Hoffman 2014, 2017).

People also have cultural explanation, often fixed, of where phenomena emanate from, sometimes in explication of their environment and sometimes not, and those explanations can quite stubbornly deflect any external explanation or advice. Eruptions may be due to an angry god, often a double-faced one like Pele. Massive waves are roused by a fickle sea spirit. Cause might be charged to an error in people's ways, lack of respect, straying from ceremony, or an individual's dire sin. The interpretations may be integral to a people's religion, overlap with their morality, or simply be part of their long adhered to cosmology and cosmology as epistemology. Whatever the foundation, the culture's own ingrained explanations can well operate like a wall of insouciance toward any newcomer's admonitions.

Other underlying aspects of culture—again aspects that may remain obscure to officials or agents—can furtively deflect any agenda as well. For one, the agents of policy often assume that official governing authorities, like a state governor, a mayor, village president, or other, holds the power to direct hazard mitigation or disaster response. However, while those officials may be part of how a local community is formally articulated to larger society, they may not actually bear effective mastery over a population. Rather, within almost every culture, there exists a set of integral organizing structures, both manifest and latent, that instead determines who truly has jurisdiction of the people. The assumption that titled administrators hold sway very likely ignores a deeper examination of where real bidding lies. Very commonly, command over a people's decisions and actions lies within their kinships—that is, family. If so, actual power may reside in the hands of a father, mother, grandfather, grandmother, oldest sibling, or other prominent relative, and not a political functionary. Outsiders also may often view a people's kinship in terms of the rather modern phenomena of small nuclear families, perhaps like their own, which acts on its own. They neglect that extended kinship groups prevail in many places. Outsiders may further assume that people recognize blood through both father and mother, as many of them do. This is not necessarily the case. Some people calculate blood relations only through their father and his line or mother and hers. Even if descent is calculated through both father and mother, frequently one side carries more weight than the other. Throughout most of European culture, for example, despite kinship seen through both parents generally the father's descent carries more weight, at least symbolically in last names and in canonic legal right. However, among other people such as a particular ethnic or tribal group, the mother's line dominates. In China, for instance, the paternal grandfather tends to be the patriarch of not just the nuclear family but also his entire descent line. It is at his decree that the whole family mitigate or endure a disaster, or leave. In contrast, in many regions of Louisiana, especially among certain ethnic enclaves, a set of related women, sisters, and aunts organize everyone and direct the comings and goings. If they, especially the maternal grandmother, says it is time to go, that is when, and not before, the family takes to the road (Browne 2015) and they further attempt to collect together wherever they are displaced. Among some people, certain kinship groups and persons within those groups hold higher status, and thus the ordinance, than others. People will act according only to the command of the genealogically designated leader, in short the chief male or female.

Disregarding the intricacies of kinship can lead to serious adversity, and the kinship system may be more complex than at first seems apparent. In Santorini, Greece, after the 1956 earthquake, the incoming reconstruction team ignored that, despite the typical Greek patronymic last names and dominant male line, underneath a hidden matrilineal arrangement persisted, to wit, village houses were owned and passed down through the women, not the men. Indeed, a mother's own house was given to her eldest daughter upon marriage, with subsequent daughters receiving new houses constructed nearby. As a result, village neighborhoods were sororal and featured collections of mothers, sisters, and aunts. Men, quite contrary to many other parts of Greece, moved about the village from their mother's house to that of their wives. In total inattention, the government builders gave house and house title to men and randomly assigned the new residences, breaking up the matrifocal neighborhoods that had previously given women solace from nearby female kin, legacy, and their one modicum of wealth. The actions proved highly disruptive to the villagers and added to their disenfranchisement and struggles. In reverse, years later, after the Athens earthquake of 2011, the native mayor of heavily impacted Ana Louisos, a Pontian Greek suburb on the outskirts of Athens, recognized that homes there belonged to women. The mayor rebuilt neighborhoods as they were, gave new homes back to women, and even nominated women to run the temporary tent communities. As a result, the reconstruction of the suburb proved quite successful (Hoffman 2017).

Among some people, rather than kinship individuals owe their land use, employment, status, and entailed obedience to a patron. A patron is generally a person from a higher class or a business owner whose position and prestige, and, most importantly, ability to grant favor, provides guardianship over those less advantaged. In such cases, it may be the various patrons of a place who possesses authority over the populace, not officials nor family. The bond between patron and client is often passed down from generation to generation. The tie may be confirmed through ceremonial means, such as by the patron sponsoring a baptism or wedding and becoming a godparent, or the tie may remain officially unofficial. It may entail formal deference in behavior or bestow on the parties quasi-siblinghood. The patron's favors may include employment, land use, protection, or promotion, but whatever the endowment provides the client's essential subsistence. Therefore, risk reduction, means of reconstruction, abandonment of place, or return, may lie at a patron's behest, and not in the control of a governor or agent. In fact, a patron's desires quite frequently come into conflict with the recommendations of outsiders.

In yet other instances, rather than a kinship coterie or a client-patron set-up, people in a particular culture might be members of a governing association. The association might be based on employment, propinquity, shared tillage, or co-operative chore arrangements like the *minga work groups* of Ecuador (Faas 2017). Membership may follow from legacy, a vote cast, or through payment. Group boundary and group rule may be rigid or flexible. Sometime confederacies follow from church affiliation and have religious stricture behind them or from ethnic faction, or communal customs. Nonetheless, just as with kinship or patronage, an association may hold decision power over members' actions and lives. All in all, whatever the underlying system, mitigation of a hazard or response to a disaster may not lie with individuals or occur in expected fashions. They may lie with the culture's decreed blocs, cliques, and divisions.

Kinship, patronage, and association also entail another sort of hold. They involve who people turn to when in need, which is not necessarily the disaster center or aid group. It is been hypothesized that where people consider blood relations to be through both parents they may have more possibilities in finding shelter. Where kinship is unilineal it forms a distinct social unit, or clan, and it or a member's associations may be able to absorb a single person's loss through collective, and binding, resources. All the underlying yet compelling systems of connectivity, kinship, patronage, and associations play a part in determining how people deal with threat and adversity and provide different kinds of agency. They also embody a drawback. Very commonly, they bring with them conventions from the past, so that in situations of peril or calamity people often return to and reaffirm traditional customs, no matter how modern they may have become. Often the old customs inhibit coping and adaptation, and those intervening do not see why, but once more it merits learning a people's abiding practices (Hoffman 1998, 1999a).

It is important to note that people in different cultures may operate altogether as an entity, and not only as an extended family, an association, or a grouping of clients, but also as the whole community. Individualism is not always admired nor sought; in fact, it quite rarely is. Rather, resolutions and actions may be determined as a cultural unit. Decisions and actions can also be constrained by the cultural whole. There may exist a subduing code of propriety. There is also the loss of social standing and, very significantly, the pressure of gossip. By and large, humans in general, no matter the culture, are askance to do what others are not doing. From hamlet to city avenue, they will not act unless others do the same, and it is critical for any expert, official, and help to remember that humans are innately communal.

Connected to kinship, association, and patronage are two matters that may endure in a culture that can drastically affect a people's recognition of danger and ability to act. Most crucial of these is land tenure. The people within a culture might have purchased land and home, or inherited their property down generations; the grasp of property cannot be overstated. The holdings might include house, fields, water, and pasture rights. They could entail distributional shares or seasonal grassland. In any case, the clutch of land tenure—both ownership and right of use—is formidable, compelling, and undeniable. It pulls people. It makes them stay where they are, as they are, and it makes them return. Within many cultures, the holdings mean more than livelihood—though that is compelling enough—they impart the identity, status, and glory of legacy. They imply enrichment, bear future prospects, and, most crucially, implicate justice. People will fight for their land no matter if faced with loss, change, and most certainly with appropriation. Loss involves rightness.

Within a culture's ethos, tenure can also extend beyond just home. It may involve objects—that is, a person's furnishings and trinkets. Cultures differ in their implicit evaluation of goods and accumulation. Some people are imbued to treasure cars, clothes, jewels, pots, pans. Other are exhorted to forgo attachment and leave trappings behind. After the 2004 tsunami in Sumatra, villagers placed flags on heaps of debris that were indistinguishable to others to mark the location of their former gardens. In Florida people will pile their cars full of the most seemingly inconsequential possessions when fleeing hurricanes. In Myanmar, people simply walk away from flooding houses. The materials they need to reconstruct when the waters dissipate are readily procurable again. It almost goes without saying that land use, ownership, inheritance, and objects also intermesh with place attachment.

How people are primed to define their cultural character is quite material to the enactment of policy or any program. Some cultures, and thus the people within them, esteem and inculcate compliance. Other cultures encourage insubordination, affront, and bravado. Quite clearly, earthquake survivors in Nepal respond very differently from survivors in Italy. For some, resistance is called for. It may be temporary or endless. Different cultures also prescribe different rules of etiquette. Adhering to or slighting the rules can enable success or lead to the failure of any intervention. In some societies, men do not address unrelated women nor do women speak to men. In others, certain family members are not allowed to speak to certain others or are only allowed to joke, and not impart anything of consequence. In some cultures, it is impolite to say "no," so that authorities and agents are continuously met with "yes," whether parties intend to follow through or not. In some, certain families claim privilege and special treatment. They believe they are entitled due to rank or religion. Occupations may be so culturally assigned that converting people to different labor is impossible.

Studies have shown that type of employment, level of education, literacy, native tongue, health, and healing systems affect a people's attitude toward the hazard they face, or their recovery and resettlement (Bolin and Stanford 1999). Class, race, ethnicity, age, and generation also greatly bear on risk acceptance and disaster survival. Among some people, age is esteemed and the authority of elders cannot be contradicted. In others, youth and the desires of the young are highly valued. The makeup of the community is of significance—that is, if all the people come from a similar background or if diverse sectors and factions exist. Is there unity in general or contentiousness? Is crime high or low? Do people trust or distrust? The issues surrounding gender and women's abilities worldwide are manifold. Phillips outlines these issues in this volume and in many other works (chapter this volume; Enarson and Morrow 1998).

Another major concern surrounds whether a people have ever been colonized, and what repercussions the colonization left behind. Among some, the structure of the imposed dominance resulted in stifling bureaucracies in which nothing can be decided or done. Any number of competing bureaus and bureaucrats must approve. In others, there remains a legacy of class rigidity with so little wiggle room left, some people suffer enduring deprivation. Others communities are heir to dogged post-colonization legacies of corruption and payoffs. In some, indigenous methods of handling risk and calamities were so uprooted and replaced with imposed and inappropriate practices, their living conditions of the people today are far more perilous than originally (Oliver-Smith 1999). And among almost all once colonized, a kernel of resistance lingers that can readily turn on any outsider, program, or decree.

Persons affected by risk or calamity everywhere may also have their own opinions, or frames of what is perilous, who is in danger, and who suffered from the disaster. Those opinions may strongly differ from those coming in, but are key to clarity of any official's or agent's actions. On top of that, every mile upon mile, village upon village, within any region faces hazards and undergoes upheaval differently. No two locations are exactly alike, and, for those entering, it is crucial to recognize how discretely disparate any particular hazard and disaster is, depending on the precise site. When the word "local" is quoted, it truly means local. Risk and disaster understanding are always local. One cannot stress that more. And, always, absolutely always, there remain the root causes of risk and disaster: poverty, hunger, illiteracy, and woebegone living conditions.

The catalogue of the many deep-seated yet profound ways cultures may differ presented here is by no means complete. In addition, different people not only have different languages, and perhaps distinct local dialects within them, or share a major world language but speak an indigenous one, they often have different body language and gestures as well. In Turkey a tip of the head down means yes, while a tip of the head up means no, not sideways movement. In short, a Western nod may imply agreement and lifting a head to note attention may mean no. Very likely there are different ideas of what constitutes an ailment, some completely unique to a culture, along with ideas of treatment and medication. Work ethics differ. On the island where I often work, a north wind means too cold to work and a south wind means the fishing is too good to work, and everyone understands. Tolerated ideas of cleanliness vary and also order. Child rearing practices drastically diverge, in some places the idea of a bedtime simply does not exist. Modesty and personal space dictates diverge, everyone together on one sofa; everyone apart in an isolate chair. Pride, and what constitutes it, differs and offence may be taken when its subtleties are even accidentally crossed. Notions of justice differ and certainly religious practices, beliefs, and celebrations do, including informal, often clandestine, long-rooted folk ones that may strongly influence what to avoid and how to manipulate fate. In some cultures competition is utterly discouraged and anyone who dares to get ahead instantly scorned; steadfast uniformity rules. In others, competition and improvement underlie daily striving and goals, sometimes to the point that anything others receive mean less for another. While these possible differences are posed as opposites here, they are actually present on a continuum. Cultures express them to various degrees.

As a final point, there is no denying that many people in today's world no longer live in what was their original milieu. That does not mean, however, that, although they have been displaced from primary environments, they have been displaced from their culture. Much of the world's migration is, after all, chain migration. Once one person from a place or family moves, others follow. As a result, many cities feature patchwork neighborhoods populated entirely by the former denizens of a certain place. With them comes their customs. Authority in the face of peril still lies with prior governance. Determining behavior patterns and cosmological precepts continue as before. Perhaps there is pressure to change. Sometimes change occurs. But culture is persistent and it behoove any official or agent to know where a people, and from what, a people came.

Conclusion

Let me give one last reason anthropology is essential to the study of disaster. It is, in fact, the basic reason. Despite years of attempts, it is clear that there are simply no physical solutions to risk and disaster—no seawall, no rechanneling the lava flows, no windbreaks, no cloud seeding, no complete retrofitting, no boron safety rods, no thinning the trees, no levees. After all, one person's levee is another person's flood, and warning systems only warn. After much examination and development in the entire risk and disaster field, encompassing seismology, climatology, city planning, meteorology, engineering, agronomy, ecology, environmental studies, and other pertinent studies, all agree on one basic premise:

disasters do not spring up like sudden happenings from some unknown sphere. All disasters are human caused in one way or another.

In short, there is no such thing as a natural disaster. It matters not if the disaster springs forth from the fundamental terrain, from the modifications that have taken place, or from the contrivances of human manufacture, all have a people part to their creation and unfolding. Therefore, they all have a cultural substance and all have a societal foundation. Without the presence of humans, earth's disruptions are merely events of environmental process, not calamities (Oliver-Smith and Hoffman 1999; Blaikie et al. 1994; Hartman and Squires 2006). To constitute a disaster, people are present. They have planted themselves in a certain place. They have done things and made things. Thus all the solutions are cultural and social, including decisions and actions to build the walls, put on hurricane straps, move the community, back away from the forest, or evacuate. To attribute a disaster to the mere ecological happening is like assigning a noise to "one hand clapping." It obfuscates the underlying hand.

In this chapter and all the others in this volume, it is also important to remember that, while we largely talk about major disasters, many people live with minor disasters—a mudslide, a flood, lack of rain—on a daily basis. So, I will end with how I always end the talks I give on risk and disaster. Not only is there no such thing as a natural disaster, there is no such thing as a small disaster. Big or small, no matter when, what, or how, all people, and places, and cultures are equally valuable.

Susanna M. Hoffman (Ph.D., Anthropology, UC Berkeley) is author, coauthor, and editor of twelve books, including *The Angry Earth*, first and second edition, and *Catastrophe and Culture*, two ethnographic films, and more than forty articles. She initiated the Risk and Disaster Thematic Interest Group for the Society for Applied Anthropology and is founder and chair of the Risk and Disaster Commission for the International Union of Anthropology and Ethnographic Sciences. She was the first recipient of the Aegean Initiative Fulbright concerning the Greek and Turkish earthquakes, and helped write the UN Statement on Women and Disasters.

References

Altman, I., and S. Low, eds. 1992. *Place Attachment: Human Behavior and the Environment,* 12th ed. Thousand Oaks, CA: Sage.
Berlin, B., and P. D. Kay. 1969. *Basic Color Terms.* Berkeley: University of California Press.

Blaikie, Piers, T. Cannon, I. Davis, and B. Wisner. 1994. *At Risk: Natural Hazards, People's Vulnerability and Disasters.* London: Routledge.

Boas, Franz. 1940. *Race, Language, and Culture.* Chicago: University of Chicago Press.

Bolin, R., and L. Stanford. 1999. "Constructing Vulnerability in the First World: The Northridge Earthquake in Southern California, 1994." In *The Angry Earth: Disaster in Anthropological Perspective,* edited by A. Oliver-Smith and S. Hoffman, 89–112. New York: Routledge.

Brightman, R. 1995. "Forget Culture: Replacement, Transcendence, Relexification." *Cultural Anthropology* 10, no. 4, 509–46.

Browne, K. 2015. *Standing in the Need: Culture, Comfort, and Coming Home After Katrina.* Austin: University of Texas Press.

Button, G. 2010. *Disaster Culture.* Walnut Creek, CA: Left Coast Press.

Cosgrove, D., and S. Daniels. 1988. "Introduction: Iconography and Landscape." In *Iconography of Landscape,* edited by D. Cosgrove and S. Daniels, 1–10. Cambridge: Cambridge University Press.

Enarson, E., and B. Morrow, eds., 1998. *The Gendered Terrain of Disasters: Through Women's Eyes.* Westport, CT: Greenwood Publishing Group.

Faas, A. J. 2017. "Enduring Cooperation: Space, Time, and Minga Practice in Disaster-Induced Displacement and Resettlement in the Ecuadorian Andes." *Human Organization* 76, no. 2, 99–108.

Feld, S., and K. Basso, eds. 1996. *Senses of Place.* Santa Fe, NM: School of American Research.

Fischer, M. 2007. "Culture and Cultural analysis as Experimental Systems." *Cultural Anthropology* 22, no. 1, 1–65.

Fiske, S., S. Crate, S. Crumley, C., K. Galvin, H. Lazrus, L. Lucero, A. Oliver-Smith, B. Orlove, S. Strauss, and R. Wilk. 2014. *Changing the Atmosphere: Anthropology and Climate Change: Final Report of the American Anthropology Association Global Climate Change Task Force.* Arlington, VA: American Anthropology Association.

Firth, R. 1959. *Social Change in Tikopia: Restudy of a Polynesian Community After a Generation.* London: Allen and Unwin.

Frake, C. 1962. *Cultural Ecology and Ethnography.* Washington, DC: American Anthropological Association.

Friedel, D., I. Schele, and J. Parker. 1993. *Maya Cosmos.* New York: Morrow.

Gibson, J. 1979. *The Ecological Approach to Visual Perception.* Boston: Houghton Mifflin.

Gill, T. 2013. "This Spoiled Place: People, Place and Community in an Irradiated Village in Fukushima Prefecture." In *Japan Copes with Calamity,* edited by T. B. Gill, B. Steger, and D. Slater, 201–34. Bern, Switzerland: Peter Lang Publishers.

Gupta, A., and J. Ferguson. 1992. "Beyond 'Culture': Space, Identity, and the Politics of Difference." *Cultural Anthropology* 7, no. 1, 6–23.

Harris, M. 1968. *The Rise of Anthropological Theory.* New York: Routledge.

Hartman, C., and G. Squires, eds. 2006. *There Is No Such Thing as a Natural Disaster: Race, Class and Hurricane Katrina.* New York: Routledge.

Hoffman, Susanna. 1998. "Eve and Adam among the Embers: Gender Patterns After the Oakland Berkeley Firestorm." In *The Gendered Terrain of Disasters: Through Women's Eyes,* edited by E. Enarson and B. H. Morrow, 55–61. Westport, CT: Greenwood.

———. 1999a. "The Regenesis of Traditional Gender Pattern in the Wake of Disaster." In *The Angry Earth: Disaster in Anthropological Perspective,* edited by A. Oliver-Smith and S. Hoffman, 173–191. New York: Routledge.

———. 1999b. "After Atlas Shrugs: Cultural Change or Persistence After a Disaster." In *The Angry Earth: Disaster in Anthropological Perspective*, edited by A. Oliver-Smith and S. Hoffman, 302–326. New York: Routledge.

———. 2002. "The Monster and the Mother: The Symbolism of Disaster." In *Catastrophe and Culture: The Anthropology of Disaster*, edited by S. Hoffman and A. Oliver-Smith, 113–42. Santa Fe, NM: School of American Research.

———. 2005. "Katrina and Rita: A Disaster Anthropologist's Thoughts." *Anthropology News* 46, no. 8, 19.

———. 2013. "The Flight of a Firebird: Becoming a Disaster Anthropologist and Practitioner." In *Handbook of Practicing Anthropology*, edited by R. Nolan, 114–23. Hoboken, NJ: Wiley and Sons.

———. 2014. "Culture: The Crucial Factor in Hazard, Risk and Disaster Recovery: The Anthropological Perspective." In *Hazards, Risks, and Disasters in Society*, edited by A. Collins, S. Jones, B. Manyena, and J. Jayawickrama, 289–394. London: Elsevier.

———. 2016. "The Question of Cultural Continuity and Change After Disaster: Further Thoughts." In *Continuity and Change in the Applied Anthropology of Risk and Disaster*, edited by A. J. Faas, *Annals of Anthropological Practice* 40, no. 1, 39–51.

———. 2017. "Disasters and Their Impact: A Fundamental Feature of Environment." In *Handbook of Environmental Anthropology*, edited by H. Kopnina and E. Shoreman-Ouimet, 193–205. London: Routledge.

Hoffman, S., and A. Oliver-Smith, eds. 2002. *Catastrophe and Culture: The Anthropology of Disaster*. Santa Fe, NM: School of American Research.

Howell, W., and G. Kenneth. 2016. "The Sixth Wave: Tlingit Cultural Responses to the Giant Tsunami." Paper presented at the Society of Applied Anthropology Meeting, Vancouver, British Columbia, 1 April.

Ingold, T. 1992. "Culture and the Perception of the Environment." In *Bush, Base: Forest Farm*, edited by E. Cross and D. Parkin, 39–55. London: Routledge.

———. 2011. *The Perception of the Environment*. New York: Routledge.

Kahneman, D. 2011. *Thinking Fast and Slow*. New York: Farrar, Strauss and Giroux.

Kawabata, H., and S. Zeki (2000). "Neural Correlates of Beauty." *Journal of Neurophysiology* 91, no. 4, 1699–705.

Kroeber, A. 1948. *Anthropology: Race, Language, Culture, Psychology, Prehistory*. New York: Harcourt Brace.

———. 1952. *The Nature in Culture*. Chicago: University of Chicago Press.

Lakoff, G., and M. Johnson. 1980. *Metaphors We Live By*. Chicago: University of Chicago Press.

Low, S., and D. Lawrence-Zuniga. 2003. *The Anthropology of Space and Place: Locating Culture*. Oxford, UK: Blackwell.

Nicholson, I. 1967. *Mexican and Central American Mythology*. London: Paul Hamlyn.

Oliver-Smith, A. 1999. "Peru's Five-Hundred Year Earthquake." In *The Angry Earth: Disaster in Anthropological Perspective*, edited by A. Oliver-Smith and S. Hoffman, 74–88. New York: Routledge.

Oliver-Smith, Anthony, and Susanna Hoffman, eds. 1999. *The Angry Earth: Disaster in Anthropological Perspective*. New York: Routledge.

Ortner, S. 1996. *Making Gender: The Polities and Erotics of Culture*. Boston: Beacon Press.

Paine, R. 2002. "Danger and the No-Risk Thesis." In *Catastrophe and Culture: The Anthropology of Disaster*, edited by S. Hoffman and A. Oliver-Smith, 67–90. Santa Fe, NM: School of American Research.

Piaget, J. 1926. *The Language and Thought of the Child*. London: Routledge and Kegan Paul.

Rajan. 1999. R. 1999. "Bhopal: Vulnerability, Routinization, and the Chronic Disaster." In *The Angry Earth: Disaster in Anthropological Perspective*, edited by A. Oliver-Smith and S. Hoffman, 257–277. New York: Routledge.

———. 2002. "Missing Expertise, Categorical Politics, and Chronic Disasters: The Case of Bhopal." In *Catastrophe and Culture: The Anthropology of Disaster*, edited by S. Hoffman and A. Oliver-Smith, 237–260. Santa Fe, NM: School of American Research.

Scarry, E. 1989. *Love and Beauty*. Princeton, NJ: Princeton University Press.

Sircell, G. 1975. *A New Theory of Beauty: Essays on the Arts, 1*. Princeton, NJ: Princeton University Press.

Steward, J. 1955. *The Theory of Culture Change*. Urbana: University of Illinois Press.

Taussig, M. 2010. *What Color Is the Sacred?* Chicago: University of Chicago Press.

Tyler, E. 1871. *The Interpretation of Cultures*. New York: Basic Books.

Zhang, Q., and R. Barrios, 2017. "Imagining Culture: The Politics of Culturally Sensitive Reconstruction and Resilience Building in Post-Wenchuan Earthquake China." In *Responses to Disaster and Climate Change: Understanding Vulnerability, Building Resilience*, edited by M. Companion and M. Chaiken, 93–102. Boca Raton, FL: CRC Press.

CHAPTER 12

Engaged

Applying the Anthropology of Disaster to Practitioner Settings and Policy Creation

KATHERINE E. BROWNE, ELIZABETH MARINO, HEATHER LAZRUS, and KEELY MAXWELL

Introduction

In the past two decades, a growing body of anthropological work on disasters has generated case studies and theoretical advancements that contribute to a more robust understanding of the human condition under extreme stress. The momentum for the rapidly expanding field of disaster anthropology owes much to early publications by Susanna Hoffman and Anthony Oliver-Smith (Hoffman and Oliver-Smith 1999; Oliver-Smith 1996). Anthropologists today routinely chronicle the human impacts of profound loss and, in the process, document policies, institutions, and communities where anthropological insight can be applied to lessen the suffering of human beings. That suffering, anthropologists have learned, often (and ironically) arises from fundamental cultural misalignments between institutions offering aid and communities receiving aid.

In this chapter we aim to make explicit some of the most critical insights to date from the anthropology of disaster. Rather than offering a comprehensive review, we will report on those insights that have resonated most with our own experiences working at the juncture of communities and policymaking institutions in the wake of disasters. We show how and why these insights have difficulty gaining traction with practitioners and policymakers. We conclude with a set of recommendations for bridging these gaps. To organize this inquiry, we ask three questions: (1) What do we currently know about the causes of disaster, about reducing their impact, and about managing impacted communities when disaster does occur? (2) What are the differences between academic anthropological work on disasters and practitioner work on disasters? and (3) What recommendations can help academics and practitioners bridge gaps between knowledge and practice? Our responses to these questions have a bearing on how we theorize disaster, how we operationalize disaster as an organizing concept, how we understand the application of anthropological knowledge in disaster contexts, how anthropological knowledge is different from practitioner knowledge, and how asking these questions gives rise to theoretical insights into the roles of both academics and practitioners.

One of the ways in which anthropological insight on disasters is distinct from other social science traditions is that it grounds disaster vulnerability, response, and risk in colonial and postcolonial histories and contemporary neoliberal political economic traditions (Barrios 2017). Anthropological insight adds needed cultural dimensions to our understanding of post-disaster trauma (Browne 2015). Yet, anthropologists can be challenged when engaging with academics and practitioners from other disciplines. By the term "engagement" we mean the collective processes of relationship building with other disaster specialists and practitioners, influencing the decision making of, or policy creation for, disaster specialists, and/or explicitly attempting to mitigate negative outcomes for communities that suffer when a disaster occurs. Engaged anthropology, as defined above, is a complex and fuzzy term that may include activist anthropology (the political alignment of anthropologists with "an organized group of people in struggle") (Hale 2006), or it can stand outside of alignment with a particular group. For purposes of this discussion, we define "engaged anthropologists of disaster" as those who reflexively and pragmatically consider how research findings can be applied to disaster scenarios in order to reduce suffering.

Engaged anthropologists recognize the challenge inherent in bringing anthropological insights to bear in disaster interventions. Hurricanes, radiological releases, earthquakes, space weather, industrial accidents, and climate change can all be considered disasters, but they vary greatly in terms of their spatial extent, chronic or acute duration, anthropogenic causation, capacity for risk management, hazard risk, and perceptions of risk. Geomagnetic storms related to space weather can damage critical infrastructure, while extreme heat poses health risks. Sociological traditions contain a rich literature on the social situatedness of hazard risk and risk management in modern industrial society (Beck 1992). Emergency responders in the United States may not be able to attend to deep social histories and diverse cultural needs using practices established under the Stafford Act and National Response Framework. Engineers, ecologists, and public health officials may point out that incidents as diverse as bioterrorism, flooding, and tornadoes present discrete physical and health risks that must be considered. Anthropologists can engage with these diverse perspectives to better parse out and communicate how anthropological insights apply to different types of disasters.

We, the authors, come from across the spectrum of practicing and academic anthropologists. Browne and Marino work within university contexts and have extensive experience collaborating with nonacademic partners including those in policy institutions. Lazrus works at a federally funded research and development center that produces fundamental science in service to society, thus effectively straddling the academic and practical arenas. Maxwell works for the research arm of a federal management agency. Our perspectives in this chapter reflect our positions in these various institutions as well as our fieldwork, research, and collaborations. We note that any differences between academic and practicing anthropologists do not form a binary, but instead fall along a continuum that defies easy categorization. We contend that theoretically informed work often has important practical implications and practice often informs and builds theory.

The unconventional, bulleted form of this piece represents a deliberate choice. Because many of these points are already embedded in the literature, we chose to crystalize and lay them out in a format that suits the urgency for applying this material in practical ways.

Question 1: What Do We Currently Know about the Causes of Disasters, about Reducing Their Impact, and about Managing Impacted Communities when Disaster Does Occur?

Margaret Mead said that anthropologists in the field "must wait for events to reveal much that must be learned. A storm, an earthquake, a fire, a famine—these are extraordinary events that sharply reveal certain aspects of a people's conception of life and the universe" (Mead and Metraux 1970, 311). To be sure, anthropologists end up studying disasters for many reasons: a catastrophic event occurs in the community where they have gone to study, they are drawn by the aftermath of an event, or they seek to understand aspects of a community that perseveres in the face of heightened risk. In any case, what anthropologists learn in the field when a disaster threatens or strikes does reveal deeper systems and structures that we learn are part of both the occurrence of disaster and the recovery from disaster (Garcia-Acosta 2002).

Causes of Disaster

Anthropologists and others who study disasters continue to learn much about the root and proximate causes of disaster. Here are five concepts we believe we understand in this regard:

- We know that there is no such thing as a natural disaster (Hoffman and Oliver-Smith 1999; O'Keefe, Westgate, and Wisner 1976). Disasters are produced by the confluence of hazards, which themselves may be socially produced, and social relationships. Furthermore, as Smith points out, "Whatever the political tampering with science, the supposed 'naturalness' of disasters here becomes an ideological camouflage for the social (and therefore preventable) dimensions of such disasters" (2006, 1–2). While recognizing that the natural world does exert pressures on human populations that have the potential to cause harm (Hoffman and Oliver-Smith 1999), this harm is inequitably distributed and often preventable and it is human decision making across time that makes this so.
- We know that the root cause of disaster, whether acute or chronic, is almost always embedded in social histories of inequality produced by power relations (colonial or otherwise) that make people unequal in the face of hazards (Barrios 2017; Laska and Morrow 2006; Marino 2012; Oliver-Smith 1996; Schuller 2016). These inequalities can include increased exposure to hazards, lack of economic resources, political oppression, and gender discrimination, among many others. In this context, we can say that the contours of disaster and the difference between who lives and who dies is to a greater or lesser extent a social calculus (Browne 2015; Marino 2015; Oliver-Smith 1996; Smith 2006). Some communities, as some geographical spaces, are—formally and informally, in codified and implicit ways—designated as sacrifice zones where risk is compiled (Buckley and Allen 2011). For example, when people around the world saw images of stranded African American communities in the deadly sewage pond of post-Katrina New Orleans, Illinois senator Barack Obama (the son of an anthropologist) said, "The people of New Orleans weren't just abandoned during the hurricane, [they were] abandoned long ago" (Obama 2005). Anthropologists have demonstrated that social and colonial histories create disasters in the wake of earthquakes (Oliver-Smith 2012), or flooding linked to climate change (Marino 2012, 2015) because they are tied to decisions that stretch back decades, even centuries.
- We know that neoliberal policies can increase risk (Buckley and Allen 2011; Tierney 2015). Shifting the burden of disaster risk reduction and disaster risk management to private companies only increases public exposure, especially among people who cannot afford private services, such as the kind Funk describes in *Windfall: The Booming Business of Global Warming* (2014). One example is the growing trend of relegating firefighting services to private companies in parts of California. Of course, only some people can afford this basic protection.

- Evidence suggests that a warming planet, or what Monbiot (2013) calls "climate breakdown," has accelerated and intensified many types of disasters including storms, flooding, wildfire, and heat waves (United States Global Change Research Program 2017). Moreover, the rising incidence of water scarcity, drought, waterborne diseases, and infectious diseases of many types, as well as potential human conflict, all exist in the wake of global warming (Barnett and Adger 2007).
- We have seen a disturbingly frequent occurrence we call a second (and completely human-induced) disaster that is caused by outside authorities' disregard for local needs, local adaptations, and local knowledge. Institutional ignorance, even if unintentional, produces culturally and locally inappropriate solutions that can cause an equally painful disaster on top of the first disaster. This second disaster can be triggered by the dismantling of preexisting social networks, the undermining of local economies, the displacing of local leaders (Schuller 2016), the building of locally inappropriate infrastructure (Barrios 2017), and the manipulating of post-disaster contexts for the purposes of land-grabbing (Leckie 2005), among many other misguided actions. As an emblematic ethnographic insight, Browne's work in post-Katrina New Orleans showed how much more significant this second disaster was than the first disaster in terms of the suffering it caused survivors (Browne 2015).

Mitigation of Disasters

Just as there are some things we know about the human causes of disaster, anthropologists also have learned a good deal about disaster mitigation. According to the Federal Emergency Management Agency (FEMA), the term "mitigation" refers to the effort to reduce loss of life and property by lessening the impact of disasters. In mitigation, we can pair what we know about causes to specific kinds of action.

- We know that social inequity is a root cause of disaster in that it affects how risks are distributed and why certain populations are more exposed to and susceptible to different types of hazards (Fothergill and Peek 2004; Hardy et al. 2018; Laska and Morrow 2006). By extension, we also know that an effective way to mitigate disaster risk is to reduce inequality, including political, social, economic, and environmental inequities (Fothergill and Peek 2004).
- We know addressing risk involves embracing the reality that disasters are a normal part of life, and not an out of the ordinary event. Thinking of disasters as outside of what is normal underscores our separation from nature. If, as Oliver-Smith (1996) and others have argued, disas-

ters are instead understood as one end of the continuum of everyday life, then our opportunity for understanding our role in producing risk and disaster becomes real and solutions become more apparent. This orientation to hazards as normative becomes even more critical since climate change and sea level rise increase the likelihood that conditions such as high water become an ecological norm and multi-annual event in many places around the world (Marino 2018).

- We know that there is an urgency to reduce risk and vulnerability, as the Sendai Framework for Disaster Risk Reduction spells out (United Nations International Strategy for Disaster Risk Reduction [UNISDR] 2015). Disasters are getting more expensive because of twin factors related to population: increasing numbers of people in the world and increasing numbers of people living in high-risk areas, especially along the coasts (de Sherbinin et al. 2012; Neumann et al. 2015).

- We also know that proactive efforts to reduce risk are more cost-effective than cleaning up the devastation of a disaster after the fact (UNISDR 2007), which suggests an imperative to fund risk reduction efforts, including climate change mitigation (O'Brien et al. 2006). According to FEMA, for every dollar ($1) spent in risk reduction, $6 is saved in costs that would otherwise be expended in recovering from disaster losses (Multihazard Mitigation Council 2017).

- We know and are seeing that the rise in risk, vulnerability, and disasters is causing a rise in noneconomic costs as James Morrissey and Anthony Oliver-Smith's (2013) innovative work on non-economic losses and damages demonstrates. To mitigate loss and damage that are non-economic in nature, we must first recognize these and then agree that we must account for them and understand their inherent value in any community. Such noneconomic losses and damages may include, for example, the rise of chronic health problems, the loss of trust among family members, the erosion of faith in democracy and citizenship— social costs that can be brought about by the second disasters noted above (Browne 2015).

- We know, based on our collective experience, that efforts to reduce risk cannot effectively be achieved without the participation of local people. The academic/practitioner network called the Culture and Disaster Action Network (CADAN) has demonstrated through a variety of case studies just how unsuccessful and unsustainable risk reduction and disaster recovery projects are that do not take local values and practices into account. That is, local knowledge and consideration of place-based cultural practices support the agency of local people and enhance a form of resilience that is collective, enduring, and sustainable (Browne 2016; CADAN et al. 2017a, 2017b).

- Cultural patterns also figure importantly in local responses to weather forecasts and warnings. Human groups develop complex systems of knowledge about their environments that shape the very concept of hazard, support identification of risks, and inform decision making (Rockman 2003). Therefore, information that can increase prepared-ness, reduce vulnerability, or mitigate disaster outcomes must draw on a mix of sources including traditional environmental knowledge, official sources of forecasts, environmental and social cues, and information transmission through social networks and social media.
- The capacity to access, understand, and use information about the potential of a disaster is an important determinant of disaster out-comes. Preparation involves being able to anticipate the potential for harm. Preparation is hindered with absent or inaccurate information, a common problem that haunts the gap between academic research and practitioner work. Access to information that is understandable, relatable, and actionable is therefore an important component of miti-gating disasters. Not only should information be provided in necessary languages through relevant channels of transmission, but it must also be culturally appropriate (Hardy et al. 2018).
- Preparedness and risk communication that leverage and extend exist-ing adaptive capacities is efficient and effective. We know that when faced with a disaster, people behave within their cultural context including seeking further information and validation through social networks (Morss et al. 2017).

Management of Disasters

The timeline of a disaster might be said to move from root causes to mitiga-tion to management. Here are some things we know about management.

- We know that management is a multifaceted endeavor involving a mul-tiplicity of actors from housing specialists to fire chiefs to floodplain experts.
- We have observed that recovery efforts often center on rebuilding infrastructure, housing, and other aspects of the visible material envi-ronment; it is more challenging to foster recovery of the invisible but profoundly important social tissue of the community (Barrios 2017; Browne 2015). To the extent that there is something we can call recov-ery, it is a process and it is nonlinear. Recovery depends not only on restoration of the built environment, but also on fulfilling social needs and recognizing affective experiences (Barrios 2017) that are rarely factored into recovery efforts. Browne describes recovery as nothing

like healing a broken bone that gets better incrementally over time. Instead, she documents the long "lurching process, sometimes involving major setbacks or complete stalls . . . [in which] different emotional and material challenges come into view." Sometimes there is no closure (Browne 2015, 1).

- We know that disasters impact communities in complex, intersecting ways that include economic, social, cultural, historical, and environmental dimensions. As intersectional studies have shown (Crenshaw 1997) race, class, and gender identities produce interacting vulnerabilities. For anthropologists, sustainable disaster management involves addressing these cumulative human vulnerabilities—the variety of structured systems of inequality, as well as place-based histories and histories of disproportionate exposure to and impacts from other hazards and risks. Future disaster recovery efforts would benefit from operational strategies to anticipate and recover from disasters that embrace this level of complexity and multicausality.
- We have seen in some cases that disaster recovery efforts are being performed by private companies, NGOs, nonprofits, and faith-based groups, not government agencies (Browne 2015; Schuller 2016). In terms of sustainability, this shift raises important concerns. NGOs, nonprofits, and faith-based groups have agendas and often quotas that drive the kind of work they do and the length of time they have to do it. The concerns about multinational corporations hired to take on the work of cleanup and rebuilding raise even bigger questions about who gets these jobs and how many layers of subcontractors are paid before the work gets done.
- We know there have been distinct efforts to address community considerations, such as the the "whole community" approach, made doctrine by FEMA in 2011, and the "long-term recovery groups" that FEMA advocates for in the aftermath of disaster, and that are based in local communities. These have not always had the intended effect, however. For example, national volunteer groups and faith-based groups are limited in their effectiveness by the demands of distinct donors with agendas that are not tailored to address survivors' needs as much as to fulfill these donors' goals (Browne 2015; Browne and Even 2018).

When disaster recovery is privatized, disaster reconstruction can reproduce social oppression and exploitation. In the aftermath of Katrina, FEMA gave no-bid contracts to companies that routinely sought the cheapest possible laborers from other areas and outside the country. The Bush administration made the recruiting of cheap labor possible by suspending the law requiring federal contractors to pay the going local wage (Gotham 2012, Petterson et al. 2006). The population in the New Orleans area and employment in the construction industry underwent a noticeable shift towards Latinos (Petterson et al. 2006). The impact on local, African-American residents desperate for work was a double whammy of exclusion (Browne 2015).

- We know that local communities want to intervene after a disaster, but that the relationships between experts and non-expert "locals" can be challenging. After Hurricane Katrina and other disasters, reports anthropologists heard from the field indicated that responders and emergency managers shunned help from untrained local groups of volunteers that spontaneously formed to help rescue people and animals. This posture of bias against non-experts contributed to failures during the response phase of Katrina when experts turned away volunteers despite their own insufficient numbers to handle the scale of need. Because of this catastrophic performance, volunteer bands of local citizens were allowed during Harvey, and many of these groups were thus able to save hundreds and perhaps thousands of lives (Phillipps 2017). How these spontaneously formed groups will be viewed by and supported by government institutions (or not) is, yet, unclear.

- We know that during and immediately following a disaster, responses from different arms of public sector agencies come together in what organizational theorists call "virtuoso teams." One problem with these virtuoso teams is that they may be unfamiliar with each other's organizational language or culture. Additionally, by relegating these collaborative efforts to extraordinary rather than ordinary circumstances, they lose efficiency and effectiveness that could be avoided if readiness were an everyday state (Fischer and Boynton 2005). Cross-agency organization is further complicated by rigid mandates, state/federal disparities in operation and budgetary timelines, and high turnover rates among agency workers that limit institutional memory (Marino 2015).

- We know that communities are not homogeneous, and there may be many different groups of people with very different worldviews at play (Rayner and Cantor 1987). Cultural worldviews represent different ideals about how society should be structured, and about how society and nature should interact. People with differing worldviews may have correspondingly different ideas about risk, blame, and how disaster risk reduction and recovery should be managed (McNeeley and Lazrus 2014). Divergent cultural worldviews within disaster-affected communities are, thus, important to understand in order to craft management responses that are culturally congruent and socially palatable (Lazrus 2016).

- We know that the timelines of management agencies and those of communities do not always align. As Marino and Lazrus (2016) demonstrate from two very different communities—one in Alaska and in Polynesia Tuvalu—notions of time, efficiency, and flexibility are deeply cultural. Thus, part of the misalignments among disaster responders and communities are expressed via misaligned needs given different senses of time and timing.
- We know that managing a post-disaster situation requires paying attention to well-being. Yet, health-care interventions may assume that the trauma imposed by disaster in one area has the same meaning across different cultural settings. The natural implication of this assumption is that trauma is trauma and can be treated with the same approaches, a problem medical anthropologists call "Kleinman's Fallacy" for the work of anthropologist and physician Arthur Kleinman. After Katrina, for example, mental health provisioning was organized around models of conventional one-on-one therapist/patient exchanges. These individualized therapies did not correspond to the cultural practices of many African American families who follow group-style talk therapy (Browne 2015). Mental health teams that swoop into disaster sites assuming the universal relevance of dominant Western models may get the treatment wrong and create bigger problems (Watters 2010).

Question 2: What Are the Differences between Academic Anthropological Work on Disasters and Practitioner Work on Disasters?

Applying anthropological insights to disaster contexts is a goal for many academics because, ultimately, they are driven and ethically bound to frame and use research for the ultimate purpose of reducing suffering. Many disaster practitioners are motivated to partner with academics because practitioner knowledge and experience is critical for creating robust theories of disaster and because practitioners are often eager to access and apply research to the contexts in which they work. However, despite the clear overlapping goal of reducing suffering, these partnerships can be confounded by substantial differences among the contexts, institutions, and habits of practitioners, academics, and those who move between these divides. At CADAN, these in-between individuals self-identify as pracademics, modeled by Laura Olson and Alessandra Jerolleman, two working pracademics. In the following section we outline work assembled largely from our lead author, who has developed these ideas with coauthors Caela O'Connell and Laura Yoder (Browne, O'Connell, and Yoder

2018). We describe both differences and the shaping influences on these differences among academic work on disasters and practitioner work on disasters.

Academics and Practitioners

- What separates us as groups of people involves different kinds of work, different time pressures, and different products to show for our efforts. These differences that Browne, O'Connell, and Yoder (2018, 421) identified from the literature may seem benign, but, as they argue, "the gaps that exist between disaster research and disaster practice are not neutral." They carry with them dire human consequences. These divides affect real people's lives: the survivors and future survivors of disasters (Browne, O'Connell, and Yoder 2018).
- Academic/practitioner divides exist in most fields, and the fact of these gaps has been studied systematically since the 1980s. These divides are commonly explained as a product of professional silos that people inadvertently get stuck inside because the everyday functioning in their worlds is constrained by certain cultural assumptions and ways of working. Yet when we talk about practitioner/academic divides, the image of grain silos to depict broad differences across two domains does not capture the variation within each of these areas of work. Variations within practitioner work and within academic work resemble something more like separate, well-established cityscapes themselves constrained by distinctive internal logics (just as a city might be constrained by a river or a railroad track). Within the practitioner cityscape, and within the academic cityscape, structural hierarchies (among other variations) can create noteworthy differences. For example, many people at high levels in NGOs, international NGOs, and nonprofits deal less with the day-to-day pragmatics of disaster response or recovery than with abstract notions of community and recovery and reconstruction. Likewise, intra-institutional structures of universities, like separate disciplines, departments, or colleges, also commonly operate as distinctive sites of knowledge and practice.
- For academics, crossing disciplinary lines into a different academic neighborhood can challenge one's professional identity (Hein et al. 2018) and raise ideological and epistemological issues (Miller et al. 2008). Similarly, practitioners may encounter difficulty in shifting their neighborhoods of practice, as when practitioner experts move from disaster risk recovery to disaster response, from small nonprofits to large nonprofits, or from nonprofits to governmental organizations. All these complexities and contextualizations tend to produce even

greater challenges when we cross from the cityscape of an academic world into a cityscape of the practitioner world or vice versa (Browne, O'Connell, and Yoder 2018).

- Practitioners often work in a variety of different national and international contexts and different countries in a short span of time. Practicing disaster social scientists (and pracademics) like Adam Koons, Laura Olson, or Mo Hamza have literally worked around the world, which differentiates them from the anthropologist conducting longitudinal research. Practitioners go where there is need. Anthropologists may have multiple field sites, but are not on call to work across the entire globe, as practitioners often are.

- The literature suggests that the most recognized source of disconnect between practitioners and academics involves issues of communication that may be divided into familiar sub-issues: (1) inaccessible language, or jargon, of academic writing, (2) length of academic articles, (3) publications in unknown outlets that are inaccessible to practitioners because of both cost and invisibility, and (4) limitations for academics in accessing practitioner knowledge. There are also communication issues rooted in problem framing and driving ideas (Browne, O'Connell, and Yoder 2018).

- Academic and practitioner logics are often incompatible, as Browne, O'Connell, and Yoder (2018) describe. Researchers are driven by academic curiosity and are trained to identify their research problem in relation to, and advancement of, existing literature. By contrast, practitioners identify their problems as the pressing concerns facing them in a given context, made evident to them as sector experts and within their institutional mandates to respond. Within that larger frame, it is possible to identify specific freedoms and constraints. For example, though constrained by funding and funding trends, academics have a degree of freedom to place themselves in geographical and community spaces of their choosing. Generally, they assign themselves.

- Academics and practitioners have different measures of success. In academia, success is measured by research publications and the quality of teaching. The reach of academic research into public life receives little institutional recognition. By contrast, practitioners are measured by the efficiency and effectiveness of their assignments to help communities and survivors, not on their contribution to the state of knowledge.

- Finally, and possibly most importantly, are time pressures. Practitioners must get it done quickly and do not have time to consult the literature or make visits to the library in the middle of a reconstruction project. Academics have an academic year with teaching and service responsibilities to consider (Browne, O'Connell, and Yoder 2018).

Question 3: What Recommendations Can Help Academics and Practitioners Bridge Gaps between Knowledge and Practice?

- Certainly academic anthropologists have made much progress in prob-
lematizing disasters by bringing to light the unnaturalness of disasters,
their deep political economic history, the consequences of neoliberal
approaches to disaster aid and humanitarian relief, and the cultural
misalignments of helping institutions and survivor communities on the
ground. Yet much of the literature stops short of problem solving. These
insights leave unanswered some important questions. What can first
responders, humanitarian organizations, state and federal agencies, or
local activists do differently? What other steps can they take to reduce
suffering and help individuals, families, and communities recover in a culturally sensitive, politically empowering, and economically just manner?
Practitioners need theory to be translated into actionable items.
- We have learned that transforming theory and anthropological insights
into recommendations for policy and practice requires a nuanced
approach, one that is informed by the insights of disaster practitioners.
Recommending to policymakers that they make radical changes to the
economy or government systems will, as one of our co-authors points
out, drive away the very audience we seek to attract. Understanding
the avenues of possibility and the language and framing used in policy
domains can help shape agendas for change. Making suggestions
related to mandated goals under presidential policy directives or
agency strategic plans, for example, are more likely to succeed.
- A general recommendation that "policymakers should" or "decision
makers should" is not likely to be implemented. Moreover, there are
many people in disaster agencies who make decisions, but probably
very few of them consider themselves policymakers. They may be more
likely to pay attention if anthropologists or other academics specify
the type of decisions to which their suggestions apply, such as regulatory decisions, contract decisions, or research-prioritizing decisions.
Recommendations that demonstrate a knowledge of institutional
structures are more likely to be heeded. FEMA, other federal agencies, and state and local governments all play different roles during
disasters. For example, improvement of building codes to improve
resilience is often recommended for localities, when in many cases it
is at the state level where building codes get set. Attention to the
science-policy nexus in the disaster world, where the uptake points are
for implementing scientific insights in disaster decision making, can be
effective. Additionally, suggestions can be couched in the language of
practitioners, such as trainings, tactical procedures, or guidelines.

- Practitioners can nudge the institutions they work in to look hard at their own institutional assumptions. For example, many disaster policies in the United States are built around individuals or the nuclear family. This deeply embedded assumption of what a family is has real consequences for who has access to risk mitigation and recovery aid (Marino 2018) and how social needs of recovery are understood (Browne 2015). Assumptions about individualism, a liberal economy, or notions of property undergird much institutional and governmental decision-making. As insiders, practitioners are uniquely positioned to help guide their institutions to recognize the fact and relevance of their own assumptions.
- We have learned that engagement between practitioners and academics requires interactions that will not necessarily be comfortable. Engagement requires all parties to move into uncomfortable conversations that exist in what workshop facilitators sometimes refer to as the "groan zone." As spaces where confrontation occurs and disagreements emerge, a groan zone will cause distress but can also provide a pathway for bridging gaps and allowing progress and idea integration to occur (Browne, O'Connell, and Yoder 2018).
- There, in the groan zone, we learn that there is the need to learn one another's jargon, places of publication, institutional assumptions, hierarchical structures, and realistic constraints. Academics cannot realistically publish only government reports and still seek tenure. No practitioner can realistically change the funding structure of FEMA simply because an article in *Human Organization* suggested it. As our colleague Heather Kirkland joked recently, "Do we really think that first responders and emergency managers got into this work to perpetuate neoliberal ideas?" (Kirkland, personal communication, 2018). We need to acquire more-grounded perspectives and work to comprehend the motivations of the day-to-day lives of the other.
- Disaster anthropologists, whether academics or practitioners, embody a wealth of understanding about the causes of disasters and how disaster impacts permeate affected communities. Our vantage from theory as well as from the ground allows us to apply what we know to ease actual or potential hardship. We work in the field among people who may have just lost more than we can imagine. What can we offer them with our understandings of disasters? Our research methods often involve listening. Our presence and our connections through our anthropological toolkit can be a form of compassion that is both rigorous theoretically and robust methodologically. In this way, knowledge and practice can come together in the enactment of anthropology to

help community members navigate and make sense of the disaster they have experienced.

Conclusions

Our experience demonstrates that the gap between practitioners and academics can be bridged. In 2016, Browne and O'Connell hosted a National Science Foundation–funded workshop at Colorado State University to bring together twelve disaster practitioners and academics to discuss how to integrate cultural factors into disaster work on the ground. In the process of this three-day facilitated workshop, participants inevitably encountered the gaps described above and detailed in their article (Browne, O'Connell, and Yoder 2018). By the end of our experience, and having pushed through the groan zone, it was our collective sense we had begun the makings of a network that had the potential to transform an important area of disaster practice. The anthropological lens could be used to scale up our concern to more systematically reduce unnecessary suffering. To that end, and as an example of one such approach, we formed CADAN (www.cultureanddisaster.org).

Today, this young network of CADAN members has carried out research projects, formal presentations, and workshops that have brought together researchers and practitioners. In 2017 at the Cancun, Mexico, site of the Global Platform on Disaster Risk Reduction (GPDRR), CADAN members publicly presented the concept they developed and called Culture-Based DRR [disaster risk reduction]. There, they also conducted a rapid ethnography, interviewing more than forty international disaster relief, reconstruction, and risk reduction specialists.

In a 2018 project in Washington, DC, funded by the FEMA Higher Education Program, CADAN members were invited to organize and lead a workshop to discuss FEMA's new strategic plan (issued in April 2018), "Building a Culture of Preparedness." The workshop brought together academics, practicing anthropologists, FEMA staff, and representatives of other federal agencies and nonprofit organizations. This workshop showed great promise in generating fresh understandings after a short period of engagement.

One small but significant shift in our FEMA collaborators occurred over the course of the workshop as people began referencing "cultures of preparedness" rather than "a culture of preparedness." At the same time, our academic colleagues were better able to understand the unique challenges and limitations FEMA workers face, including their own experiences of stress and trauma during disaster events. The CADAN-led FEMA workshop took the subsequent form of a September 2018 nationwide Webinar and in January 2019, the team released a formal, forty-page report for FEMA Higher Education entitled "Building Cultures of Preparedness." It may be impossible to measure the outcomes of this kind of engagement. However, it clearly represents the kind of gap-bridging that CADAN and other organizations seek to achieve in the way academics and practitioners frame the concept of disaster and preparedness itself. As CADAN members and founders we are simultaneously hopeful and overwhelmed by what our self-appointed task has become: to formalize and systematize practices, relationships, and institutions that can move information and action among those who are trying to stem the suffering caused by disasters. How CADAN will move into the future is not fully clear, but what is absolutely certain is that this work is needed and is resonating with the very people we hope to engage.

We would be remiss not to point out that since the late 1980s there have been numerous efforts to create networks that straddle academic and practitioner divides. La Red, for example, is one such group that is considered foundational in the development of the vulnerability approach to disasters. Its founding members are based in academic and disaster risk reduction institutions in Mexico, South America, and Europe. Additionally, the Society for Applied Anthropology houses a Risk and Disaster Thematic Interest Group that connects academics of numerous disciplines, academic practitioners, and full-time field practitioners who work all over the world. In Europe, the European Association of Social Anthropologists is the umbrella organization for DICAN, the Disaster and Crisis Anthropology Network (https://www.easaonline.org/networks/dican/) and functions much the same way, uniting anthropologists and other concerned academics with both practitioners and academic practitioners. The International Union of Anthropological and Ethnographical Sciences (IUAES; https://www.iuaes.org/), recognizing the growing risk and disaster issue worldwide, has formed a Risk and Disaster Commission directed by Susanna M. Hoffman and Virginia Garcia-Acosta that brings together academics and policymakers from many countries to confer with one another. The Integrated Research on Disaster Risk (IRDR; housed in Beijing, China) puts together internationally represented scientific and practical committees and alliances that work to alleviate risk creation and promote risk reduction. The examples above do not form an exhaustive list, but are examples of efforts by anthropological researchers and practitioners to come together to attempt solutions.

Despite these many valuable attempts to bridge divides between practitioners and academic anthropologists in disaster work, we have found that persistent obstacles remain. The CADAN mission may offer an innovative space for progress on this front. CADAN was founded as an action network. The network is focused on changing disaster policy and practice to integrate cultural awareness in disaster work. For that reason, CADAN is vigilant about the bi-directionality of practitioner/academic exchanges including not only the immersion of practitioners in academic settings (which is not unusual in the networks listed above), but also the immersion of academics in practitioner settings (which, we have so far found, is more unusual). As academics and practitioners interested in working together, the present is an exciting time for innovation and hybrid collaborations and we look forward to seeing where this work will take us.

Disclaimer

This manuscript has been subjected to the Environmental Protection Agency's review and has been approved for publication. Note that approval does not signify that the contents necessarily reflect the views of the Agency. Mention of trade names, products, or services does not convey official Agency approval, endorsement, or recommendation.

Katherine E. Browne is University Distinguished Professor at Colorado State University and professor of anthropology. Her research examines how social history, place-based cultural identities, and sense of agency shape trajectories of well-being and resilience after disaster. Her public-facing books and films argue for a paradigm nudge to integrate sociocultural factors into practices of mitigation and recovery. She cofounded CADAN as a collaborative space for academics and practitioners. In 2018 the American Anthropological Association named Browne recipient of its highest honor, the Franz Boas Award for Exemplary Service to Anthropology.

Elizabeth Marino is assistant professor of anthropology and sustainability at OSU-Cascades, and co-director of CADAN. She conducts research on issues of repetitive flooding, climate change displacement, risk perception, and the intersection of culture and the environment. Marino is an author in the Fourth National Climate Assessment; and is a U.S. delegate of the Arctic Science Ministerial. Her book, *Fierce Climate, Sacred Ground: An Ethnography of Climate Change in Shishmaref, Alaska* was published in 2015.

Heather Lazrus is an environmental anthropologist and a project scientist at the National Center for Atmospheric Research in Boulder, Colorado. Using the theories and methods in the anthropological toolkit, she investigates the cultural mechanisms through which all weather and climate risks are perceived, experienced, and addressed. Lazrus focuses on the interface between extreme weather and climate change and works closely with colleagues from diverse disciplines. Her research contributes to improving the utility of weather forecasts and warnings, reducing social vulnerability to atmospheric and related hazards, and understanding community and cultural adaptations to climate change.

Keely Maxwell is a general anthropologist in the U.S. Environmental Protection Agency's Office of Research and Development, National Homeland Security Research Center. An environmental anthropologist and ecologist by training, she researches community resilience to disasters and the social context of environmental cleanups.

References

Barnett, Jon, and W. Neil Adger. 2007. "Climate Change, Human Security and Violent Conflict." *Political Geography* 26, no. 6, 639–55.

Barrios, Roberto E. 2017. *Governing Affect: Neoliberalism and Disaster Reconstruction.* Lincoln: University of Nebraska Press.

Beck, Ulrich. (1992). *Risk Society: Towards a New Modernity,* 1st ed. London: Sage.

Browne, Katherine E. 2015. *Standing in the Need: Culture, Comfort, and Coming Home After Katrina.* Austin: University of Texas Press.

———. 2016. "Roux and Resilience: Eleven Years After Hurricane Katrina." *SAPIENS.* 31 August. Retrieved from https://www.sapiens.org/culture/hurricane-katrina-af termath-roux-resilience/.

Browne, Katherine E., and Trevor Even. 2018. "The 'Culture of Disaster' Student Immersion Project: First-Hand Research to Learn about Disaster Recovery after a Colorado Flood." *International Journal of Mass Emergencies and Disasters.* 36, no. 3, 264–86.

Browne, Katherine E., Caela O'Connell, and Laura Yoder. 2018. "Journey through the Groan Zone with Academics and Practitioners: Conflict, Difference, and the Process of Bridging to Strengthen Disaster Risk and Recovery Work." *International Journal of Disaster Risk Science* 9, no. 3, 421–28.

Buckley, Geoffrey L., and Laura Allen. 2011. "Stories about Mountaintop Removal in the Appalachian Coalfields." In *Mountains of Injustice: Social and Environmental Justice in Appalachia,* edited by Michele Morrone, Geoffrey L. Buckley, and Jedidiah Purdy, 161–80. Athens: Ohio University Press.

Crenshaw, Kimberle. 1997. "Intersectionality and Identity Politics: Learning from Violence Against Women of Color." In *Reconstructing Political Theory: Feminist Perspectives,* edited by M. L. Shanley and U. Narayan, 178–93. State College: Pennsylvania State University Press.

Culture and Disaster Action Network (CADAN), Marino, E., A. Koons, L. Olson, K. E. Browne, A. J. Faas, J. Maldonado. 2017a. "A Helping Hand." *The Mark News*, 10 July.

————. 2017b. "Local Resilience." *The Mark News*, 17 July.

de Sherbinin, Alex, Marc Levy, Susana Adamo, Kitt MacManus, Greg Yetman, Valentina Mara, Liana Liana Razafindrazay, Benjamin Goodrich, Tanja Srebotnjak, Cody Aichele. 2012. "Migration and Risk: Net Migration in Marginal Ecosystems and Hazardous Areas." *Environmental Research Letters* 7, no. 4, 045602.

Fischer, Bill, and Andy. Boynton 2005. "Virtuoso Teams." *Harvard Business Review* 83, 7/8, 116–23.

Fleming, E., J. Payne, W. Sweet, M. Craghan, J. Haines, J.F. Hart, H. Stiller, and A. Sutton-Grier. 2018. "Coastal Effects." In *Impacts, Risks, and Adaptation in the United States: Fourth National Climate Assessment, Volume II*, edited by D.R. Reidmiller, C.W. Avery, D.R. Easterling, K.E. Kunkel, K.L.M. Lewis, T.K. Maycock, and B.C. Stewart, 322–352. Washington, DC: U.S. Global Change Research Program. doi:10.7930/NCA4.2018.CH8.

Fothergill, Alice, and Lori A. Peek. 2004. "Poverty and Disasters in the United States: A Review of Recent Sociological Findings." Natural Hazards 32, no. 1, 89–110.

Funk, McKenzie. 2014. *Windfall: The Booming Business of Global Warming*. New York, New York: Penguin Books.

Garcia-Acosta, Virginia 2002. "Historical Disaster Research." In *Catastrophe and Culture: The Anthropology of Disaster*, edited by S. Hoffman and A. Oliver-Smith, 49–66. Santa Fe, NM: School of American Research.

Gotham, Kevin Fox. 2012. "Disaster Inc.: Privatization and Post-Katrina Rebuilding in New Orleans." *Perspectives on Politics* 10, no. 3, 633–46.

Güneralp, Burak, Inci Güneralp, and Ying Liu. 2015. "Changing Global Patterns of Urban Exposure to Flood and Drought Hazards." *Global Environmental Change* 31, 217–225.

Hale, Charles R. 2006. "Activist Research v. Cultural Critique: Indigenous Land Rights and the Contradictions of Politically Engaged Anthropology." *Cultural anthropology* 21, no. 1, 96–120.

Hardy, Dean, Heather Lazrus, Michael Mendez, Ben Orlove, Isabel Rivera-Collazo, J. Timmons Roberts, Marcy Rockman, Kimberley Thomas, Benjamin P. Warner, Rob Winthrop. 2018. *Social Vulnerability: Social Science Perspectives on Climate Change*, Part 1. Washington, DC: USGCRP Social Science Coordinating Committee. https://www.globalchange.gov/content/social-science-perspectives-climate-change-workshop.

Hein, Christopher, John ten Hoeve, Sathya Gopalakrishnan, Ben Liveh, Henry Adams, Elizabeth Marino, and C. Susan Weiler. 2018. "Overcoming Early Career Barriers to Interdisciplinary Climate Change Research." *WIREs Climate Change* 9, e530.

Hoffman, Susanna M., and Anthony Oliver-Smith. 1999. "Anthropology and the Angry Earth: An Overview." In *The Angry Earth*, edited by A. Oliver-Smith and S. Hoffman, 1–16. New York: Routledge.

Laska, Shirley, and Betty H. Morrow. 2006. "Social Vulnerabilities and Hurricane Katrina: An Unnatural Disaster in New Orleans." *Marine Technology Society Journal* 40, F16–26.

Lazrus, Heather 2016. "'Drought Is a Relative Term': Drought Risk Perceptions and Water Management Preferences among Diverse Community Members in Oklahoma, USA." *Human Ecology* 44, no. 5, 595–605.

Leckie, Scott. 2005. "The Great Land Theft." *Forced Migration Review* 25, 15–16.

Marino, Elizabeth. 2012. "The Long History of Environmental Migration: Assessing Vulnerability Construction and Obstacles to Successful Relocation in Shishmaref, Alaska." *Global Environmental Change* 22, no. 2, 374–81.

———. 2015. *Fierce Climate, Sacred Ground: An Ethnography of Climate Change in Shishmaref, Alaska.* Fairbanks: University of Alaska Press.

———. 2018. "Adaptation Privilege and Voluntary Buyouts: Perspectives on Ethnocentrism in Sea Level Rise Relocation and Retreat Policies in the US." *Global Environmental Change* 49, 10–13.

Marino, Elizabeth, and Heather Lazrus. 2016. "'We are Always Getting Ready': How Diverse Notions of Time and Flexibility Build Adaptive Capacity in Alaska and Tuvalu." In *Contextualizing Disaster,* edited by Gregory V. Button and Mark Schuller, 153–70. New York: Berghahn Books.

McNeeley, Shannon M. and Heather Lazrus. 2014. "The Cultural Theory of Risk for Climate Change Adaptation." *Weather, Climate, and Society* 6, no. 4, 506–19.

Mead, Margaret and Rhoda Metraux. 1970. *A Way of Seeing.* New York: McCall.

Miller, Thaddeus, Timothy Baird, Caitlin Littlefield, Gary Kofinas, F. Stuart Chapin III, and Charles Redman. 2008. "Epistemological Pluralism: Reorganizing Interdisciplinary Research." *Ecology and Society* 13, 2, 46. Retrieved 6 July 2019 from https://pdxscholar.library.pdx.edu/cgi/viewcontent.cgi?article=1097&context=usp_fac.

Monbiot, George. 2013. "Climate Breakdown." *The Guardian,* 4 October. Retrieved from https://www.monbiot.com/2013/10/04/climate-breakdown/.

Morrissey, James, and Anthony Oliver-Smith. 2013. "Perspectives on Non-Economic Loss and Damage: Understanding Values at Risk from Climate Change." Report by the Loss and Damage in Vulnerable Countries Initiative, edited by Koko Warner and Sönke Kreft. Bonn: United Nations University Institute for Environment and Human Security.

Morss, Rebecca E., Julie L. Demuth, Heather Lazrus, Leysia Palen, C. Michael Barton, Christoper A. Davis, Chris Snyder, Olga V. Wilhelmi, Kenneth A. Anderson, David A. Ahijevych, JenningsAnderson, Melissa Bica, Kathryn R. Fossell, Jennifer Henderson, Marina Kogan, Kevin Stowe, and Joshua Watts. 2017. "Hazardous Weather Prediction and Communication in the Modern Information Environment." *Bulletin of the American Meteorological Society* 98, 2653–2674.

Multihazard Mitigation Council. 2017. *Natural Hazard Mitigation Saves. 2017 Interim Report: An Independent Study.*" Retrieved May 2018 from https://www.fema.gov/natural-hazard-mitigation-saves-2017-interim-report.

Neumann, Barbara, Athanasios T. Vafeidis, Juliane Zimmermann, and Robert J. Nicholls. 2015. "Future Coastal Population Growth and Exposure to Sea-Level Rise and Coastal Flooding—A Global Assessment." *PLOS ONE* 10, no. 3, e0118571.

O'Brien, Geoff, Phil O'Keefe, Joanne Rose, and Ben Wisner. 2006. "Climate Change and Disaster Management." *Disasters* 30, no. 1, 64–80.

O'Keefe Phil, Ken Westgate, and Ben Wisner. 1976. "Taking the Naturalness out of Natural Disasters." *Nature* 260, 566–67.

Obama, Barack. 2005. "Hurricane Katrina Relief Efforts." Best Speeches of Barack Obama through his 2009 Inauguration. Retrieved from http://obamaspeeches.com/029-Statement-on-Hurricane-Katrina-Relief-Efforts-Obama-Speech.htm.

Oliver-Smith, Anthony. 1996. "Anthropological Research on Hazards and Disasters." *Annual Review of Anthropology* 25, no. 1, 303–28.

————. 2012. "Peru's Five-Hundred-Year Earthquake: Vulnerability in Historical Context." In *The Angry Earth: Disaster in Anthropological Perspective*, edited by A. Oliver-Smith and S. Hoffman, 88–102. New York: Routledge.

Oliver-Smith, Anthony, and Susanna Hoffman, eds. 1999. *The Angry Earth: Disaster in Anthropological Perspective*. New York: Routledge.

Petterson, John S., Laura D. Stanley, Edward Glazier, James Philipp. 2006. "A Preliminary Assessment of Social and Economic Impacts Associated with Hurricane Katrina." *American Anthropologist* 108, 643–670.

Phillipps, Dave. 2017. "Seven Hard Lessons Federal Responders to Harvey Learned from Katrina." *New York Times*, 7 September. Retrieved October 2017 at https://www.nytimes.com/2017/09/07/us/hurricane-harvey-katrina-federal-responders.html?_r=0.

Rayner, Steve, and Robin Cantor. 1987. "How Fair Is Safe Enough? The Cultural Approach to Societal Technology Choice." *Risk Analysis* 7, no. 1, 3–9.

Rockman, Marcy. 2003. "Knowledge and Learning in the Archaeology of Colonization." In *The Colonization of Unfamiliar Landscapes: The Archaeology of Adaptation*, edited by M. Rockman and J. Steele, 3–24. London: Routledge.

Schuller, Mark 2016. *Humanitarian Aftershocks in Haiti*. New Brunswick, NJ: Rutgers University Press.

Smith, Neil. 2006. "There's No Such Thing as a Natural Disaster." *Understanding Katrina: Perspectives from the Social Sciences* 11. Retrieved May 2019 from http://understandingkatrina.ssrc.org/Smith/.

Tierney, Kathleen. 2015. Resilience and the Neoliberal Project: Discourses, Critiques, Practices—And Katrina. *American Behavioral Scientist* 59, no. 10, 1327–1342.

United Nations International Strategy for Disaster Risk Reduction (UNISDR). 2007. *Hyogo Framework for Action 2005–2015: Building the Resilience of Nations and Communities to Disasters*. Final Report of the World Conference on Disaster Reduction. United Nations, Geneva. Retrieved July 2018 from https://www.unisdr.org/we/inform/publications/1037.

————. 2015. *Sendai Framework for Disaster Risk Reduction 2015–2030*. Retrieved July 2018 from http://www.wcdrr.org/uploads/Sendai_Framework_for_Disaster_Risk_Reduction_2015-2030.pdf.

United States Global Change Research Program. 2017. *Climate Science Special Report: Fourth National Climate Assessment, Volume I* [Wuebbles, D.J., D.W. Fahey, K.A. Hibbard, D.J. Dokken, B.C. Stewart, and T.K. Maycock (eds.)]. U.S. Global Change Research Program, Washington, DC, USA, 470 pp.

Watters, Ethan. 2010. *Crazy Like Us: The Globalization of the American Psyche*. New York: Simon and Schuster.

CHAPTER 13

Future Matter Matters

Disasters as a (Potential) Vehicle for Social Change— It's About Time

ANN BERGMAN

Introduction

In this theoretical and conceptual chapter I argue, from a sociological perspective, that the future is an important dimension to consider when understanding disasters and vulnerability in relation to power and inequality.

The future is especially critical if one views both structure and agency as important factors for social change. Individuals, groups, organizations, and nations act in relation to the past, the present, and the future. Social structures generated from the past condition the practices in the present (Adam 2008). Therefore, present practices tend to, but are not determined to, reproduce existing conditions since transforming them often demands a struggle. Social structures are transformed in some cases by laws, regulations, policies, and elections, and in others by resistance or subversive actions. Various critical events in society, where the existing social order is shaken and challenged, can also be potential transformative factors. Disasters are examples of events that create temporary societal cracks and social and cultural dizziness. In the aftermath of these events, there might be an unintended and unpredicted opportunity to act differently since, as argued by Hoffman and Oliver-Smith (2002, 9), "Disasters unmask the nature of a society's social structure" and thereby the "distribution of power reveals itself." However, powerful groups tend to tighten their grip on resources and positions, ensuring that things are drawn back to the "normal" state. The reproduction of vulnerability and social inequality is made possible by durable concentrations of power and privilege.

In the rest of this chapter I will discuss disasters and vulnerability in relation to the social context and power. Thereafter the focus is on disasters, causality, social change, and time. The belief in the art of management as the best way to solve problems related to disasters is then contrasted to the concepts of participation and solidarity. In the final sections attention is given to how future matters matter, and various ways to relate to and approach the future—such as predictions, utopias, and dystopias.

Disasters as the New Normal

Definitions of disasters are often based on assumptions of a linear causality. For example, Chang (2004, 2) defined a disaster as a consequence of an event where the event is defined as a hazard. Other definitions include the social context; for example, Wisner and Luce (1993, 130) stated that a disaster occurs when a physical hazard meets a vulnerable population. Another definition was given by Lavell et al. (2012, 31), who stated that disasters are "severe alterations in the normal functioning of a community or a society due to hazardous physical events interacting with vulnerable social conditions, leading to widespread adverse human, material, economic, or environmental effects that require immediate emergency response to satisfy critical human needs and that may require support for recovery." Although the last definition shows an awareness of the complexity of causal relations and the situation from which a disaster emerges, it is still caught in a framework in which the pre-disaster state is the normal and thus the norm. However, the awareness of the risk of being vulnerable and a possible victim of a disaster is increasing with regard to people and societies. According to the United Nations Office for Disaster Risk Reduction (UNISDR 2012, 2–3) in *Towards a Post-2015 Framework for Disaster Risk Reduction,* there is a growing exposure to disasters and risk in general in society. That document states that both developing and developed countries face increasing natural and man-made risks due to unsustainable development, poverty and segregation, and financial and ecological collapses. Consequently, we live in a society where risks, hazards, and potential disasters are always present, emerging, creeping, and ambiguous (Kouzmin 2008). This means that a general conception of what is normal to some extent has shifted from being the absence of disasters, to being the presence of potential or manifested disasters.

Since disasters are regarded as bad and as things that we have to handle in the best possible way, countries—mainly developed ones—have invested in methods and techniques to track and predict disasters. Models have been developed to carry out rescue work and to reestablish what has been destroyed. Thus, some potential disasters can be avoided or altered and the negative consequences of disasters can be reduced. As a consequence of the threats of disasters, voices from politics and science have stressed the relationship between risks, vulnerability, disasters, and sustainability, and there is an increasing awareness that we need to take social, environmental, political, and economic responsibility if we are to take the call for sustainability seriously.

The relations between vulnerability, disasters, sustainability, and equality are discernable. Hence, inequality is a major challenge when trying to achieve a sustainable society by reducing vulnerability. Segregation, marginalization, misrecognition, and powerlessness are neither sustainable nor desirable, and these patterns exist today in all known societies— more or less. Capitalism and industrialization have brought advantages for some by decreasing poverty and increasing human choice, the standard of living, and the availability of knowledge and science. However, damage to the environment and humankind can largely, but not solely, be blamed on industrialization and capitalism. Due to capitalism, markets are still growing globally and the actions that rest on finances do not stop at national borders. As long as states are committed to the same ideology of growth as large corporations, they will underestimate the very risk this growth causes to workers, citizens, and the environment. In many cases, the state itself is a risk and a threat since it distributes vulnerability and precariousness unevenly. Working conditions, waste products, and the exploitation of humans and of natural resources are all risks from which disasters can emerge. The happy marriage between capitalism and globalization and the unending strife for growth by producing more and more, faster and faster, intensified the West's predatory activities in other parts of the world. As pointed out by Nandy (2013), the urge for growth has led to a globally mobile exploitation of so-called underdeveloped countries. In line with this, Augé (2014, 72) argued that today we face a "globalization [that] is more perverse than colonization in that its actors are less easily identifiable even though it is imposed on everyone. We all have the feeling that we are colonized but we do not know who by." Therefore, according to Augé (2014), the local context, with its specific characteristics, is now also global in nature. Thus, the local context being global in nature means, among other things, that it is part of a global greedy capitalist system—a system without consciousness.

There is growing awareness that disasters are unintended outcomes of a causal cocktail of historical and present social, economic, geographical, and natural conditions (Oliver-Smith and Hoffman 2002; Ullberg 2013). A future without disasters is hard to imagine since we simply cannot control and predict everything as if in a laboratory. Even if we could, there is still a lack of local and worldwide responsibility, solidarity, and equality.

A part of the root system of disasters is connected to social structures. For instance, a lack of power and control is among the most important conditions from which vulnerability emerges (Fothergill 1998). Hoffman (2002, 136) used the term "societal auto-da-fé" to depict how a phenomenon, such as disasters, that emerges from society and culture "turns into a public execution." Therefore, fighting disasters is not about fighting nature, it is about fighting ourselves. On a meta level, we are all vulnerable and so are our systems and institutions. Disasters, their causes, and their effects are a part of our society and simply reveal how we make society work. This revelation indicates that there is a price to pay for reducing vulnerability and that this price requires some people to give up some of their power and privileges. In order to create a less vulnerable society perhaps one way forward is a decent society for all (Nandy 2013, 13).

Vulnerability: A Power-Drenched Concept

As mentioned above, a central component in a discussion about disasters is the existence of vulnerable groups. Blaikie et al. (1994, 9) defined vulnerability as "the characteristics of a person's or a group's capacity to anticipate, cope with, resist, and recover from the impact of a natural hazard." Today, scholars in the field would add "man-made" hazards and feminists would emphasize the word "man." By using the term "vulnerability," the interactions, conditions, resources, and contexts are stressed in order to extend the understanding to a dynamic concept beyond the more passive concept of being a victim of a specific event called "disaster." Although simply being human makes us all vulnerable to hazards in different ways, less privileged groups tend to be more exposed and defenseless than others. Fothergill and Peek (2004) showed that when disasters are geographically and socioeconomically widespread, a society or nation tends to mobilize its resources to a higher degree to lessen their consequences and their likelihood of happening again than when they only hit segments and groups in society that are already on the margin and less privileged. In the latter cases, the focus tends to be on emergency rescue work and preventing effects that are more widespread. Furthermore, high-income sufferers and other powerful groups tend to seek and receive aid to a higher degree than low-income sufferers, for example, from federal aid and relief programs. Fothergill (1998) argued that we can understand disaster settings as a strategic site since they strip away the veil that otherwise hides social conditions. García-Acosta (2002, 57) described disasters as "revealers," for the same reason. As mentioned above, a disaster setting and the disaster's aftermath unmask invisible power relations (Oliver-Smith and Hoffman 2002). This, in turn, affects the interaction patterns and the strategies used by both the holders of authority and power and their counterparts. Since individuals interact in different structural positions, social relations are drenched by power, facilitating for some and hindering for others (Scott 2001). For example, Fothergill (1998) stresses that class, ethnicity, age, and gender stratify power and resources in society as a whole, as well as in relation to vulnerability.

Thus, a lack of power, resources, and control in specific situations reduces agency, which in turn, is a driver for vulnerability. Lavell et al. (2012, 32–33) argued that the relationship between vulnerability and capacity is crucial since vulnerable groups often lack certain resources and capacities to act before, during, and after a disaster. To define someone as vulnerable is to make a statement about his or her capacity to meet hazards. This statement can be a relevant description of the situation, but it can also have stereotyping and stigmatizing effects and thus push some vulnerable groups farther into powerlessness—making it a self-ful-filling prophecy. An important aspect of being in a position of power is to develop strategies for staying in that position. Although disasters make the invisible power relations visible, groups that benefit from an unequal social and material order tend to overlook this or try to legitimize it.

When discussing power using a Weberian perspective, power is defined as an agent's capacity to impose his or her will on the other part—even against their will and resistance (Weber 1978). However, power is not only domination (i.e., power over), it is also a causal capacity (i.e., power to) (Morriss 2002) and a capacity that is conditioned by materiality, social structure, and discourse. Power is about capabilities and abilities to act in certain ways; therefore, vulnerability is tightly associated with power. Thus, an inability and incapability to act in relation to a hazard indicates a lack of power. In some cases, vulnerable groups are given the chance to influence in a certain situation or context, but influencing is not, if lending an ear to Morriss (2002), the same as having power as a disposition. Power as a disposition means that one can have it and decide whether to exercise it. A person cannot have influence, on the other hand, without exercising it. Besides, influence is often something that is given to someone by some-one with power. An inability and incapability to act in relation to a hazard can thus be a lack of both influence and power.

Disasters as Causal Forces

As argued above, disasters emerge as a consequence of, among other things, the accumulated decisions and practices that form a society.

However, causality goes the other way around as well. Disasters are forceful phenomena that affect society both as potential and as manifest full-fledged disasters. Regardless of whether it is a potential, a risk, or a manifested disaster, a disaster can trigger social change since it carries with it a possibility for society to, through people's actions, develop in new directions. Nevertheless, the literature contains evidence of a tendency for society to return to a pre-disaster state. This process and capacity to bounce back to "normal" is often defined by the contested concept of resilience. According to the United Nations Office for Disaster Risk Reduction (UNISDR 2017), resilience is "the ability of a system, community or society exposed to hazards to resist, absorb, accommodate, adapt to, transform and recover from the effects of a hazard in a timely and efficient manner." Some view resilience as a positive aspect and worth striving for, as a potential for long-term maintenance of well-being and safety (Hollnagel 2009; Petersen and Johansson 2008). However, Kelman, Gaillard, and Mercer (2015, 22), among others, argued that the core idea of resilience could be bouncing back to poverty, segregation, vulnerability, and so on. Instead of bouncing back to normal, some have talked about the ability to "bounce forward" (Manyena et al. 2011) or to "build back better" (Kennedy et al. 2008). Thus, vulnerability and resilience are related to each other and the question is whether some groups' resilience relies on other groups' vulnerabilities.

As pointed out above, power relations need to be considered in order to make it possible to transform the existing social order. McEntire (2004, 193) argued that there is now an understanding that the relationship between disasters and development is close and complex and that disasters can have both positive and negative effects on vulnerability and the future development of society. To be able to generate an inclusive and holistic view of disasters, vulnerability, and society, McEntire stressed the need for both a radical Marxist view focusing on capitalism and inequality and a more liberal Weberian view focusing on culture and rational management. On the one hand, McEntire was referring to Hewitt (1983), who claimed that disasters should be understood as a reflection of the existing unequal and exploitative social order. Accordingly, McEntire stressed that radical social changes have to take place in order to create a more sustainable and safer society. On the other hand, he also referred to Mileti (1999), who argued that culture, beliefs, norms, and people's preferences are to blame for disasters and vulnerability. In this latter Weberian-influenced perspective, it is people's choices that will determine their vulnerability and losses in the future. So, from a holistic perspective, it is important to consider both structure and agency. Social structures cannot act, but people can. Social structures do not have intentions, hopes, and fears, but people do. However, structures condition people's lives by facilitating or impeding their agency. Although society is not deterministic, privileged interests are very successful at reproducing existing structural orders, rather than transforming them.

Disasters are potential vehicles for social change, not only after a disaster when a society is trying to recover, but also as future potential disasters that impinge on how societies are developed and changed in certain directions—both away from something and toward something. When understanding how disasters are causally interrelated with structure and agency, the temporal dimension is central. In relation to the future, one can distinguish between two causal time orientations (Adam and Groves 2007). One is being oriented toward the future—that is, from the present and directed forward. Given the past and present situation, we can expect the emergence of various disasters in the future and therefore act to prevent them. The other is guided by the future—that is, its time orientation is the opposite and thus backwards from the future toward the present. For example, if we desire a sustainable and equal society in the future, we have to act in certain ways now. Thus, disasters and vulnerability can be seen as both a past and a present fact. They can also be seen and used as predictions about the future (given the past and the present) or an idea about what a preferable or non-preferable future would be. These thoughts are not new. In *Philosophy of the Present,* Mead (2002[1932]) stated that all our actions are a result of the three temporal dimensions the past, the present, and the future, and that actions are future-oriented. Hence, it is not only the past and the present that condition social activity, but also the future. Therefore, even though the future has not yet happened, it is real since it is causal and it makes things happen in the present. In other words, social life consists of a dynamic mix of experience of the past and expectations of the future. In some cases the mixture is not only dynamic but contradictory. As Barrios (2017, chap. 6) shows, disaster-affected groups can be more prone to accentuate the past to detect what went wrong in order to find paths to recovery, while experts and political and socioeconomic elites tend to be using the future as guidance for the recovery work. That is a future defined by those experts and driven by their hopes and anxieties. Once again, the future can be used as a tool to oppress and control certain groups in the present (Nandy 2013, 3).

Future orientations go hand in hand with visions, plans, goals, ideological views, and achievements. Consequently, the future—the not yet—is seen as a possibility and a resource for some and as a threat for others. Powerful agents use claims about the future as a way to achieve desired outcomes by defining the future in in a certain way. Resistance and emancipation are also driven by ideas and hopes related to a future that is different from the past and the present situation, as in the feminist vision of a gender-equal society. Although the feminist movement is far from homogenous, it has been a powerful driving force in society (Walby 2011). Therefore, the future and statements and visions about the future can be used to influence and impinge on the actions of individuals and collectives, thereby transforming social structures. In order to influence social change in a specific direction, the future can be used as a dystopian warning signal in order to try to prevent certain things from happening. An example is the environmental movement, where research, politics, media, and popular culture often use the future to depict a worst-case scenario in order to change how people and organizations behave regarding sustainability. Hence, considerations about the future spark emotions of both anxiety and hope. Also in the field of disaster, one could argue that the future tends to be seen in the light of a constant flashing warning lamp, which in turn influences the perception of how to understand, plan, organize, and carry out work in relation to potential and actual disasters. As argued, it is not only what to avoid that guides our actions: perceptions about what is a preferable future tend also to play an important role in people's and communities' everyday practices and in the policies and practices on local, national, and global levels. However, the power of definition appears to be a double-edged sword since what is defined as both a non-preferable (what to avoid) and a preferable (what to achieve) future tends to be defined by the powerful and the privileged. Nevertheless, the power to define is a necessary but not a sufficient condition to have impact on the social order. Another crucial factor is making people act accordingly, and here is where the art of management comes in.

The Art of Management

Management is often seen as the solution to a number of social, economic, and environmental concerns. Together with the art of predicting, the art of management holds a strong position in modern societies when they try to meet and govern challenges such as hazards and disasters (e.g., Barrios 2017). A belief in the art of management is a belief in management as a system or a function that provides solutions to societal problems. When it comes to disaster management and organizational arrangements, management has been dominated by a functionalist view (today with a neoliberal twist) where disasters are seen as something disruptive for society and where the main concern is to return to normality and stability (Barnshaw and Trainor 2007). Suparamaniam and Dekker (2003) pointed out there is an overwhelming risk that disaster relief work will be planned and carried out in an instrumental way, as in a military operation. However, as mentioned above, there is a growing awareness that concepts such as stability and normality are problematic and that they disguise more than they reveal (Bankoff 2001; Ullberg 2013). What is defined as abnormal or as normal is an outcome of power relations, and thus an outcome of the capability to be in charge of plans and actions. For example, to set up a recovery plan is to make a claim about the future; in order to realize that future, people must act accordingly—whether forced or of their own free will. Management is a way of trying to make the future predictable according to specific interests or ideas turned into goals and visions. As such, management is a way of coordinating differentiated tasks (and the often-neglected coordination by nonhierarchal collaboration is yet another way).

If knowledge guiding the work related to disasters emanates from homogenous and powerful groups that are carrying out the work in a mechanic top-down managerial manner, there might be a lack of knowledge about the local situation, which can create a gap between formal power and authority and the local knowledge and practice (e.g., Haraway 1988). In situations that are characterized by insecurity, unpredictability, and a lack of shared understanding and knowledge, these characteristics tend to be compensated for by even more emphasis on the art of management—manifested as more control, plans, and directives. More weight on management seldom results in getting authority and resources to where the local knowledge is, or vice versa. The lack of success or progress is another aspect that tends to result in more planning, more management, more experts, and more directives—and a wider gap. It can also set in motion false assumptions, misdirected efforts, and a vicious blame game.

With management and expertise comes responsibility. If management and experts (including scientists) are too isolated in their own perspectives, beliefs, and visions of what is best to do before, during, and after a disaster, their isolation can increase vulnerability. The power of management and the power of expertise tend to be based on trust in and/ or acceptance of management's capacities to solve a problem. Trust in and/or acceptance of the powers of others can undermine trust in the knowledge and capacities of oneself—or even result in a belief that one lacks knowledge and capacity (Scott 2001). Although there are resistance and alternative voices, it is important to consider whether those voices are recognized at all. Lack of recognition is a durable cause of inequality, vulnerability, and reproduced power relations and, as Roy (2004, 1) put it, "there is really no such thing as the 'voiceless.' There are only the deliberately silenced, or the preferably unheard." Disaster research has shown how, for example, women lack decision making power, how they are dismissed as hysterical, how they are denied leadership positions in emergency management, how their work throughout a disaster process is devalued or made invisible at the same time as they suffer more than men (Phillips, this volume; Wisner and Luce 1993). This is one side of vulnerability that shows that power and the art of management together can result in a system that preserves stupidity.

If there is an ambition to change and develop society in a more equal and sustainable direction, then what do we need to consider and do differently so that we do not rely too heavily on experts and the art of management? As pointed out above, a more equal distribution of resources and recognition of difference is important (e.g., Fraser 1995). However, there is also a need for action in order to transform society; therefore, action through participation, for example, is vital. In most cases, redistribution along with recognition and participation must be fought for, which makes it even more difficult for the vulnerable and powerless to achieve them. Being able to participate is important in any type of development of a democratic society; at least, this is what is stated in regulations, policy documents, and the rhetoric. However, participation and the participation of interest groups can, in reality, be situations where the voices that have impact and are heard are the voices of a few. In some cases, participation on conditional terms increases the knowledge gap and power distance between different groups instead of decreasing it.

White (1996) described four different modes of participation, only one of which is about empowering people. In the first mode, the participants are viewed as objects and their participation is more of a functional exercise to secure compliance, minimize dissent, and gain legitimacy. In the other mode, the participation is instrumental, where the participants are viewed as instruments to make projects more efficient by delegating responsibilities and assignments. A third example is where the participants are viewed as actors and the mode of participation is consultative in order to get in tune with public views and values, to collect good ideas and defuse opposition, and to enhance responsiveness. A fourth mode of participation is transformative and the participants are viewed as agents. Here, the purpose of the participation is to build political capabilities, critical consciousness, and confidence, and to enable recognition and demand rights. In the transformative mode of participation, people's own experiences are recognized and they are enabled to articulate and analyze their own situations.

Marginalized groups' chances to actually participate and have impact on decision making in a transformative way is not secured just by including them in certain core groups. There is always the pitfall of nominally including a group that tends to satisfy equality or democratic goals in a tokenistic and non-transformative way. This raises questions about how that group represents its own interests. Is it possible for everyone to have a voice and, if so, does anyone listen, and do they actually have an impact? As Barrios (2017) points out, the questions of what defines a good life are too important to be defined, implemented, and evaluated by specific ethno- and sociocentric interests.

Besides recognition, reflection is another crucial aspect of disaster work and management. This involves reflecting on and evaluating plans, actions, and consequences, and asking whether we are doing things right, whether we are doing the right things, and what the right things to do are (Börjesson et al. 2006). While these questions might appear trivial, they could be fruitful if one reflects on them, the norms and values that guide the answers, and the conditions generating the actual outcomes. Organizational and managerial conditions that continuously facilitate a broad spectrum of insights to develop and to be recognized are important in critical situations such as disasters (Blackmand and Sadler-Smith 2009). Broad and heterogeneous wisdom emerges from the past and present experiences of various groups. We need not only common knowledge, rational thinking, and abstract scientific models, but also futuristic thoughts about possible, probable, and preferable futures and courses of action (Amara 1981). Complex phenomena need multifaceted knowledge, from insiders and outsiders, experts and common people, and the knowledge needed is both general and local. An absence of reflexivity in organizations and management feeds what Alvesson and Spicer (2012) defined as a functional stupidity, which is the more or less conscious rejection of using intellect as something other than a tool to avoid justifications and refuse open-mindedness. Since organizations and managerial practices are part of the broad field of disaster work, one can expect that functional stupidity can also occur in them. Functional stupidity makes life easier—at least for the moment and for the management.

Future Matters Matter

I have argued elsewhere that the future is real and important (Bergman 2015). The future is not only imagined (Adam and Groves 2007, xiii), but also made and created all the time, everywhere. This creation is done in our everyday practice at work and elsewhere, through policies and politics, through technology and medicine, and through education and science. According to one of the most famous futurists, Wendell Bell (2003, 117), "All of us here on earth are time travelers. We are constantly mowing with the arrow of time on a one-way trip out of the past and in on [*sic*] ongoing present toward the future." This might sound like something from a science fiction movie and rather fatalistic, but the important point is that we are agents creating the future and that we, including researchers, are therefore morally responsible for the "not yet" (Adam 2008, 111–12). Some disciplines and researchers are more prone to take active part in this creation, while others prefer to make claims and predictions about it.

From a historical perspective, the future has been looked into by various experts who have claimed to have a direct channel into it, such as prophets, oracles, witches, druids, priests, fortune tellers, and astrologers (Adam and Groves 2007). In many of these historical ways of understanding the future, the lesson is that you cannot escape your fate even if you know it. It is a deterministic way of looking at the future. Today, we may ridicule the use of tools such as cards, bird bones, or the contents of a dead fish's stomach to say something about the future. Nonetheless, Adam and Groves (2007) argued that we still rely on tools such as advanced statistical modeling and trend analysis for essentially the same purpose—that is, to try and control the future and to make it less threatening and uncertain. Examples of methods that rely on the logic of past-based causality and continuity are prediction, prognosis, forecast, and trend analysis—all of which are calculations and claims about the future. These rather mechanistic ways of understanding society, which relies on calculations and estimations of repetition and continuity, is strong in the Western world. Facts and their operational domain are the past and the present, and these facts are presented as evidence for the future as well. One example is a prediction about the societal order in democratic countries based on a rather advanced and secure situation in the 1980s, when it was formulated. In an article entitled *The Irreversible Welfare State,* Therborn and Roebroek (1986, 332) predicted that "the welfare state is an irreversible major institution of advanced capitalist countries, as long as democracy prevails. The building of a majoritarian anti-welfare state coalition seems impossible for the foreseeable future." Today, we are witnessing the demolition of the welfare state in the democratic Nordic countries, due to deregulation and neoliberal influences such as New Public Management. The lesson learned is that predictions and statements about the future need to be under constant inquiry—and that includes predictions made by AI.

Predictions about hazards, vulnerability, and disasters are problematic due to the unpredictability of nature and society. However, this does not mean that we should not engage in a serious discussion of what can happen in the future. Thus, in order to handle unpredictable situations, it is important to have an awareness of what can happen, about possible and probable scenarios, and a readiness and capacity to reduce the negative consequences if it happens. However, there is always the risk, especially with the currently available technology, that disaster work will turn into a predicting game with a focus on trying to predict and prevent disasters as specific events with consequences for vulnerable groups. On the other hand, another problem is that predictions that actually has been able to identify risks and emerging disasters tend to be neglected. For example, Fekete (2012, 1175) notes that no one paid attention to the predictions made about Hurricane Katarina or the events in Japan in 2011 until the aftermath of these events. Which knowledge makes it all the way through to action and gets the chance to make a difference is also a matter of priorities made by agents in powerful positions.

De Jouvenel (1967, 5) argued that knowledge about the future is a contradiction in itself. This is because, at the same time as you state something about the future, you might alter it one way or another, and conditions and interdependencies will affect the predicted outcome. De Jouvenel continued, "Our knowledge of the future is inversely proportional to the rate of progress" (1967, 275). In other words, in a society that is complex and has a great capacity to produce innovation and change, one should be wary about relying on predictions based on past facts in order to gain knowledge about future consequences, since they are best suited to stable contexts (e.g., Adam and Groves 2007).

As an alternative way of dealing with the future, futurologists argue—in line with de Jouvenel (1967) and Bell (2003)—that there are no future facts, only future possibilities for things to happen. Instead of predictions, techniques used include scenario planning, foresight, back-casting, weak signal scanning, horizon scanning, and Delphi methods. These techniques also rest on past- and present-based knowledge about what is likely to happen, as well as on conjecture about what could happen. Some of them also include normative visions about what is a desirable or preferable future. In other words, what do we want and, given existing conditions, how do we get there? These techniques regard, according to Adam and Groves (2007, 37), the future as empty and open, ready to be filled with content. Even though the past and the present are conditioning the future, it is subject to human intervention and is therefore ours to make. Progressive research projects have focused on what researchers see as a preferable future by exploring various theories and models for social change, driven by the idea of emancipatory agency. For example, between 2014–2018 there was a research project based at the KTH Royal Institute of Technology in Sweden, called "Beyond GDP Growth: Scenarios for a Sustainable Building and Planning." Another example is the Real Utopias Project, which started in the United States in 1991 in order to create a wide range of proposals and models for radical social change. Although the objectives of these projects are mainly to create radical alternatives to capitalism, their concerns and awareness include how risks and hazards generated by the greedy forces of capitalism are related to a lack of equality, and sustainability.

As mentioned above, one can approach the future from the past looking forward, or from the future looking backwards. Both these time horizons are necessary when understanding the future, since it is always in the making and in between the past, the present, and the not yet manifested. Adam and Groves (2007, 37) suggested that the future is not an empty territory, a tabula rasa waiting to be filled, but is instead occupied by visions, plans, decision, and actions—in the process of being materialized into empirically accessible facts. Therefore, it is important to formulate alternatives about preferable futures and how to get there, as well as non-preferable futures and how to avoid them. This is more progressive than the adaptive resilience approach, which rests on a readiness to meet and cope with the disasters.

There have been dystopias and utopias that have projected a critique toward contemporary society to an imagined future or imagined "no place" as More (1965[1516]) defined utopia in his book with the same name. Perhaps the "no place" analogy is more relevant than we can think of today, but with a twist of meaning referring not to an imagined place or future but to a real place, earth, where "no one" is safe "no where."

Our Responsibility for the Not Yet

Globalization and the imperatives of competitiveness and growth on a worldwide market make existing and future generations vulnerable. The same applies for hazards and disasters: they do not stop at borders (Fekete 2012, 1173). However, we are not predestined to experience rapidly increasing hazards, vulnerability, and disasters as the bitter but necessary fruits of social, economic, and technological development. Then again, economic growth cannot be trusted as the main driver for societal change. Nor can the modern witchcraft of the science of economics, which has been so successful in conceptualizing and shaping our Western society (Nandy 2013), remain unquestioned. Experiences and knowledge about disasters and vulnerability need to be considered when raising fundamental questions about how to deal with efficient production systems and global sustainability. Today, we tend to sacrifice human safety, well-being, freedom, and welfare for economic growth. Here, there is an inevitable moral dimension where we need to develop solidarity and responsibility—even among strangers—to be able to alter this priority.

As stated, visions about the future are likely to be created by influential vested interests (Williams 2007). Powerful groups, stakeholders, organizations, and institutions try to make the future predictable and controllable through promises, obligations, contracts, laws, or other ways of managing things in their desired direction. As Abraham Lincoln's famous quote unveils, "[t]he best way to predict your future is to create it." Hence, powerful groups and interests need to be challenged and alternatives need to be developed based on questions that have to do with justice, equality, and minimized harm and neglect. Our society requires broad transformative participation and future-oriented emancipatory agency in order to reach sustainability and equality. So what about the role of the scientific community? Well, there is an unequal distribution of scientific knowledge as a consequence of the cost related to research—nationally and globally. Therefore, as Augé (2014, 76) suggested, the concentrations of accumulated capital—not only financial but also intellectual—need to be redistributed and circulated. This is yet another area in which researchers can prove their solidarity and responsibility. Therefore, scientists in the field of critical and radical disasters research need to pay attention to and engage in the future as a potential subject for moral and normative interference. In other words, researchers in this field can help improve society through their knowledge and reasoning—together with potentially and actually vulnerable groups and with scientists from other fields. Otherwise, there is a risk that society might continue to be strongly influenced by economic and political interests that do not hesitate to articulate statements about the future as if they were true, by using the art of predicting or the art of management (or governance).

Theories about disaster do not prevent disasters, but they might have an impact on social consciousness and practice, and reveal different hopes and fears. I would like to stress the value of emancipatory agency and the possibility for disaster researchers to be part of the development of our society. If these researchers do not take action, and instead regard the future as decontextualized and depersonalized, the future be can used and abused by others without them feeling guilt, remorse, or responsibility. That is problematic since our future is coming generations' present (Adam and Groves 2007, 13). Disaster researchers need to be explicit about the possible, probable, and preferable futures—both inside and outside academia—based on their knowledge of disasters, vulnerability, and the causes and consequences of disasters. Adam (2008, 113) stated that the future is a contemporary social condition for our actions at the same time as we are creating it. This means researchers are a part of this creation process, whether or not we want to be. Even more importantly—since we are a part of it—we can also be held responsible for it.

To Conclude

Despite this chapter's focus on the future in a Western modernist manner, the intention is not to argue that the Western understanding of time and the importance of future is the blueprint for all or that it is universal across all cultures. Neither is the quality of the predictability of the future, or of utopias or dystopias, the main point. Instead, their potential to encourage and open up for criticism, reflexion, hope, anxiety, and alternative futures is of importance. Utopias and dystopias are visions that create a gap between reality and hope or fear. Nandy (2013, 3) argues that it is this very gap that "becomes a source of cultural criticism and a standing condemnation of the oppression of everyday life to which we otherwise tend to get reconciled." When utopias are taken seriously, the outcome has often been devastating for some since what is a utopia for some can be a dystopia for others. For that reason, dialogue and collaboration are essential.

Disaster research and work need to be seen from a wider perspective in relation to the possible futures that we ourselves are busy creating right now. Thus, one important aspect is to make use of the individual, collective, and organizational capacity to learn, reflect, and change when necessary. Power and functional stupidity together make up a vicious combination; we have to fight such stupidity collectively by using a wide range of referent groups. This is a shared responsibility and perhaps our only chance to reduce vulnerability in the long run. Our actions or absence thereof have consequences for the future. Therefore, paradoxically and as Adam and Groves (2007, 165) declare, we are all responsible for something that has not yet happened and might not even happen. The future is not ours, it is theirs—the coming generations'—and our practices today will condition their present. In our future, we are history. Equality and sustainability involve the responsibility and actions to improve the quality of life—for all, and even for the ones not born yet and whose present is our future.

Ann Bergman is a professor of working life science at Karlstad University, Sweden. Her research and teaching interest lies within the field of gender, work and management, gender segregation and inequality, and futures studies. Currently she is doing research on the digitalization of work, work-life boundaries and the future of work. Her recent interest in disaster research draws on the connection between organization, risk management, mitigation strategies and social groups' exposure to risks and vulnerability in relation to the futures.

Acknowledgments

As a total outsider in the field of disaster studies I am grateful beyond measure for valuable collegial support given by members of the Risk and Disasters Topical Interest Group who commented on earlier drafts of this text. The editors Susanna M. Hoffman and Roberto E. Barrios have definitely earned a gold star each. Some parts of sections "Disasters as Causal Forces," "Future Matters Matter," and "Our Responsibility for the Not Yet" are based on a keynote speech published as a commentary (Bergman 2015).

References

Adam, B. 2008. "Future Matters: Futures Known, Created and Minded." *Twenty-First Century Society* 3, no. 2, 111–16.

Adam, B., and Groves, C. 2007. *Future Matters: Action, Knowledge, Ethics.* Leiden, Netherlands: Brill.

Amara, R. 1981. "The Futures Field: Searching for Definitions and Boundaries." *The Futurist* 15, no. 1, 25–29.

Alvesson, M., and Spicer, A. 2012. "A Stupidity-Based Theory of Organizations." *Journal of Management Studies* 49, no. 7, 1194–220.

Augé, M. 2014. *The Future.* London: Verso Books.

Bankoff, G. 2001. "Rendering the World Unsafe: 'Vulnerability' as Western Discourse." *Disasters 25, no. 1,* 19–35.

Barnshaw, J., and Trainor, J. 2007. "Race, Class, and Capital amidst the Hurricane Katrina Diaspora." In *The Sociology of Katrina: Perspectives on a Modern Catastrophe,* edited by D. L. Brunsma, D. Overfelt, and J. S. Picou, 91–105. Lanham, MD: Rowman and Littlefield.

Barrios, R. E. 2017. *Governing Affect: Neoliberalism and Disaster Reconstruction.* Lincoln: University of Nebraska Press.

Bell, W. 2003. *Foundations of Futures Studies: Human Science for a New Era.* Vol. 1, *History, Purposes, and Knowledge.* New Brunswick, NJ: Transaction Publishers.

Bergman, A. 2015. "Back to the Future: Not looking into the Future but at Futures." *Nordic Journal of Working Life Studies* 5, no. 1, 3–8.

Blackman, D., and E. Sadler-Smith. 2009. "The Silent and the Silenced in Organizational Knowing and Learning." *Management Learning* 40, no. 5, 569–85.

Blaikie, P., T. Cannon, I. Davis, and B. Wisner. 1994. *At Risk: Natural Hazards, People's Vulnerability and Disasters.* New York: Routledge.

Börjesson, L., M. Höjer K-H., Dreborg, T. Ekvall, and G. Finnveden. 2006. "Scenario Types and Techniques: Towards A User's Guide." *Futures* 38, no. 7, 723–39.

de Jouvenel, B. 1967. *The Art of Conjecture.* New York: Basic Books.

Fekete, A. 2012. "Spatial Disaster Vulnerability and Risk Assessments: Challenges in Their Quality and Acceptance." *Natural Hazards* 61, no. 3, 1161–78.

Fothergill, A. 1998. "The Neglect of Gender in Disaster Work: An Overview of the Literature." In *The Gendered Terrain of Disaster: Through Women's Eyes,* edited by E. Enarson and B. H. Morrow, 63–84. Westport, CT: Praeger.

Fothergill, A., and Peek, L. A. 2004. "Poverty and Disasters in the United States: A Review of Recent Sociological Findings." *Natural Hazards* 32, no. 1, 89–110.

Fraser, N. 1995. From Redistribution to Recognition? Dilemmas of Justice in a 'Post-Socialist' Age. *New Left Review* 212, 68–93.

García-Acosta, V. 2002. "Historical Disaster Research." In *Catastrophe and Culture*, edited by S. Hoffman and A. Oliver-Smith, 49–166. Santa Fe, NM: School of American Research.

Haraway, D. 1988. "Situated Knowledges: The Science Question in Feminism and the Privilege of Partial Perspective." *Feminist Studies* 14, no. 3, 575–99.

Hewitt, K. 1983. *Interpretations of Calamity from the Viewpoint of Human Ecology.* Boston: Allen & Unwin.

Hoffman, S. M. 2002. "The monster and the mother: The symbolism of disaster." In *Catastrophe and Culture*, edited by S. Hoffman and A. Oliver-Smith, 113–42. Santa Fe, NM: School of American Research.

Hoffman S. M., and A. Oliver-Smith, eds. 2002. *Catastrophe and Culture: The Anthropology of Disaster.* Santa Fe, NM: School of American Research Press and James Currey.

Hollnagel, E. 2009. "The Four Cornerstones of Resilience Engineering." *Resilience Engineering Perspectives*, vol. 2, *Preparation and Restoration*, 117–33. Farnham, UK: Ashgate.

Kelman, I., Gaillard, J. C., and Mercer, J. 2015. "Climate Change's Role in Disaster Risk Reduction's Future: Beyond Vulnerability and Resilience." *International Journal of Disaster Risk Science* 6, no. 1, 21–27.

Kennedy, J., Ashmore, J., Babister, E., and Kelman, I. 2008. "The Meaning Of 'Build Back Better': Evidence from Post-Tsunami Aceh and Sri Lanka." *Journal of Contingencies and Crisis Management* 16, no. 1, 24–36.

Kouzmin, A. 2008. "Crisis management in crisis?" *Administrative Theory & Praxis* 30, no. 2, 155–83.

Lavell, A., M. Oppenheimer, C. Diop, J. Hess, R. Lempert, R. Li, R. Muir-Wood, and S. Myeong. 2012. "Climate Change: New Dimensions in Disaster Risk, Exposure, Vulnerability, and Resilience." In *Managing the Risks of Extreme Events and Disasters to Advance Climate Change Adaptation*, edited by C. B. Field, V. Barros, T. F. Stocker, D. Qin, D. J. Dokken, K. L. Ebi, M. D. Mastrandrea, K. J. Mach, G. K. Plattner, S. K. Allen, M. Tignor, and P. M. Midgley, 25–64. Cambridge: Cambridge University Press.

Manyena, S. B., O'Brien, G., O'Keefe, P., and Rose, J. 2011. "Disaster Resilience: A Bounce Back or Bounce Forward Ability." *Local Environment* 16, no. 5, 417–24.

McEntire, D. A. 2004. "Development, Disasters and Vulnerability: A Discussion of Divergent Theories and The Need for Their Integration." *Disaster Prevention and Management: An International Journal* 13, no. 3, 193–98.

Mead, G. H. 2002[1932]. *The Philosophy of the Present.* Amherst, NY: Prometheus Books.

Mileti, D. 1999. *Disasters by Design: A Reassessment of Natural Hazards in the United States.* Washington DC: Joseph Henry Press.

More, T. 1965[1516]. *The Complete Works of St. Thomas More. Utopia.* New Haven, CT: Yale University Publishers.

Morriss, P. 2002. *Power: A Philosophical Analysis*, 2nd ed. New York: Manchester University Press.

Nandy, A. 2013. *Traditions, Tyranny and Utopias: Essays in the Politics of Awareness*, 4th ed. New Delhi: Oxford University Press.

Oliver-Smith, A. and Hoffman, S. M. 2002. "Introduction." In *Catastrophe and Culture,* edited by S. Hoffman and A. Oliver-Smith, 3–22. Santa Fe, NM: School of American Research.

Petersen, K., and Johansson, H. 2008. "Designing Resilient Critical Infrastructure Systems Using Risk and Vulnerability Analysis." In *Resilience Engineering: Remaining Sensitive to the Possibility of Failure,* edited by E. Hollnagel, C. P. Nemeth, and S. Dekker, 160–171. Aldershot, UK: Ashgate.

Roy, A. 2004. "Peace and the New Corporate Liberation Theology." *City of Sydney Peace Price Lecture,* CPACS Occasional Paper No. 04/2, 3 November 2004. Sydney: University of Sydney, 1–9.

Scott, J. 2001. *Power.* Cambridge, UK: Polity Press.

Suparamaniam, N., and Dekker, S. 2003. "Paradoxes of Power: The Separation of Knowledge and Authority in International Disaster Relief Work." *Disaster Prevention and Management: An International Journal* 12, no. 4, 312–18.

Therborn, G., and J. Roebroek. 1986. "The Irreversible Welfare State: Its Recent Maturation, Its Encounter with the Economic Crisis, And Its Future Prospects." *International Journal of Health Services* 16, no. 3, 319–38.

United Nations Office for Disaster Risk Reduction. 2012. "Towards a Post-2015 Framework for Disaster Risk Reduction." Retrieved 24 April 2019 from https://www.unisdr.org/files/25129_towardsapost2015frameworkfordisaste.pdf.

———. 2017. "Terminology: Letter R." Retrieved 26 March 2019 from https://www.unisdr.org/we/inform/terminology#letter-r.

Ullberg, S. 2013. *Watermarks: Urban Flooding and Memoryscape in Argentina.* PhD diss., Institute of Latin American Studies, Stockholm University, Sweden.

Walby, S. 2011. *The Future of Feminism.* Cambridge: Polity.

Weber, M. 1978. "Basic Sociological Terms." In *Economy and Society,* edited by G. Roth and C. Wittich, 3–62. Berkeley: University of California Press.

White, S. C. 1996. "Depoliticising Development: The Uses and Abuses of Participation." *Development in Practice* 6, no. 1, 6–15.

Williams, C. C. 2007. *Rethinking the Future of Work: Directions and Visions.* Basingstoke, UK: Palgrave Macmillan.

Wisner, B., and Luce, H. R. 1993. "Disaster Vulnerability: Scale, Power and Daily Life." *GeoJournal* 30, no. 2, 127–40.

Index

www.ingramcontent.com/pod-product-compliance
Lightning Source LLC
Chambersburg PA
CBHW070903030426
42336CB00014BA/2302

* 9 7 8 1 7 8 9 2 0 6 4 8 7 *